Classic Edition

Sources

Environmental Studies

Classic Edition

Sources

Environmental Studies

Edited by

THOMAS A. EASTON
Thomas College, Waterville, ME

CLASSIC EDITION SOURCES: ENVIRONMENTAL STUDIES, FOURTH EDITION

Published by McGraw-Hill, a business unit of The McGraw-Hill Companies, Inc., 1221 Avenue of the Americas, New York, NY 10020. Copyright © 2012 by The McGraw-Hill Companies, Inc. All rights reserved. Previous editions © 2009 and 2007. No part of this publication may be reproduced or distributed in any form or by any means, or stored
in a database or retrieval system, without the prior written consent of The McGraw-Hill Companies, Inc., including, but not limited to, in any network or other electronic storage or transmission, or broadcast for distance learning.

Some ancillaries, including electronic and print components, may not be available to customers outside the United States.

Classic Edition Sources® is a registered trademark of the McGraw-Hill Companies, Inc.

Classic Edition Sources is published by the **Contemporary Learning Series** group within the McGraw-Hill Higher Education division.

1 2 3 4 5 6 7 8 9 0 QDB/QDB 1 0 9 8 7 6 5 4 3 2 1

ISBN 978-0-07-352764-2
MHID 0-07-352764-5
ISSN 2159-0605 (print)
ISSN 2159-0613 (online)

Managing Editor: *Larry Loeppke*
Developmental Editor: *Dave Welsh*
Permissions Coordinator: *DeAnna Dausener*
Marketing Specialist: *Alice Link*
Project Manager: *Robin A. Reed*
Design Coordinator: *Margarite Reynolds*
Buyer: *Nicole Baumgartner*
Media Project Manager: *Sridevi Palani*

Compositor: Laserwords Private Limited

Advisory Board

Members of the Advisory Board are instrumental in the final selection of articles for each edition of CLASSIC EDITION SOURCES. Their review of articles for content, level, currentness, and appropriateness provides critical direction to the editor and staff. We think that you will find their careful consideration well reflected in this volume.

Editor

Thomas A. Easton
Thomas College

Adboard Members

THOMAS A. EASTON
Thomas College

THOMAS A. EASTON is a professor of science at Thomas College in Waterville, Maine, where he has been teaching environmental science, science, technology and society, and computer science since 1983. He received a BA. in biology from Colby College in 1966 and a PhD. in theoretical biology from the University of Chicago in 1971. He writes and speaks frequently on scientific and futuristic issues. His books include *Focus on Human Biology*, 2nd ed., coauthored with Carl E. Rischer (HarperCollins, 995), *Careers in Science*, 4th ed. (VGM Career Horizons, 2004), *Taking Sides: Clashing Views on Controversial Issues in Science, Technology and Society* (McGraw-Hill, 10th ed., 2011), and *Taking Sides: Clashing Views on Controversial Environmental Issues* (McGraw-Hill, 14th ed., 2011). Dr. Easton is also a well-known writer and critic of science fiction.

Dedication

I dedicate this book to the next generation in hope that by understanding the journey they may also understand the goal.

Preface

*U*ndesirable human impacts on the environment have been noted for centuries, but it has been only a few decades since anything that could be called an "environmental movement" took form and even less time since the field of "environmental science" coalesced from those of ecology, chemistry, medicine, politics, public policy, law, economics, and all the other areas that must come together and interact in order to identify environmental problems, decide what to do about them, and implement the decisions. The birth of the movement was heralded by the first Earth Day in 1970, but that was possible only because of the momentum initiated by Rachel Carson, whose book, *Silent Spring*, appeared in 1962 and condemned the use of pesticides because of their toxic effects on wildlife and humans. In a pattern that was to be repeated many times with future issues, there was an initial violent reaction from the chemical and agribusiness industries, but the general public's response was more positive.

The situation before Rachel Carson and *Silent Spring* is nicely captured by Judge Richard Cudahy, who in "Coming of Age in the Environment," *Environmental Law,* Winter 2000, writes, "It doesn't seem possible that before 1960 there was no 'environment'—or at least no environmentalism. I can even remember the Thirties, when we all heedlessly threw our trash out of car windows, burned coal in the home furnace (if we could afford to buy any), and used a lot of lead for everything from fishing sinkers and paint to no-knock gasoline. Those were the days when belching black smoke meant a welcome end to the Depression and little else."

The 1960s saw the passage of the U.S. Air Quality Act, which set standards for air pollution; the U.N. Biosphere Conference, which discussed global environmental problems; and the U.S. National Environmental Policy Act, which required (among other things) that federal agencies prepare environmental impact statements for their projects. By 1972 the use of DDT was banned, and other pesticides were much more carefully regulated. In 1992, the first "Earth Summit" was held in Rio de Janeiro, Brazil; it became famous for generating agreement on addressing climate change, biodiversity loss, sustainability, global poverty, and more. Since then, a great deal more has happened. More problems have been identified,

basic principles (such as sustainability) have been formulated, there has been an immense amount of debate, and a few problems (such as ozone depletion) have seen significant progress. In 2001, the National Research Council's Committee on Grand Challenges in Environmental Sciences published *Grand Challenges in Environmental Sciences* (National Academy Press, 2001) in an effort to reach "a judgment regarding the most important environmental research challenges of the next generation—the areas most likely to yield results of major scientific and practical importance if pursued vigorously now." These areas include biogeochemical cycles (the cycling of plant nutrients, the ways human activities affect them, and the consequences for ecosystem functioning, atmospheric chemistry, and human activities); biological diversity; climate variability; hydrologic forecasting (groundwater, droughts, floods, etc.); infectious diseases; resource use; land use; and reinventing the use of materials (e.g., recycling). Many of these issues have corresponding selections in this book. They are also discussed in my *Taking Sides: Clashing Views on Controversial Environmental Issues* (McGraw-Hill, 14th ed., 2011).

In 1992, that first Earth Summit grew from the recognition that the world faced many environmental problems. It generated much agreement that solutions had to be found, and in its wake it left a great deal of optimism that those solutions could be found. Unfortunately, that optimism has weakened. The second Earth Summit (Johannesburg, 2002) focused on sustainability but generated very little in the way of action plans. As the world approaches the third Earth Summit (Rio de Janeiro, 2012), the mood is bleak. As many of the selections in this book show, problems identified years ago are worsening despite international agreements to find solutions. Planners and supporters are hoping that "Rio +20" will repeat the success of the 1992 Summit and generate real international commitment to solve the world's environmental problems. Such success will, however, be difficult to achieve, for environmental issues are infamous for the conflicts of interest that surround them.

No one should be surprised that "environmental science" or "environmental studies" has become a common program or major in colleges and universities. Yet in order for that to happen, there had to be an accompanying recognition that not all subjects worth studying fit neatly into the traditional academic departmental structure. Environmental science started in biology departments as a human-oriented version of ecology, but it soon became apparent that it required resources from other departments, both in the sciences and in the humanities. To study environmental problems and solutions effectively requires studying biology and ecology, chemistry and physics, geology and geography, economics and politics, law and public policy. The result has been a host of mutidisciplinary, interdisciplinary, and cross-disciplinary programs, most aimed at producing graduates equipped to deal with the most important real-world problems, whether they will work as scientists, teachers, lawyers, or public servants. Today there are even entire colleges that center their curricula on environmental studies, endeavoring to unify human knowledge under one of the most inclusive of labels. At the College of the Atlantic in Bar Harbor, Maine, for instance, "All students major in Human Ecology, the study of our relationship with our environment. This major gives you the flexibility to design your own course of study. It's all about creativity, investigation, engagement, and community."

This fourth edition of *Classic Edition Sources: Environmental Studies* brings together 43 selections—essays, book excerpts, journal articles, even an opinion from a Supreme Court case—of enduring relevance to the field. They illuminate history, discovery, debate, and solutions. The book is rich enough to serve as the sole text for an introductory environmental studies course. It may also serve as a companion to an environmental science text.

Many of the selections in the book have been retained from previous editions. More have been chosen to emphasize the historical background, current problems (such as global warming), and future prospects. My work with *Taking Sides: Clashing Views on Controversial Environmental Issues* has been a great help in cultivating insight and perspective.

ORGANIZATION OF THE BOOK. The selections are organized topically around several major areas of environmental studies. Part 1 provides An Overview of Environmental Studies; Part 2 includes selections related to Energy; Part 3, Environmental Degradation; Part 4, Human Health and the Environment; Part 5, Environment and Society. Many of the selections could easily have been placed in a different Part. The placement is not entirely arbitrary, however. The divisions are intended to provide clusters of roughly equal size that lend themselves to teaching as a unit. However, instructors are free to rearrange the sequence to suit their own preferences. Each selection is preceded by a short introductory headnote that provides biographical information about the author and briefly describes the relevance of the topic and the value of the selection.

New to this edition are Learning Outcomes and Critical Thinking study questions to help students better understand what they have read.

ON THE INTERNET. For each chapter in this book, we have provided several Internet site addresses relevant to the chapter, as well as a brief description of each site. See the "Internet References" section on p. x-ref.

A WORD TO THE INSTRUCTOR. An *Instructor's Manual with Test Questions* (multiple-choice and essay) is available through the publisher for instructors using *Classic Edition Sources: Environmental Studies*, 4th ed., in the classroom.

ACKNOWLEDGMENTS. I wish to thank McGraw-Hill's David Welsh and Larry Loepke for offering me the opportunity to carry on the work of Theodore D. Goldfarb.

Comments, criticisms, and especially suggestions of selections to be considered for future editions are welcome from all readers or users—both instructors and students—of this book.

Thomas A. Easton
Thomas College Waterville, ME

Contents

Fundamental Causes of Environmental Problems

14

Ecosystems and Ecosystem Services

29

Part 2: Energy

4 Energy and Ecosystems — 48

5 Renewable and Nonrenewable Energy — 57

Part 3: Environmental Degradation

6 Forests, Wilderness, and Wildlife — 62

Part 4: Human Health and the Environment

Chemicals   *129*

Part 5: Environment and Society

Political and Economic Issues   *143*

Correlation Guide

*E*ach volume in the *Classic Edition Sources* series brings together over 35 selections of enduring intellectual value—classic articles, book excerpts, and research studies—that have shaped a discipline of study. Edited for length and level, the selections are organized topically around the major areas of study within the discipline. For more information on *Classic Edition Sources* and other *McGraw-Hill Contemporary Learning Series* titles, visit www.mhhe.com/cls.

This convenient guide matches the Parts in **Classic Edition Sources: Environmental Studies, 4/e** with the corresponding chapters in two of our best-selling McGraw-Hill Environmental Science textbooks by Cunningham/Cunningham.

Classic Edition Sources: Environmental Studies, 4/e	Principles of Environmental Science, 6/e by Cunningham/Cunningham	Environmental Science: A Global Concern, 11/e by Cunningham/Cunningham
Part 1: An Overview of Environmental Studies	**Chapter 1:** Understanding Our Environment **Chapter 2:** Environmental Systems **Chapter 5:** Biomes and Biodiversity **Chapter 6:** Environmental Conservation	**Chapter 1:** Understanding Our Environment **Chapter 2:** Principles of Science and Systems **Chapter 3:** Matter, Energy, and Life **Chapter 4:** Evolution, Biological Communities, and Species Interactions Chapter 5: Biomes: Global Patterns of Life
Part 2: Energy	**Chapter 12:** Energy	**Chapter 14:** Geology and Earth Resources **Chapter 15:** Air, Weather, and Climate **Chapter 19:** Conventional Energy **Chapter 20:** Sustainable Energy
Part 3: Environmental Degradation	**Chapter 2:** Environmental Systems **Chapter 5:** Biomes and Biodiversity **Chapter 6:** Environmental Conservation **Chapter 9:** Air: Climate and Pollution **Chapter 13:** Solid and Hazardous Waste	**Chapter 11:** Biodiversity: Preserving Species **Chapter 12:** Biodiversity: Preserving Landscapes **Chapter 13:** Restoration Ecology **Chapter 16:** Air Pollution **Chapter 17:** Water Use and Management **Chapter 18:** Water Pollution

Classic Edition Sources: Environmental Studies, 4/e	Principles of Environmental Science, 6/e by Cunningham/ Cunningham	Environmental Science: A Global Concern, 11/e by Cunningham/ Cunningham
Part 4: Human Health and the Environment	**Chapter 1:** Understanding Our Environment **Chapter 4:** Human Populations	**Chapter 8:** Environmental Health and Toxicology **Chapter 9:** Food and Hunger **Chapter 10:** Farming: Conventional and Sustainable Practices
Part 5: Environment and Society	**Chapter 4:** Human Populations **Chapter 14:** Economics and Urbanization **Chapter 15:** Environmental Policy and Sustainability	**Chapter 6:** Population Biology **Chapter 7:** Human Populations **Chapter 21:** Solid, Toxic, and Hazardous Waste **Chapter 22:** Urbanization and Sustainable Cities **Chapter 23:** Ecological Economics **Chapter 24:** Environmental Policy, Law, and Planning **Chapter 25:** What Then Shall We Do?

Internet References

Chapter 1

The Association for the Study of Literature and Environment is an organization dedicated to promoting an interest in literature that considers the relationship between human beings and the natural world.
www.asle.org

Environment and History is an interdisciplinary journal containing articles by scholars in the humanities and the biological sciences. These articles examine current environmental problems from an historical perspective.
www.erica.demon.co.uk/EH.html

The International Institute for Sustainable Development advances sustainable development policy and research by providing information and engaging in partnerships worldwide. It says it "promotes the transition toward a sustainable future. We seek to demonstrate how human ingenuity can be applied to improve the well-being of the environment, economy and society."
www.iisd.org/default.asp

The Wildlife Conservation Society protects wildlife and wild lands through careful science, international conservation, and education. It also manages the world's largest system of urban wildlife parks.
www.wcs.org

Since 1935, the U.S. Department of Agriculture's Natural Resources Conservation Service (originally called the Soil Conservation Service) has provided leadership in a partnership effort to help America's private land owners and managers conserve their soil, water, and other natural resources.
www.nrcs.usda.gov

The U.S. Department of Agriculture's Forest Service manages public lands in national forests and grasslands. Gifford Pinchot, the first Chief of the Forest Service, summed up its mission as "to provide the greatest amount of good for the greatest amount of people in the long run."
www.fs.fed.us

Chapter 2

The mission of the Environmental Protection Agency is to protect human health and to safeguard the natural environment—air, water, and land—upon which life depends. This home page is a major resource for information on a wide variety of current environmental issues.
www.epa.gov

EnviroLink is a nonprofit organization dedicated to providing comprehensive, up-to-date information about available environmental resources.
www.envirolink.org

The Natural Resources Defense Council is one of the most active environmental research and advocacy organizations. Its home page lists its concerns as clean air and water, energy, global warming, toxic chemicals, nuclear waste, and much more.
www.nrdc.org

The Earth Council Alliance aims to foster increased awareness of sustainability achievements, problems and pragmatic solutions around the world.
www.earthcouncilalliance.org/en

The United Nation's Environment Programme "works to encourage sustainable development through sound environmental practices everywhere. Its activities cover a wide range of issues, from atmosphere and terrestrial ecosystems, the promotion of environmental science and information, to an early warning and emergency response capacity to deal with environmental disasters and emergencies."
www.unep.org

The Club of Rome is a global think tank that brings together scientists, economists, businessmen, international high civil servants, heads of state, and former heads of state from all five continents who are convinced that the future of humankind is not determined once and for all and that each human being can contribute to the improvement of our societies. Its mission is to act as a global catalyst of change and contribute to the solution of the complex set of the most crucial problems—political, social, economic, technological, environmental, psychological and cultural—facing humanity.

The Club of Rome is an independent, not-for-profit organization with members from all regions of the world, from different cultures and histories, from different fields of science and public policy, and from academia, civil society, and the corporate sector. It promotes interdisciplinary analysis, dialogue and action on the fundamental, systemic challenges that are determining the future of humanity and is recognized worldwide for its early work on the relations between economic growth and the environment (see *The Limits to Growth*).
www.clubofrome.org/eng/home

The Earth Day Network (EDN) promotes environmental citizenship and helps activists in their efforts to change local, national, and global environmental policies. It also makes the Ecological Footprint quiz (www.myfootprint.org/) available.
www.earthday.net

Chapter 3

Many state Departments of Environmental Protection maintain web pages that provide information about local ecosystems. Forida is one example that provides information on the state's salt marshes.
www.dep.state.fl.us/coastal/habitats/saltmarshes.htm

The United Nations initiated the Millennium Ecosystem Assessment in 2001. Its objective was "to assess the consequences of ecosystem change for human well-being and the scientific basis for action needed to enhance the conservation and sustainable use of those systems and their contribution to human well-being." It concluded that "human actions are depleting Earth's natural capital, putting such strain on the environment that the ability of the planet's ecosystems to sustain future generations can no longer be taken for granted."
www.maweb.org/en/index.aspx

The Ecosystems Services Project studies the services people obtain from their environments, the economic and social values inherent in these services, and the benefits of considering these services more fully in shaping land management policies.
www.ecosystemservicesproject.org

Chapter 4

The Center for Limnology is a multidisciplinary center within the College of Letters and Science at the University of Wisconsin-Madison.
http://limnology.wisc.edu

Water on the Web (WOW) offers aquatic ecology (limnology) curricula for high-school and college students. Its "Water Science" is designed for second-year technical students or undergraduates in water resource management, water science, or environmental resource management programs.
http://waterontheweb.org/under/lakeecology/index.html

JULES is the Joint UK Land Environment Simulator. It is based on MOSES (Met Office Surface Exchange System), the land surface model used in the Unified Model of the UK Meteorological Service. Among other things, it can be used to study inflows and outflows of energy.
www.jchmr.org/jules/science/energy.html

NASA's Earth Observatory's mission is to share with the public the images and discoveries about climate and the environment that emerge from NASA research, including its satellite missions, in-the-field research, and climate models.
earthobservatory.nasa.gov/Topics/heat.html

The Environmental and Energy Study Institute (EESI) is a nonprofit organization dedicated to promoting environmentally sustainable societies, in part by educating policy-makers.
www.eesi.org

Chapter 5

The Rocky Mountain Institute, founded by Amory Lovins, promotes the benefits of efficiency in energy and resource use.
www.rmi.org

The National Renewable Energy Laboratory (NREL) is the leading center for renewable energy research in the United States.
www.nrel.gov

The International Network for Sustainable Energy (INFORSE) is a global network of independent non-governmental organizations working for sustainable energy solutions to reduce poverty and protect the environment.
www.inforse.org

The Federal Energy Regulatory Commission regulates and oversees energy industries in the economic and environmental interest of the American public.
www.ferc.gov

The International Partnership for the Hydrogen Economy is an international institution established to accelerate the transition to a hydrogen economy.
www.iphe.net

Chapter 6

The National Wildlife Federation is dedicated to wilderness preservation and the protection of endangered species. The organization's home page contains links to information about these issues.
www.nwf.org

The Sierra Club calls itself "America's oldest, largest and most influential grassroots environmental organization." It is dedicated to protecting "the wild places of the earth" through education and lobbying.
www.sierraclub.org

The Wilderness Society applies scientific analysis to wilderness issues and advocated at the highest levels to save, protect, and restore America's wilderness areas.
www.wilderness.org

People for the Ethical Treatment of Animals (PETA) is often in the news for protesting mistreatment of domestic animals but its concerns are broader than that. PETA believes that animals have rights and deserve to be treated well without regard to whether they are useful to humans.
www.peta.org

Chapter 7

In 2002, world leaders attending the World Summit on Sustainable Development in Johannesburg agreed to achieve, by 2010, a significant reduction in the rate of loss of biodiversity. The resulting Convention on Biological Diversity represents the international community's commitment to conserving biodiversity.
www.biodiv.org/default.shtml

The Biodiversity Project works to make people aware of the importance of biodiversity and to enlist their participation in preserving it.
www.biodiversityproject.org

The Convention on International Trade in Endangered Species aims to ensure that international trade in wild animals and plants (for fur, pets, food, zoo specimens, research, etc.) and their parts (such as ivory) does not threaten their survival.
www.cites.org

The Rainforest Action Network works to protect rainforests and the rights of their inhabitants by promoting sustainable logging and "green" investment practices.
www.ran.org

The World Rainforest Information Portal provides access to a wealth of rainforest-related information.
www.rainforestweb.org

The United Nation's Secretariat of the Convention on Biological Diversity was established to support the goals of the Convention on Biological Diversity (CBD), signed at the Earth Summit in Rio de Janeiro, Brazil, in 1992. The CBD was the first global agreement to cover all aspects of biological diversity: the conservation of biological diversity, the sustainable use of its components and the fair and equitable sharing of benefits arising from the use of genetic resources.
www.cbd.int/secretariat

Chapter 8

The Natural Resources Defense Council is one of the most active environmental research and advocacy organizations. Its home page lists its concerns as clean air and water, energy, global warming, toxic chemicals, nuclear waste, and much more.
www.nrdc.org

The Pesticide Action Network North America (PANNA) challenges the global proliferation of pesticides, defends basic rights to health and environmental quality, and works to ensure the transition to a just and viable society.
www.panna.org

The Silicon Valley Toxics Coalition (SVTC) was formed in 1982 to engage in research, advocacy, and organizing associated with environmental and human problems caused by the rapid growth of the high-tech electronics industry, to advance environmental sustainability and clean production in the industry, and to improve health, promote justice, and ensure democratic decision making for affected communities and workers in the United States and the world.
www.svtc.org

The Center for Health, Environment, and Justice, founded by Love Canal activist Lois Gibbs, has the overarching goal of preventing harm, "particularly among vulnerable populations such as children. If a safer process, material or product exists it should be used."
www.chej.org

The U.S. Environmental Protection Agency provides a great deal of information on the Superfund program, including material on environmental justice.
www.epa.gov/superfund

RiverRestoration.org provides planning, engineering, and technical support for river restoration projects.
www.riverrestoration.org

American Rivers, founded in 1973, protects and promotes rivers as valuable assets that are vital to human health, safety and quality of life. Among its activities, it promotes rubbish cleanup campaigns and the removal of dams.
www.americanrivers.org

The United States Army Corps of Engineers serves the armed forces and the nation by providing vital engineering services and capabilities in support of national interests in the areas of water resources, environment, infrastructure, homeland security, and war.
www.usace.army.mil

The U.S. Department of Commerce's National Oceanic and Atmospheric Administration's Office of Ocean and Coastal Resource Management works to protect, restore, and manage ocean and coastal resources through ecosystem-based management.
http://coastalmanagement.noaa.gov

Chapter 9

The Intergovernmental Panel on Climate Change (IPCC) was formed by the World Meteorological Organization (WMO) and the United Nations Environment Programme (UNEP) to assess the scientific, technical, and socioeconomic information relevant for the understanding of the risk of human-induced climate change. The fourth assessment report was released in 2007 and is available at this site.
www.ipcc.ch

The United Nations Environmental Program maintains this site as a central source for substantive work and information resources with regard to climate change.
http://climatechange.unep.net

For a tutorial on stratospheric ozone depletion, see this NASA page:
www.nas.nasa.gov/About/Education/Ozone

The Global Warming International Center (GWIC) is an international body disseminating information on global warming science and policy, serving governmental and non-governmental organizations, and industries, in more than 145 countries.
http://globalwarming.net

World View of Global Warming offers numerous reports and photo essays documenting the effects of global warming.
www.worldviewofglobalwarming.org

350.org, founded by journalist and environmentalist Bill McKibben, is an international campaign that aims "to inspire the world to rise to the challenge of the climate crisis—to create a new sense of urgency and of possibility for our planet." The 350.org name comes from the level of atmospheric carbon dioxide—350 ppm—that we must get below, for scientists warn that staying above that level means dire consequences for the world as a whole.
www.350.org

Chapter 10

The Institute for Food and Development Policy (also known as Food First) analyzes the root causes of global hunger, poverty, and ecological degradation, develops solutions in partnership with movements working for social change, and publishes newsletters, reports, and books. It seeks to eliminate the injustices that cause hunger.
www.foodfirst.org

The International Food Policy Research Institute (IFPRI) has a vision of "a world free of hunger and malnutrition. The vision is based on the human right to adequate food and nutrition and recognition of the inherent dignity of all members of the human family. It is a world where every person has secure access to sufficient and safe food to sustain a healthy and productive life and where decisions related to food are made transparently and with the participation of consumers and producers."
www.ifpri.org

The Food Policy Institute (FPI) of Rutgers University aims to focus academic expertise on issues of food policy in ways "responsive to the needs of government, industry and the consumer."
www.foodpolicyinstitute.org

The Agriculture Network Information Center is a guide to quality agricultural (including biotechnology) information on the Internet as selected by the National Agricultural Library, Land-Grant Universities, and other institutions.
www.agnic.org

Chapter 11

The Pesticide Action Network North America (PANNA) challenges the global proliferation of pesticides, defends basic rights to health and environmental quality, and works to ensure the transition to a just and viable society.
www.panna.org

e.hormone is hosted by the Center for Bioenvironmental Research at Tulane and Xavier Universities in New Orleans. It provides accurate, timely information about environmental hormones and their impacts.
http://e.hormone.tulane.edu

The Silicon Valley Toxics Coalition (SVTC) was formed in 1982 to engage in research, advocacy, and organizing associated with environmental and human problems caused by the rapid growth of the high-tech electronics industry, to advance environmental sustainability and clean production in the industry, and to improve health, promote justice, and ensure democratic decision making for affected communities and workers in the United States and the world.
www.svtc.org

The U.S. Environmental Protection Agency provides a great deal of information on the Superfund program, including material on environmental justice.
www.epa.gov/superfund

Scorecard, founded by Environmental Defense in 1998 and now operated by Green Media Toolshed, is a rich source of information about pollution problems and toxic chemicals. Use it to learn about pollution problems in your community and see which geographic areas and companies have the worst pollution records.
www.scorecard.org

Chapter 12

The Ecumenical EcoJustice Network provides educational information on environmental and ecological issues from a Christian faith perspective.
www.ecojusticenetwork.org

ActionPA.org is a Pennsylvania-based research, networking, and organizing center for the grassroots environmental justice movement. It educates and trains student and community activists, among other things.
www.actionpa.org

The Community Coalition for Environmental Justice works "to achieve environmental and economic justice in low-income communities and communities of color. We believe that everyone, regardless of race or income, has the right to a clean and healthy community."
www.ccej.org

The Union of Concerned Scientists describes itself as "the leading science-based nonprofit working for a healthy environment and a safer world. UCS combines independent scientific research and citizen action to develop innovative, practical solutions and to secure responsible changes in government policy, corporate practices, and consumer choices."
www.ucsusa.org

The Worldwatch Institute is a nonprofit public policy research organization dedicated to informing policymakers and the public about "the interactions among key environmental, social, and economic trends. Our work revolves around the transition to an environmentally sustainable and socially just society—and how to achieve it."
http://worldwatch.org

Chapter 13

The Population Council is an international, nonprofit, non-governmental organization that seeks to improve the well-being and reproductive health of current and future generations around the world and to help achieve a humane, equitable, and sustainable balance between people and resources. Many of the Council's documents are available online.
www.popcouncil.org

For nearly 40 years, Population Connection (formerly Zero Population Growth) has advocated for progressive action to stabilize world population at a level that can be sustained by Earth's resources.
www.populationconnection.org

The United Nations Population Division is responsible for monitoring and appraising a broad range of population topics. This site offers a wealth of recent data and links.
www.un.org/esa/population/unpop.htm

The United Nations Population Fund (UNFPA) is an international development agency that supports countries in using population data for policies and programmes to reduce poverty and to ensure that every pregnancy is wanted, every birth is safe, every young person is free of HIV/AIDS, and every girl and woman is treated with dignity and respect.
www.unfpa.org/index.htm

The Population Reference Bureau provides data about population, health, and the environment to help advance the well-being of current and future generations.
www.prb.org

The Population Index is the primary reference tool to the world's population literature.
http://popindex.princeton.edu

Chapter 14

The United Nations's Division for Sustainable Development (in the Department of Economic and Social Affairs) coordinates international and national efforts (especially developing countries

and countries with economies in transition) to achieve sustainable development. Reports such as *Trends in Sustainable Development* are available for downloading.
www.un.org/esa/sustdev

The Foundation for Deep Ecology (FDE) believes that true ecological sustainability may require a rethinking of our values as a society and that if we are to achieve ecological sustainability. Nature can no longer be viewed only as a commodity; it must be seen as a partner and model in all human enterprise. It supports education and advocacy on behalf of wild Nature, primarily through its publications and public programs.
www.deepecology.org

Navdanya started as a program of the Research Foundation for Science, Technology and Ecology (RFSTE) in India, a participatory research initiative founded by world-renowned scientist and environmentalist Dr. Vandana Shiva, to provide direction and support to environmental activism.
www.navdanya.org

The Online Ethics Center for Engineering, established in 1995 with National Science Foundation funding, provides engineers, computer scientists, and engineering students with resources useful for understanding and addressing work-related ethical issues. It became an activity of the National Academy of Engineering (NAE) in March 2007 as part of the NAE's new Center for Engineering Ethics. This page contains cases, essays, codes of ethics, and links about the issues facing engineers and scientists in relation to the environment and sustainable development.
www.onlineethics.org/Topics/Enviro.aspx

Part 1: An Overview of Environmental Studies

Preservation vs. Conservation

Selection 1
GEORGE PERKINS MARSH, from *Man and Nature* (Charles Scribner, 1864)

Selection 2
JOHN MUIR, from *The Mountains of California* (Houghton Mifflin, 1916)

Selection 3
GIFFORD PINCHOT, from *The Fight for Conservation* (Doubleday, 1910)

Selection 4
ALDO LEOPOLD, from *A Sand County Almanac* (Oxford University Press, 1977)

Learning Outcomes

After reading this unit, you should be able to:

1. Describe how environmental problems are caused by human activity.
2. Explain the difference between conservationism and preservationism.
3. Explain the relationship between sustainability and the long-term prospects for human civilization.
4. Understand the value of including the well-being of nature in human decision making.

Man and Nature

George Perkins Marsh

As a diplomat in Constantinople and Greece between 1849 and 1854, Vermont lawyer, linguist, and scholar George Perkins Marsh (1801–1882) traveled across Europe and the Mediterranean area. Later, from 1861 to 1882, he was the American ambassador to the Kingdom of Italy. *Man and Nature* was first published in 1864; Perkins continued working on the book until his death; the third edition appeared in 1885.

Marsh has been called a "prophet of modern conservation," and his book, coming as it did well before the conservation movement, was enormously influential. It had its roots in Marsh's recognition that human activity had greatly altered the Vermont landscape, removing trees and leading to erosion and flooding. He showed early concern with methods for preventing damage and restoring the land. When he went to Europe, he found that similar changes had been happening for much longer, with more profound results, and that efforts to prevent damage and restore the land had considerable history.

It was never Marsh's aim to encourage the restoration or preservation of wilderness. The Earth, he said, was given to man for "usufruct" (to use the fruit of), not for consumption or waste. Resources should remain to benefit future generations. Stewardship was the point, a concept which later took a central place in the conservation movement and which lies at the root of sustainability.

As David Lowenthal wrote in his introduction to the centenary edition of *Man and Nature* (Belknap/Harvard University Press, 1965), "Anyone wielding a hoe or an axe knows what he is doing, but before Marsh no one had assessed the cumulative effect of all axes and hoes. For him the conclusion was inescapable. Man depends upon soil, water, plants, and animals. But in obtaining them he unwittingly destroys the fabric of nature. Therefore man must learn to understand his environment and how he affects it. For his own sake, not for nature's alone, man must restore and maintain it as long as he tenants the earth."

Key Concept: The value to humans of protecting and restoring the environment

---

The Roman Empire, at the period of its greatest expansion, comprised the regions of the earth most distinguished by a happy combination of physical advantages. The provinces bordering on the principal and the secondary basins of the Mediterranean enjoyed a healthfulness and an equability of climate, a fertility of soil, a variety of vegetable and mineral products, and natural facilities for the transportation and distribution of exchangeable commodities, which have not been possessed in an equal degree by any territory of like extent in the Old World or the New. . . .

If we compare the present physical condition of the countries of which I am speaking, with the descriptions that ancient historians and geographers have given of their fertility and general capability of ministering to human uses, we shall find that more than one half of their whole extent—including the provinces most celebrated for the profusion and variety of their spontaneous and their cultivated products, and for the wealth and social advancement of their inhabitants—is either deserted by civilized man and surrendered to hopeless desolation, or at least greatly reduced in both productiveness and population. Vast forests have disappeared from mountain spurs and ridges; the vegetable earth accumulated beneath the trees by the decay of leaves and fallen trunks, the soil of the alpine pastures which skirted and indented the woods, and the mould of the upland fields, are washed away; meadows, once fertilized by irrigation, are waste and unproductive, because the cisterns and reservoirs that supplied the ancient canals are broken, or the springs that fed them dried up; rivers famous in history and song have shrunk to humble brooklets; the willows that ornamented and protected the banks of the lesser watercourses are gone, and the rivulets have ceased to exist as perennial currents, because the little water that finds its way into their old channels is evaporated by the droughts of summer, or absorbed by the parched earth, before it reaches the lowlands; the beds of the brooks have widened into broad expanses of pebbles and gravel, over which, though in the hot season passed dryshod, in winter sea-like torrents thunder; the entrances of navigable streams are obstructed by sandbars, and harbors, once marts of an extensive commerce, are shoaled by the deposits of the rivers at whose mouths they lie; the elevation of the beds of estuaries, and the consequently diminished velocity of the streams which flow into them, have converted thousands of leagues of shallow sea and fertile lowland into unproductive and miasmatic morasses.

Besides the direct testimony of history to the ancient fertility of the regions to which I refer—Northern Africa, the greater Arabian peninsula, Syria, Mesopotamia, Armenia, and many other provinces of Asia Minor, Greece, Sicily, and parts of even Italy and Spain—the multitude and extent of yet remaining architectural ruins, and of decayed works of internal improvement, show that at former epochs a dense population inhabited those now lonely districts. Such a population could have been sustained only by a productiveness of soil of which we at present discover but slender traces; and the abundance derived from that fertility serves to explain how large armies, like those of the ancient Persians, and of the Crusaders and the Tartars in later ages, could, without an organized commissariat, secure adequate supplies in long marches through territories which, in our times, would scarcely afford forage for a single regiment....

The decay of these once flourishing countries is partly due, no doubt, to that class of geological causes, whose action we can neither resist nor guide, and partly also to the direct violence of hostile human force; but it is, in a far greater proportion, either the result of man's ignorant disregard of the laws of nature, or an incidental consequence of war, and of civil and ecclesiastical tyranny and misrule....

Destructiveness of Man

Man has too long forgotten that the earth was given to him for usufruct alone, not for consumption, still less for profligate waste. Nature has provided against the absolute destruction of any of her elementary matter, the raw material of her works; the thunderbolt and the tornado, the most convulsive throes of even the volcano and the earthquake, being only phenomena of decomposition and recomposition. But she has left it within the power of man irreparably to derange the combinations of inorganic matter and of organic life, which through the night of æons she had been proportioning and balancing, to prepare the earth for his habitation, when, in the fulness of time, his Creator should call him forth to enter into its possession.

Apart from the hostile influence of man, the organic and the inorganic world are, as I have remarked, bound together by such mutual relations and adaptations as secure, if not the absolute permanence and equilibrium of

both, a long continuance of the established conditions of each at any given time and place, or at least, a very slow and gradual succession of changes in those conditions. But man is everywhere a disturbing agent. Wherever he plants his foot, the harmonies of nature are turned to discords. The proportions and accommodations which insured the stability of existing arrangements are overthrown. Indigenous vegetable and animal species are extirpated, and supplanted by others of foreign origin, spontaneous production is forbidden or restricted, and the face of the earth is either laid bare or covered with a new and reluctant growth of vegetable forms, and with alien tribes of animal life. These intentional changes and substitutions constitute, indeed, great revolutions; but vast as is their magnitude and importance, they are, as we shall see, insignificant in comparison with the contingent and unsought results which have flowed from them.

The fact that, of all organic beings, man alone is to be regarded as essentially a destructive power, and that he wields energies to resist which, nature—that nature whom all material life and all inorganic substance obey—is wholly impotent, tends to prove that, though living in physical nature, he is not of her, that he is of more exalted parentage, and belongs to a higher order of existences than those born of her womb and submissive to her dictates.

There are, indeed, brute destroyers, beasts and birds and insects of prey—all animal life feeds upon, and, of course, destroys other life,—but this destruction is balanced by compensations. It is, in fact, the very means by which the existence of one tribe of animals or of vegetables is secured against being smothered by the encroachments of another; and the reproductive powers of species, which serve as the food of others, are always proportioned to the demand they are destined to supply. Man pursues his victims with reckless destructiveness; and, while the sacrifice of life by the lower animals is limited by the cravings of appetite, he unsparingly persecutes, even to extirpation, thousands of organic forms which he cannot consume.

The earth was not, in its natural condition, completely adapted to the use of man, but only to the sustenance of wild animals and wild vegetation. These live, multiply their kind in just proportion, and attain their perfect measure of strength and beauty, without producing or requiring any change in the natural arrangements of surface, or in each other's spontaneous tendencies, except such mutual repression of excessive increase as may prevent the extirpation of one species by the encroachments of another. In short, without man, lower animal and spontaneous vegetable life would have been constant in type, distribution, and proportion, and the physical geography of the earth would have remained undisturbed for indefinite periods, and been subject to revolution only from possible, unknown cosmical causes, or from geological action.

But man, the domestic animals that serve him, the field and garden plants the products of which supply him with food and clothing, cannot subsist and rise to the full development of their higher properties, unless brute and unconscious nature be effectually combated, and, in a great degree, vanquished by human art. Hence, a certain measure of transformation of terrestrial surface, of suppression of natural, and stimulation of artificially modified productivity becomes necessary. This measure man has unfortunately exceeded. He has felled the forests whose network of fibrous roots bound the mould to the rocky skeleton of the earth; but had he allowed here and there a belt of woodland to reproduce itself by spontaneous propagation, most of the mischiefs which his reckless destruction of the natural protection of the soil has occasioned would have been averted. He has broken up the mountain reservoirs, the percolation of whose waters through unseen channels supplied the fountains that refreshed his cattle and fertilized his fields; but he has neglected to maintain the cisterns and the canals of irrigation which a wise antiquity had constructed to neutralize the consequences of its own imprudence. While he has torn the thin glebe which confined the light earth of extensive plains, and has destroyed the fringe of semi-aquatic plants which skirted the coast and checked the drifting of the sea sand, he has failed to prevent the spreading of the dunes by clothing them with artificially propagated vegetation. He has ruthlessly warred on all the tribes of animated nature whose spoil he could convert to his own uses, and he has not protected the birds which prey on the insects most destructive to his own harvests.

Purely untutored humanity, it is true, interferes comparatively little with the arrangements of nature, and the destructive agency of man becomes more and more energetic and unsparing as he advances in civilization, until the impoverishment, with which his exhaustion of the natural resources of the soil is threatening him, at last awakens him to the necessity of preserving what is left, if not of restoring what has been wantonly wasted. The wandering savage grows no cultivated vegetable, fells no forest, and extirpates no useful plant, no noxious weed. If his skill in the chase enables him to entrap numbers of the animals on which he feeds, he compensates this loss by destroying also the lion, the tiger, the wolf, the otter, the seal, and the eagle, thus indirectly protecting the feebler quadrupeds and fish and fowls, which would otherwise become the booty of beasts and birds of prey. But with stationary life, or rather with the pastoral state, man at once commences an almost indiscriminate warfare upon all the forms of animal and vegetable existence around him, and as he advances in civilization, he gradually eradicates or transforms every spontaneous product of the soil he occupies. . . .

The ravages committed by man subvert the relations and destroy the balance which nature had established between her organized and her inorganic creations; and she avenges herself upon the intruder, by letting loose upon her defaced provinces destructive energies hitherto kept in check by organic forces destined to be his best auxiliaries, but which he has unwisely dispersed and driven from the field of action. When the forest is gone, the great reservoir of moisture stored up in its vegetable mould is evaporated, and returns only in deluges of rain to wash away the parched dust into which that mould has been converted. The well-wooded and humid hills are turned to ridges of dry rock, which encumbers the low grounds and chokes the watercourses with its debris, and—except in countries favored with an equable distribution of rain through the seasons, and a moderate and regular inclination of surface—the whole earth, unless rescued by human art from the physical degradation to which it tends, becomes an assemblage of bald mountains, of barren, turfless hills, and of swampy and malarious plains. There are parts of Asia Minor, of Northern Africa, of Greece, and even of Alpine Europe, where the operation of causes set in action by man has brought the face of the earth to a desolation almost as complete as that of the moon; and though, within that brief space of time which we call "the historical period," they are known to have been covered with luxuriant woods, verdant pastures, and fertile meadows, they are now too far deteriorated to be reclaimable by man, nor can they become again fitted for human use, except through great geological changes, or other mysterious influences or agencies of which we have no present knowledge, and over which we have no prospective control. The earth is fast becoming an unfit home for its noblest inhabitant, and another era of equal human crime and human improvidence, and of like duration with that through which traces of that crime and that improvidence extend, would reduce it to such a condition of impoverished productiveness, of shattered surface, of climatic excess, as to threaten the depravation, barbarism, and perhaps even extinction of the species.

Physical Improvement

True, there is a partial reverse to this picture. On narrow theatres, new forests have been planted; inundations of flowing streams restrained by heavy walls of masonry and other constructions; torrents compelled to aid, by depositing the slime with which they are charged, in filling up lowlands, and raising the level of morasses which their own overflows had created; ground submerged by the encroachments of the ocean, or exposed to be covered by its tides, has been rescued from its dominion by diking, swamps and even lakes have been drained, and their beds brought within the domain of agricultural industry; drifting coast dunes have been checked and made productive by plantation; seas and inland waters have been repeopled with fish, and even the sands of the Sahara have been fertilized by artesian fountains. These achievements are more glorious than the proudest triumphs of war, but, thus far, they give but faint hope that we shall yet make full atonement for our spendthrift waste of the bounties of nature. . . .

I have remarked that the effects of human action on the forms of the earth's surface could not always be distinguished from those resulting from geological causes, and there is also much uncertainty in respect to the precise influence of the clearing and cultivating of the ground, and of other rural operations, upon climate. It is disputed whether either the mean or the extremes of temperature, the periods of the seasons, or the amount or distribution of precipitation and of evaporation, in any country whose annals are known, have undergone any change during the historical period. It is, indeed, impossible to doubt that many of the operations of the pioneer settler tend to produce great modifications in atmospheric humidity, temperature, and electricity; but we are at present unable to determine how far one set of effects is neutralized by another, or compensated by unknown agencies. This question scientific research is inadequate to solve, for want of the necessary data; but well conducted observation, in regions now first brought under the occupation of man, combined with such historical evidence as still exists, may be expected at no distant period to throw much light on this subject.

Australia is, perhaps, the country from which we have a right to expect the fullest elucidation of these difficult and disputable problems. Its colonization did not commence until the physical sciences had become matter of almost universal attention, and is, indeed, so recent that the memory of living men embraces the principal epochs of its history; the peculiarities of its fauna, its flora, and its geology are such as to have excited for it the liveliest interest of the votaries of natural science; its mines have given its people the necessary wealth for procuring the means of instrumental observation, and the leisure required for the pursuit of scientific research; and large tracts of virgin forest and natural meadow are rapidly passing under the control of civilized man. Here, then, exist greater facilities and stronger motives for the careful study of the topics in question than have ever been found combined in any other theatre of European colonization.

In North America, the change from the natural to the artificial condition of terrestrial surface began about the period when the most important instruments of meteorological observation were invented. The first settlers in the territory now constituting the United States and the British American provinces had other things to do than to tabulate barometrical and thermometrical readings, but there remain some interesting physical records from the early days of the colonies, and there is still an immense extent of North American soil where the industry and the folly of man have as yet produced little appreciable change. Here, too, with the present increased facilities for scientific observation, the future effects, direct and contingent, of man's labors, can be measured, and such precautions taken in those rural processes which we call improvements, as to mitigate evils, perhaps, in some degree, inseparable from every attempt to control the action of natural laws. . . .

The geological, hydrographical, and topographical surveys which almost every general and even local government of the civilized world is carrying on, are making yet more important contributions to our stock of geographical and general physical knowledge, and, within a comparatively short space, there will be an accumulation of well established constant and historical facts, from which we can safely reason upon all the relations of action and reaction between man and external nature.

But we are, even now, breaking up the floor and wainscoting and doors and window frames of our dwelling, for fuel to warm our bodies and seethe our pottage, and the world cannot afford to wait till the slow and sure progress of exact science has taught it a better economy. Many practical lessons have been learned by the common observation of unschooled men; and the teachings of simple experience, on topics where natural philosophy has scarcely yet spoken, are not to be despised.

Critical Thinking

1. What does it mean to say that the earth was given to man for usufruct alone?
2. Compare the concept of "usufruct" to that of sustainability.
3. What, according to George Perkins Marsh, accounts for the decline of human population and activity over time in many regions?

Hetch Hetchy Valley

John Muir

John Muir (1838–1914) was, perhaps, America's most influential naturalist. In 1849 he and his family came to the United States from Scotland, where he was born. After abandoning plans to attend medical school, he worked at odd jobs and established a reputation as an inventor. In 1867 he decided to tramp the wilderness as an avocation. He took a 1,000-mile walk from Indiana to Florida, sailed to Cuba and then through the Isthmus of Panama, and arrived in San Francisco, California, in 1868. He eventually settled in the summits of the Sierra Nevada.

Muir referred to his years of wandering across meadows, around lakes, and up mountains as a lifelong education in the "University of the Wilderness." The detailed observations that he recorded daily were the source of both the rapturous, poetic descriptions and the scientific theories that he published. In 1871 Muir proposed that Yosemite Valley had been formed by glacial action. This theory, now widely accepted, was ridiculed by the geologists of the time. The establishment of Yosemite National Park in 1890 was largely a result of Muir's inspirational descriptions of this remarkable wonderland.

Muir was a vociferous proponent of an ecocentric rather than an ethnocentric philosophical perspective. He challenged the prevailing view of most of the conservationists of his time, who tempered their respect for nature with the multiple-use concept that gave primacy to human needs and appetites. The preservationist movement, supported by Muir, saw the need to set aside wilderness areas where no commercial or industrial activity would be permitted. The first such "primitive areas" were established by an administrative fiat of the U.S. Forest Service in the 1920s, a decade after Muir's death, but they were not officially protected by federal law until the passage of the Wilderness Act in 1964.

The most bitter disappointment of Muir's life was his failure to prevent the flooding and submerging of the Hetch Hetchy Valley in Yosemite National Park by a dam built to supply water and electricity to San Francisco. Muir focuses on the Hetch Hetchy Valley in the following selection from his book *The Mountains of California* (Houghton Mifflin, 1916). In it, Muir displays his mastery in conveying an image of the wonders of nature and his animosity toward those who would sacrifice such wonders in the interest of commercialism.

Key Concept: the primacy of nature and the case for wilderness preservation

*Y*osemite is so wonderful that we are apt to regard it as an exceptional creation, the only valley of its kind in the world; but Nature is not so poor as to have only one of anything. Several other yosemites have been discovered in the Sierra that occupy the same relative positions on the range and were formed by the same forces in the same kind of granite. One of these, the Hetch Hetchy Valley, is in the Yosemite National Park, about twenty miles from Yosemite, and is easily accessible to all sorts of travelers by a road and trail that leaves the Big Oak Flat road at Bronson Meadows a few miles below Crane Flat, and to mountaineers by way of Yosemite Creek basin and the head of the middle fork of the Tuolumne.

It is said to have been discovered by Joseph Screech, a hunter, in 1850, a year before the discovery of the great Yosemite. After my first visit to it in the autumn of 1871, I have always called it the "Tuolumne Yosemite," for it is a wonderfully exact counterpart of the Merced Yosemite, not only in its sublime rocks and waterfalls but in the gardens, groves and meadows of its flowery park-like floor. The floor of Yosemite is about four thousand feet above the sea; the Hetch Hetchy floor about thirty-seven hundred feet. And as the Merced River flows through Yosemite, so does the Tuolumne through Hetch Hetchy. The walls of both are of gray granite, rise abruptly from the floor, are sculptured in the same style and in both every rock is a glacier monument.

Standing boldly out from the south wall is a strikingly picturesque rock called by the Indians, Kolana, the outermost of a group twenty-three hundred feet high, corresponding with the Cathedral Rocks of Yosemite both in relative position and form. On the opposite side of the Valley, facing Kolana, there is a counterpart of El Capitan that rises sheer and plain to a height of eighteen hundred feet, and over its massive brow flows a stream which makes the most graceful fall I have ever seen. From the edge of the cliff to the top of an earthquake talus it is perfectly free in the air for a thousand feet before it is broken into cascades among talus boulders. It is in all its glory in June, when the snow is melting fast, but fades and vanishes toward the end of summer. The only fall I know with which it may fairly be compared is the Yosemite Bridal Veil; but it excels even that favorite fall both in height and airy-fairy beauty and behavior. Lowlanders are apt to suppose that mountain streams in their wild career over cliffs lose control of themselves and tumble in a noisy chaos of mist and spray. On the contrary, on no part of their travels are they more harmonious and self-controlled. Imagine yourself in Hetch Hetchy on a sunny day in June, standing waist-deep in grass and flowers (as I have often stood), while the great pines sway dreamily with scarcely perceptible motion. Looking northward across the Valley you see a plain, gray granite cliff rising abruptly out of the gardens and groves to a height of eighteen hundred feet, and in front of it Tueeulala's silvery scarf burning with irised sun-fire. In the first white outburst at the head there is abundance of visible energy, but it is speedily hushed and concealed in divine repose, and its tranquil progress to the base of the cliff is like that of a downy feather in a still room. Now observe the fineness and marvelous distinctness of the various sun-illumined fabrics into which the water is woven; they sift and float from form to form down the face of that grand gray rock in so leisurely and unconfused a manner that you can examine their texture, and patterns and tones of color as you would a piece of embroidery held in the hand. Toward the top of the fall you see groups of booming, comet-like masses, their solid, white heads separate, their tails like combed silk interlacing among delicate gray and purple shadows, ever forming and dissolving, worn out by friction in their rush through the air. Most of these vanish a few hundred feet below the summit, changing to varied forms of cloud-like drapery. Near the bottom the width of the fall has increased from about twenty-five feet to a hundred feet. Here it is composed of yet finer tissues, and is still without a trace of disorder—air, water and sunlight woven into stuff that spirits might wear.

So fine a fall might well seem sufficient to glorify any valley; but here, as in Yosemite, Nature seems in nowise moderate, for a short distance to the eastward of Tueeulala booms and thunders the great Hetch Hetchy Fall, Wapama, so near that you have both of them in full view from the same standpoint. It is the counterpart of the Yosemite Fall, but has a much greater volume of water, is about seventeen hundred feet in height, and appears to be nearly vertical, though considerably inclined, and

is dashed into huge outbounding bosses of foam on projecting shelves and knobs. No two falls could be more unlike—Tueeulala out in the open sunshine descending like thistledown; Wapama in a jagged, shadowy gorge roaring and thundering, pounding its way like an earthquake avalanche.

Besides this glorious pair there is a broad, massive fall on the main river a short distance above the head of the Valley. Its position is something like that of the Vernal in Yosemite, and its roar as it plunges into a surging trout-pool may be heard a long way, though it is only about twenty feet high. On Rancheria Creek, a large stream, corresponding in position with the Yosemite Tenaya Creek, there is a chain of cascades joined here and there with swift flashing plumes like the one between the Vernal and Nevada Falls, making magnificent shows as they go their glacier-sculptured way, sliding, leaping, hurrahing, covered with crisp clashing spray made glorious with sifting sunshine. And besides all these a few small streams come over the walls at wide intervals, leaping from ledge to ledge with bird-like song and watering many a hidden cliff-garden and fernery, but they are too unshowy to be noticed in so grand a place. . . .

It appears. . . that Hetch Hetchy Valley, far from being a plain, common, rockbound meadow, as many who have not seen it seem to suppose, is a grand landscape garden, one of Nature's rarest and most precious mountain temples. As in Yosemite, the sublime rocks of its walls seem to glow with life, whether leaning back in repose or standing erect in thoughtful attitudes, giving welcome to storms and calms alike, their brows in the sky, their feet set in the groves and gay flowery meadows, while birds, bees, and butterflies help the river and waterfalls to stir all the air into music—things frail and fleeting and types of permanence meeting here and blending, just as they do in Yosemite, to draw her lovers into close and confiding communion with her.

Sad to say, this most precious and sublime feature of the Yosemite National Park, one of the greatest of all our natural resources for the uplifting joy and peace and health of the people, is in danger of being dammed and made into a reservoir to help supply San Francisco with water and light, thus flooding it from wall to wall and burying its gardens and groves one or two hundred feet deep. This grossly destructive commercial scheme has long been planned and urged (though water as pure and abundant can be got from sources outside of the people's park, in a dozen different places), because of the comparative cheapness of the dam and of the territory which it is sought to divert from the great uses to which it was dedicated in the Act of 1890 establishing the Yosemite National Park.

The making of gardens and parks goes on with civilization all over the world, and they increase both in size and number as their value is recognized. Everybody needs beauty as well as bread, places to play in and pray in, where Nature may heal and cheer and give strength to body and soul alike. This natural beauty-hunger is made manifest in the little window-sill gardens of the poor, though perhaps only a geranium slip in a broken cup, as well as in the carefully tended rose and lily gardens of the rich, the thousands of spacious city parks and botanical gardens, and in our magnificent national parks—the Yellowstone, Yosemite, Sequoia, etc.—Nature's sublime wonderlands, the admiration and joy of the world. Nevertheless, like anything else worth while, from the very beginning, however well guarded, they have always been subject to attack by despoiling gain-seekers and mischief-makers of every degree from Satan to Senators, eagerly trying to make everything immediately and selfishly commercial, with schemes disguised in smug-smiling philanthropy, industriously, shampiously crying, "Conservation, conservation, panutilization," that man and beast may be fed and the dear Nation made great. Thus long ago a few enterprising merchants utilized the Jerusalem temple as a place of business instead of a place of prayer, changing money, buying and selling cattle and sheep and doves; and earlier still, the first forest reservation, including only one tree, was likewise despoiled. Ever since the establishment of the Yosemite National Park, strife has been going on around its borders and I suppose this will go on as part of the universal battle between right and wrong, however much its boundaries may be shorn, or its wild beauty destroyed.

The first application to the Government by the San Francisco Supervisors for the commercial use of Lake Eleanor and the Hetch Hetchy Valley was made in 1903, and on December 22 of that year it was denied by the Secretary of the Interior, Mr. Hitchcock, who truthfully said:

> Presumably the Yosemite National Park was created such by law because of the natural objects of varying degrees of scenic importance located within its boundaries, inclusive alike of its beautiful small lakes, like Eleanor, and its majestic wonders, like Hetch Hetchy and Yosemite Valley. It is the aggregation of such natural scenic features that makes the Yosemite Park a wonderland which the Congress of the United States sought by law to reserve for all coming time as nearly as practicable in the condition fashioned by the hand of the Creator—a worthy object of national pride and a source of healthful pleasure and rest for the thousands of people who may annually sojourn there during the heated months.

. . . That any one would try to destroy such a place seems incredible; but sad experience shows that there are people good enough and bad enough for anything. The proponents of the dam scheme bring forward a lot of bad arguments to prove that the only righteous thing to do with the people's parks is to destroy them bit by bit as they are able. Their arguments are curiously like those of the devil, devised for the destruction of the first garden—so much of the very best Eden fruit going to waste; so much of the best Tuolumne water and Tuolumne scenery going to waste. Few of their statements are even partly true, and all are misleading.

Thus, Hetch Hetchy, they say, is a "low-lying meadow." On the contrary, it is a high-lying natural landscape garden, as the photographic illustrations show.

"It is a common minor feature, like thousands of others." On the contrary it is a very uncommon feature; after Yosemite, the rarest and in many ways the most important in the National Park.

"Damming and submerging it one hundred and seventy-five feet deep would enhance its beauty by forming a crystal-clear lake." Landscape gardens, places of recreation and worship, are never made beautiful by destroying and burying them. The beautiful sham lake, forsooth, would be only an eyesore, a dismal blot on the landscape, like many others to be seen in the Sierra. For, instead of keeping it at the same level all the year, allowing Nature centuries of time to make new shores, it would, of course, be full only a month or two in the spring, when the snow is melting fast; then it would be gradually drained, exposing the slimy sides of the basin and shallower parts of the bottom, with the gathered drift and waste, death and decay of the upper basins, caught here instead of being swept on to decent natural burial along the banks of the river or in the sea. Thus the Hetch Hetchy dam-lake would be only a rough imitation of a natural lake for a few of the spring months, an open sepulcher for the others.

"Hetch Hetchy water is the purest of all to be found in the Sierra, unpolluted, and forever unpollutable." On the contrary, excepting that of the Merced below Yosemite, it is less pure than that of most of the other Sierra streams, because of the sewerage of camp-grounds draining into it, especially of the Big Tuolumne Meadows camp-ground, occupied by hundreds of tourists and mountaineers, with their animals, for months every summer, soon to be followed by thousands from all the world.

These temple destroyers, devotees of ravaging commercialism, seem to have a perfect contempt for Nature, and, instead of lifting their eyes to the God of the mountains, lift them to the Almighty Dollar.

Dam Hetch Hetchy! As well dam for watertanks the people's cathedrals and churches, for no holier temple has ever been consecrated by the heart of man.

Critical Thinking

1. How does John Muir's ecocentric approach to conservation differ from anthropocentric and "multiple use" approaches?
2. What is the difference between preservationism and conservationism?

Principles of Conservation

Gifford Pinchot

Conservation, or the wise use of natural resources to meet human needs and desires, has always had more public and political support than more restrictive forms of nature preservation. Gifford Pinchot (1865–1946) was one of the founders and one of the most effective leaders of the U.S. conservation movement that began at the end of the nineteenth century. He served as chief of the U.S. Forest Service from 1898 to 1910. In 1902 President Theodore Roosevelt directed Pinchot to lead a campaign to restore the forest and water resources, which had been ravaged by the environmentally destructive practices of large lumbering, mining, and agricultural enterprises. In 1905 Roosevelt established a Bureau of Forestry within the U.S. Department of Agriculture, enhancing Pinchot's ability to defend public lands against the destructive plans of private interests. Pinchot also established Yale University's School of Forestry, where he was a professor from 1903 to 1936. An effective politician, Pinchot helped found the Progressive Party, and he served two terms as governor of Pennsylvania, 1923–1927 and 1931–1935.

In the following selection from his book *The Fight for Conservation* (Doubleday, 1910), Pinchot details the principles that guided his actions. It is clear that his goal was to maximize the value of resources for human exploitation rather than to preserve wilderness.

Key Concept: the conservation of resources for human use

The principles which the word Conservation has come to embody are not many, and they are exceedingly simple. I have had occasion to say a good many times that no other great movement has ever achieved such progress in so short a time, or made itself felt in so many directions with such vigor and effectiveness, as the movement for the conservation of natural resources.

Forestry made good its position in the United States before the conservation movement was born. As a forester I am glad to believe that conservation began with forestry, and that the principles which govern the Forest Service in particular and forestry in general are also the ideas that control conservation.

The first idea of real foresight in connection with natural resources arose in connection with the forest. From it sprang the movement which gathered impetus until it culminated in the great Convention of Governors at Washington in May, 1908. Then came the second official meeting of the National Conservation movement, December, 1908, in Washington. Afterward came the various gatherings of citizens in convention, come together to express their judgment on what ought to be done, and to contribute, as only such meetings can, to the formation of effective public opinion.

The movement so begun and so prosecuted has gathered immense swing and impetus. In 1907 few knew what Conservation meant. Now it has become a household word. While at first Conservation was supposed to apply only to forests, we see now that its sweep extends even beyond the natural resources.

The principles which govern the conservation movement, like all great and effective things, are simple and easily understood. Yet it is often hard to make the simple, easy, and direct facts about a movement of this kind known to the people generally.

The first great fact about conservation is that it stands for development. There has been a fundamental misconception that conservation means nothing but the husbanding of resources for future generations. There could be no more serious mistake. Conservation does mean provision for the future, but it means also and first of all the recognition of the right of the present generation to the fullest necessary use of all the resources with which this country is so abundantly blessed. Conservation demands the welfare of this generation first, and afterward the welfare of the generations to follow.

The first principle of conservation is development, the use of the natural resources now existing on this continent for the benefit of the people who live here now. There may be just as much waste in neglecting the development and use of certain natural resources as there is in their destruction. We have a limited supply of coal, and only a limited supply. Whether it is to last for a hundred or a hundred and fifty or a thousand years, the coal is limited in amount, unless through geological changes which we shall not live to see, there will never be any more of it than there is now. But coal is in a sense the vital essence of our civilization. If it can be preserved, if the life of the mines can be extended, if by preventing waste there can be more coal left in this country after we of this generation have made every needed use of this source of power, then we shall have deserved well of our descendants.

Conservation stands emphatically for the development and use of water-power now, without delay. It stands for the immediate construction of navigable waterways under a broad and comprehensive plan as assistants to the railroads. More coal and more iron are required to move a ton of freight by rail than by water, three to one. In every case and in every direction the conservation movement has development for its first principle, and at the very beginning of its work. The development of our natural resources and the fullest use of them for the present generation is the first duty of this generation. So much for development.

In the second place conservation stands for the prevention of waste. There has come gradually in this country an understanding that waste is not a good thing and that the attack on waste is an industrial necessity. I recall very well indeed how, in the early days of forest fires, they were considered simply and solely as acts of God, against which any opposition was hopeless and any attempt to control them not merely hopeless but childish. It was assumed that they came in the natural order of things, as inevitably as the seasons or the rising and setting of the sun. Today we understand that forest fires are wholly within the control of men. So we are coming in like manner to understand that the prevention of waste in all other directions is a simple matter of good business. The first duty of the human race is to control the earth it lives upon.

We are in a position more and more completely to say how far the waste and destruction of natural resources are to be allowed to go on and where they are to stop. It is curious that the effort to stop waste, like the effort to stop forest fires, has often been considered as a matter controlled wholly by economic law. I think there could be no greater mistake. Forest fires were allowed to burn long after the people had means to stop them. The idea that men were helpless in the face of them held long after the time had passed when the means of control were fully within our reach. It was the old story that "as a man thinketh, so is he"; we came to see that we could stop forest fires, and we found that the means had long been at hand. When at length we came to see that the control of logging in certain directions was profitable, we found it had long been possible. In all these matters of waste of natural resources, the education of the people to understand that they can stop the leakage comes before the actual stopping and after the means of stopping it have long been ready at our hands.

In addition to the principles of development and preservation of our resources there is a third principle. It is this: The natural resources must be developed and preserved for the benefit of the many, and not merely for the profit of a few. We are coming to understand in this country that public action for public benefit has a very much wider field to cover and a much larger part to play than was the case when there were resources enough for every one, and before certain constitutional provisions had given so tremendously strong a position to vested rights and property in general. . . .

The conservation idea covers a wider range than the field of natural resources alone. Conservation means the greatest good to the greatest number for the longest time. One of its great contributions is just this, that it has added to the worn and well-known phrase, "the greatest good to the greatest number," the additional words "for the longest time," thus recognizing that this nation of ours must be made to endure as the best possible home for all its people.

Conservation advocates the use of foresight, prudence, thrift, and intelligence in dealing with public matters, for the same reasons and in the same way that we each use foresight, prudence, thrift, and intelligence in dealing with our own private affairs. It proclaims the right and duty of the people to act for the benefit of the people. Conservation demands the application of common-sense to the common problems for the common good.

The principles of conservation thus described—development, preservation, the common good—have a general application which is growing rapidly wider. The development of resources and the prevention of waste and loss, the protection of the public interests, by foresight, prudence, and the ordinary business and home-making virtues, all these apply to other things as well as to the natural resources. There is, in fact, no interest of the people to which the principles of conservation do not apply.

The conservation point of view is valuable in the education of our people as well as in forestry; it applies to the body politic as well as to the earth and its minerals. A municipal franchise is as properly within its sphere as a franchise for water-power. The same point of view governs in both. It applies as much to the subject of good roads as to waterways, and the training of our people in citizenship is as germane to it as the productiveness of the earth.

The application of common-sense to any problem for the Nation's good will lead directly to national efficiency wherever applied. In other words, and that is the burden of the message, we are coming to see the logical and inevitable outcome that these principles, which arose in forestry and have their bloom in the conservation of natural resources, will have their fruit in the increase and promotion of national efficiency along other lines of national life.

Critical Thinking

1. In what sense, according to Gifford Pinchot, is conservation related to development?
2. Is Gifford Pinchot's version of conservation compatible with strip mining? Cutting old-growth forests for timber? Tourism?

A Sand County Almanac

Aldo Leopold

The struggle between those who advocate the conservationist multiple-use concept of environmental protection and proponents of preserving nature for its own sake continues to this day. One of the most influential American preservationists of the first half of the twentieth century was Aldo Leopold (1887–1948). As an officer of the U.S. Forest Service in the 1920s, Leopold became concerned about the failure of management practices in the national forests to adequately protect the natural environment from the effects of commercial activities. Along with fellow officer Robert Marshall, he helped establish 70 "primitive areas" where all development was prohibited. In 1933 Leopold was appointed to chair of game management, a position that was created for him at the University of Wisconsin. In 1935 Leopold and Marshall founded the Wilderness Society, which was to lead the protracted struggle for wilderness preservation. Success was finally achieved—16 years after Leopold's death—with the approval of the Federal Wilderness Act in 1964. Although Leopold was a vociferous advocate of the intrinsic worth of all living things, he supported a strong role for human beings in the management and protection of wild lands.

  A Sand County Almanac: And Sketches Here and There (Oxford University Press, 1949), from which the following selection was taken, is a collection of Leopold's lyrical, philosophical writings about nature. In the section under the heading "Thinking Like a Mountain," Leopold reflects on his early enthusiasm for killing wolves as indicative of human ignorance about the ecological interdependence that sustains a mountain ecosystem. Leopold's land ethic, which he discusses in the final section of *A Sand County Almanac*, is his most quoted and influential writing. Leopold held that ethical considerations, which historically had encompassed only the relationships among human beings, must be extended to include interactions of humans with the animate and inanimate components of the natural world.

Key Concept: an ethical relationship between humans and the land

Thinking Like a Mountain

A deep chesty bawl echoes from rimrock to rimrock, rolls down the mountain, and fades into the far blackness of the night. It is an outburst of wild defiant sorrow, and of contempt for all the adversities of the world.

  Every living thing (and perhaps many a dead one as well) pays heed to that call. To the deer it is a reminder of the way of all flesh, to the pine a forecast of midnight scuffles and of blood upon the snow, to the coyote a promise of gleanings to come, to the cowman a threat of red ink at the bank, to the hunter a challenge of fang against bullet. Yet behind these obvious and immediate hopes and fears there lies a deeper meaning, known only to the mountain itself. Only the mountain has lived long enough to listen objectively to the howl of a wolf.

  Those unable to decipher the hidden meaning know nevertheless that it is there, for it is felt in all wolf country, and distinguishes that country from all other land.

It tingles in the spine of all who hear wolves by night, or who scan their tracks by day. Even without sight or sound of wolf, it is implicit in a hundred small events: the midnight whinny of a pack horse, the rattle of rolling rocks, the bound of a fleeing deer, the way shadows lie under the spruces. Only the ineducable tyro can fail to sense the presence or absence of wolves, or the fact that mountains have a secret opinion about them.

  My own conviction on this score dates from the day I saw a wolf die. We were eating lunch on a high rimrock, at the foot of which a turbulent river elbowed its way. We saw what we thought was a doe fording the torrent, her breast awash in white water. When she climbed the bank toward us and shook out her tail, we realized our error: it was a wolf. A half-dozen others, evidently grown pups, sprang from the willows and all joined in a welcoming mêlée of wagging tails and playful maulings. What was literally a pile of wolves writhed and tumbled in the center of an open flat at the foot of our rimrock.

In those days we had never heard of passing up a chance to kill a wolf. In a second we were pumping lead into the pack, but with more excitement than accuracy: how to aim a steep downhill shot is always confusing. When our rifles were empty, the old wolf was down, and a pup was dragging a leg into impassable slide-rocks.

We reached the old wolf in time to watch a fierce green fire dying in her eyes. I realized then, and have known ever since, that there was something new to me in those eyes—something known only to her and to the mountain. I was young then, and full of trigger-itch; I thought that because fewer wolves meant more deer, that no wolves would mean hunters' paradise. But after seeing the green fire die, I sensed that neither the wolf nor the mountain agreed with such a view.

Since then I have lived to see state after state extirpate its wolves. I have watched the face of many a newly woffless mountain, and seen the south-facing slopes wrinkle with a maze of new deer trails. I have seen every edible bush and seedling browsed, first to anaemic desuetude, and then to death. I have seen every edible tree defoliated to the height of a saddlehorn. Such a mountain looks as if someone had given God a new pruning shears, and forbidden Him all other exercise. In the end the starved bones of the hoped-for deer herd, dead of its own too-much, bleach with the bones of the dead sage, or molder under the high-lined junipers.

I now suspect that just as a deer herd lives in mortal fear of its wolves, so does a mountain live in mortal fear of its deer. And perhaps with better cause, for while a buck pulled down by wolves can be replaced in two or three years, a range pulled down by too many deer may fail of replacement in as many decades.

So also with cows. The cowman who cleans his range of wolves does not realize that he is taking over the wolf's job of trimming the herd to fit the range. He has not learned to think like a mountain. Hence we have dustbowls, and rivers washing the future into the sea.

We all strive for safety, prosperity, comfort, long life, and dullness. The deer strives with his supple legs, the cowman with trap and poison, the statesman with pen, the most of us with machines, votes, and dollars, but it all comes to the same thing: peace in our time. A measure of success in this is all well enough, and perhaps is a requisite to objective thinking, but too much safety seems to yield only danger in the long run. Perhaps this is behind Thoreau's dictum: In wildness is the salvation of the world. Perhaps this is the hidden meaning in the howl of the wolf, long known among mountains, but seldom perceived among men. . . .

The Land Ethic

When god-like Odysseus returned from the wars in Troy, he hanged all on one rope a dozen slave-girls of his household whom he suspected of misbehavior during his absence.

This hanging involved no question of propriety. The girls were property. The disposal of property was then, as now, a matter of expediency, not of right and wrong.

Concepts of right and wrong were not lacking from Odysseus' Greece: witness the fidelity of his wife through the long years before at last his black-prowed galleys clove the wine-dark seas for home. The ethical structure of that day covered wives, but had not yet been extended to human chattels. During the three thousand years which have since elapsed, ethical criteria have been extended to many fields of conduct, with corresponding shrinkages in those judged by expediency only.

The Ethical Sequence

This extension of ethics, so far studied only by philosophers, is actually a process in ecological evolution. Its sequences may be described in ecological as well as in philosophical terms. An ethic, ecologically, is a limitation on freedom of action in the struggle for existence. An ethic, philosophically, is a differentiation of social from anti-social conduct. These are two definitions of one thing. The thing has its origin in the tendency of interdependent individuals or groups to evolve modes of co-operation. The ecologist calls these symbioses. Politics and economics are advanced symbioses in which the original free-for-all competition has been replaced, in part, by co-operative mechanisms with an ethical content.

The complexity of co-operative mechanisms has increased with population density, and with the efficiency of tools. It was simpler, for example, to define the anti-social uses of sticks and stones in the days of the mastodons than of bullets and billboards in the age of motors.

The first ethics dealt with the relation between individuals; the Mosaic Decalogue is an example. Later accretions dealt with the relation between the individual and society. The Golden Rule tries to integrate the individual to society; democracy to integrate social organization to the individual.

There is as yet no ethic dealing with man's relation to land and to the animals and plants which grow upon it. Land, like Odysseus' slave-girls, is still property. The land-relation is still strictly economic, entailing privileges but not obligations.

The extension of ethics to this third element in human environment is, if I read the evidence correctly, an evolutionary possibility and an ecological necessity. It is the third step in a sequence. The first two have already been taken. Individual thinkers since the days of Ezekiel and Isaiah have asserted that the despoliation of land is not only inexpedient but wrong. Society, however, has not yet affirmed their belief. I regard the present conservation movement as the embryo of such an affirmation.

An ethic may be regarded as a mode of guidance for meeting ecological situations so new or intricate, or involving such deferred reactions, that the path of social expediency is not discernible to the average individual. Animal instincts are modes of guidance for the individual in meeting such situations. Ethics are possibly a kind of community instinct in-the-making.

The Community Concept

All ethics so far evolved rest upon a single premise: that the individual is a member of a community of interdependent parts. His instincts prompt him to compete for his place in that community, but his ethics prompt him also to co-operate (perhaps in order that there may be a place to compete for).

The land ethic simply enlarges the boundaries of the community to include soils, waters, plants, and animals, or collectively: the land. . . .

The Outlook

It is inconceivable to me that an ethical relation to land can exist without love, respect, and admiration for land, and a high regard for its value. By value, I of course mean something far broader than mere economic value; I mean value in the philosophical sense.

Perhaps the most serious obstacle impeding the evolution of a land ethic is the fact that our educational and economic system is headed away from, rather than toward, an intense consciousness of land. Your true modern is separated from the land by many middlemen, and by innumerable physical gadgets. He has no vital relation to it; to him it is the space between cities on which crops grow. Turn him loose for a day on the land, and if the spot does not happen to be a golf links or a 'scenic' area, he is bored stiff. If crops could be raised by hydroponics instead of farming, it would suit him very well. Synthetic substitutes for wood, leather, wool, and other natural land products suit him better than the originals. In short, land is something he has 'outgrown.'

Almost equally serious as an obstacle to a land ethic is the attitude of the farmer for whom the land is still an adversary, or a taskmaster that keeps him in slavery. Theoretically, the mechanization of farming ought to cut the farmer's chains, but whether it really does is debatable.

One of the requisites for an ecological comprehension of land is an understanding of ecology, and this is by no means co-extensive with 'education'; in fact, much higher education seems deliberately to avoid ecological concepts. An understanding of ecology does not necessarily originate in courses bearing ecological labels; it is quite as likely to be labeled geography, botany, agronomy, history, or economics. This is as it should be, but whatever the label, ecological training is scarce.

The case for a land ethic would appear hopeless but for the minority which is in obvious revolt against these 'modern' trends.

The 'key-log' which must be moved to release the evolutionary process for an ethic is simply this: quit thinking about decent land-use as solely an economic problem. Examine each question in terms of what is ethically and esthetically right, as well as what is economically expedient. A thing is right when it tends to preserve the integrity, stability, and beauty of the biotic community. It is wrong when it tends otherwise.

It of course goes without saying that economic feasibility limits the tether of what can or cannot be done for land. It always has and it always will. The fallacy the economic determinists have tied around our collective neck, and which we now need to cast off, is the belief that economics determines *all* land-use. This is simply not true. An innumerable host of actions and attitudes, comprising perhaps the bulk of all land relations, is determined by the land-users' tastes and predilections, rather than by his purse. The bulk of all land relations hinges on investments of time, forethought, skill, and faith rather than on investments of cash. As a land-user thinketh, so is he.

I have purposely presented the land ethic as a product of social evolution because nothing so important as an ethic is ever 'written.' Only the most superficial student of history supposes that Moses 'wrote' the Decalogue; it evolved in the minds of a thinking community, and Moses wrote a tentative summary of it for a 'seminar.' I say tentative because evolution never stops.

The evolution of a land ethic is an intellectual as well as emotional process. Conservation is paved with good intentions which prove to be futile, or even dangerous, because they are devoid of critical understanding either of the land, or of economic land-use. I think it is a truism that as the ethical frontier advances from the individual to the community, its intellectual content increases.

The mechanism of operation is the same for any ethic: social approbation for right actions: social disapproval for wrong actions.

By and large, our present problem is one of attitudes and implements. We are remodeling the Alhambra with a steam-shovel, and we are proud of our yardage. We shall hardly relinquish the shovel, which after all has many good points, but we are in need of gentler and more objective criteria for its successful use.

Critical Thinking

1. What is the basic lesson of Aldo Leopold's "Thinking Like a Mountain"?
2. How is the "land ethic" related to ordinary ethical systems?

Fundamental Causes of Environmental Problems

Selection 5
PAUL S. MARTIN, from "Prehistoric Overkill," in Paul S. Martin and Richard G. Klein, *Quaternary Extinctions: A Prehistoric Revolution* (University of Arizona Press, 1984)

Selection 6
LYNN WHITE, JR., from "The Historical Roots of Our Ecological Crisis," *Science* (March 10, 1967)

Selection 7
GARRETT HARDIN, from "The Tragedy of the Commons," *Science* (Vol. 162, 1968)

Learning Outcomes
After reading this unit, you should be able to:

1. Explain how even very primitive technologies can be used to cause environmental problems.
2. Explain how human attitudes toward the environment are shaped by human culture.
3. Explain why unrestrained population growth is inadvisable.
4. Explain how computer simulations can be used to forecast future possibilities.
5. Explain how computer simulations can be validated by historical data.

Prehistoric Overkill: The Global Model

Paul S. Martin

A great many people view today's environmental problems as a product of modern, industrial civilization, and certainly industry, technology, urbanization, and population growth are associated with air and water pollution, ozone depletion, global warming, fisheries exhaustion, and other problems. But environmental historians have tracked problems of desertification, erosion, soil salinization, and more as far back as the dawn of agriculture and the first cities. Clearly humans do not need modern technology to damage their environment. Just how much technology they do need is a different question, as is just how far back into the mists of time we can look and still see humans harming the environment that supports them.

Paul S. Martin is Emeritus Professor of Geosciences at the Desert Laboratory of the University of Arizona in Tucson. He is a major contributor to the "prehistoric overkill" hypothesis that human beings were the chief cause of extinctions of large animals during the last 50,000 years. The basic argument is that soon after human beings arrived in areas such as North America or Australia, the large game animals disappeared, either because humans killed (and presumably ate) them or because humans killed their prey. Critics, apparently unwilling to grant "primitive" people armed with stone-tipped spears and arrows enough potency to wipe out whole species, have argued that the disappearance of large animals was just coincidence, or due to diseases brought by humans and their domestic animals, or due to changes in climate. One critic, says Martin, seems to be motivated in his rejection of the idea by the belief that it makes Native Americans "look bad." Since Martin first broached the "prehistoric overkill" hypothesis in the 1960s, the evidence in its favor has accumulated, but it has remained controversial.

Key Concept: Human responsibility for extinctions

Toward the end of the ice age, in the last interglacial before the last great ice advance, the continents were much richer in large animals than they are today. Mammoth, mastodonts, giant ground sloths, and seventy other genera, a remarkable fauna of large mammals, ranged the Americas. Australia was home to more than a dozen genera of giant marsupials. Northern Eurasia featured woolly mammoth, straight-tusked elephant, rhinoceros, giant deer, bison, hippopotamus, and horse. New Zealand harbored giant flightless birds, the moas, rivaled by the elephant birds of Madagascar. Even Africa, whose modem fauna is often viewed as saturated with large mammals, was richer before the end of the last glacial episode by 15 to 20 percent. In America, the large mammals lost in the ice age did not disappear gradually. They endured the last as well as the previous glaciations only to disappear afterward, the last "revolution in the history of life." Armed with modern techniques of analysis, one turns to deposits of the late Pleistocene and the prospect of determining the cause of this prehistoric revolution.

Late Pleistocene sediments yielding bones of extinct megafauna are widespread. Outcrops of late Pleistocene age are commonly more accessible than those of earlier times; as a result many more fossil mammoths have been found than brontosaurs. Dating of the late Pleistocene fossil faunas can be attempted by various geochemical techniques, especially radiocarbon analysis, an essential tool in attempting correlations, now improved by the use of accelerators. Under favorable circumstances the ages of late Pleistocene sedimentary units can be determined to within less than one hundred years. Fossil pollen and associated plant and animal remains of living species disclose the natural environment at the time of the extinctions. In a few favorable cases frozen paunch contents or ancient coprolites preserved in dry caves reveal what some extinct beasts actually ate. Mummified carcasses and cave paintings may show what others looked like. In the Old World the bones found in Paleolithic hunters' camps suggest the importance of large mammals, living and extinct, as prey. Taphonomy, the study of post-mortem processes affecting fossils, may disclose the cause of local mortality. While complete life histories of the extinct animals are unattainable, helpful clues may be found in the habits of living relatives. Many opportunities to study the late Pleistocene extinctions remain unexploited.

Given the great interest among paleontologists in earlier geological revolutions, one might expect to find an especially keen interest in the last one. For various reasons, this has not been the case, perhaps because of the peculiar attributes of the late Pleistocene extinctions. Certainly the extinctions at the end of the Pleistocene are quantitatively dwarfed by those at the Cretaceous/Tertiary boundary and the Permian/Triassic boundary. Late Pleistocene sediments, deposited during the time of the last major episode of extinctions, are neglected by many geologists who seek older events. Conversely, and for the opposite reason, ecologists have often ignored the late Pleistocene megafauna, so unlike that of historic time. While archaeologists have pioneered the study of late Pleistocene deposits and chronology, they may abandon the field at the most interesting moment when ancient artifacts give way to culturally "sterile" sediments that yield only the bones of an extinct fauna. Fortunately, the disciplinary fences that "Balkanized" late Pleistocene research earlier in this century are crumbling and some superior modern studies of megafaunal bone beds have appeared in recent years, such as those of Missouri's mastodont deposits, South Dakota's mammoth Hot Springs, Wyoming's Natural Trap, New Zealand's moa deposits, and California's tar pits. . . .

A realistic global extinction model must account for differences in extinction intensity between continents of the Southern Hemisphere—Africa, Australia, and South America. Heavy losses occurred in Australia predating by thousands of years the time of Termination I. In contrast South American extinctions, also very heavy, coincided with Termination I. African generic extinctions were unusual in being heavier early rather than late in the Pleistocene; only relatively minor losses occurred from 10,000 to 12,000 years ago.

The point I raise is that the proxy data from ocean sediments indicate that climatic changes accompanying late Pleistocene deglaciation were not unique but occurred repeatedly throughout the Ice Age. Apart from Africa, continental extinctions of large land mammals are unremarkable except during the time of Termination I. Then they coincide with heavy losses in the New World and not in Africa, while postdating massive extinctions in Australia. The extinction pattern does not obviously track changes in climate throughout the Pleistocene. Furthermore, the extinctions vary in time and intensity within the late Pleistocene.

Unlike changes in climate, the global spread of prehistoric *Homo sapiens* was confined to the late Pleistocene. Prehistoric man's role in changing the face of the earth can conceivably include major depletions of fauna, an anthropogenic "overkill." Overkill is taken to mean human destruction of native fauna either by gradual attrition over

many thousands of years or suddenly in as little as a few hundred years or less. Sudden extinction following initial colonization of a land mass inhabited by animals especially vulnerable to the new human predator represents, in effect, a prehistoric faunal "blitzkrieg." A close look at the late Pleistocene extinction pattern will help establish ways in which the extinctions might be conceived of happening. A look at various land masses will establish whether extinction coincides with human invasion.

The Late Pleistocene
Extinction Pattern

The following eight attributes of late Pleistocene extinction seem especially noteworthy:

1. *Large mammals were decimated.* In North America thirty-three genera of large mammals disappeared in the last 100,000 years (or less), while South America lost even more. Not since the late Hemphillian, several million years before the Pleistocene, was there a comparable loss of large terrestrial mammals in America. Africa lost about eight genera in the last 100,000 years while Australia lost a total of perhaps nineteen genera of large vertebrates, not all of them mammals. Europe lost three genera by extinction and nine more by range shrinkage.

2. *Continental rats survived; island rats did not.* In the Pliocene and early Pleistocene many genera of small mammals were lost in continental North America. In the late Pleistocene there were significant changes in range and in body size of small mammals. However, the episode of late Pleistocene large mammal extinction was unaccompanied by equally heavy loss of small mammals. Oceanic islands are another matter. On the West Indies and oceanic islands in the Mediterranean small endemic genera of mammals experienced very heavy prehistoric loss.

3. *Large mammals survived best in Africa.* From continent to continent the loss of megafauna was highly variable. Many more genera of large mammals were lost from the late Pleistocene in America and Australia than from Africa. No late Pleistocene families were lost from Asia or Africa to match the New World loss of families and orders.

4. *Extinctions could be sudden.* In parts of New Zealand at the end of the Holocene giant birds, the moas, disappeared within 300 years or less. Radiocarbon dates on Shasta ground sloth dung from caves in the southwestern United States indicate extinction of different populations of ground sloths around 11,000 yr B.P. Evidently North American mammoth disappeared quite suddenly around the same time. Those extinct North American genera that can be dated by radiocarbon disappeared roughly between 15,000 and 8,000 years ago. It is not certain that any one genus actually died out before another.

5. *Regional extinctions were diachronous.* Moas in New Zealand and elephant birds in Madagascar were still alive eight millenia after the mammoths and mastodonts were gone from America. The latter were predeceased

by the mammoths of northern Europe and China. The European mammoths in turn survived the diprotodonts of Australia by many thousands of years. Since late Pleistocene extinctions occur on one land mass at a different time and intensity than on another, they are unlikely to be explained by a sudden extraterrestrial catastrophe *à la* the asteroid impact model of Alvarez et al. Late Pleistocene extinctions were time transgressive. . . .

6. *Extinctions occurred without replacement.* The terrestrial extinctions seen throughtout the late Pleistocene were not necessarily preceded or accompanied by significant faunal invasion. There was no mixing of long-separated or isolated faunas, sometimes claimed as the cause of geologic extinctions, unless man is considered to be the crucial element in the new community.

7. *Extinctions followed man's footsteps.* In America, Australia, and on oceanic islands, very few megafaunal extinctions of the late Pleistocene can be definitely shown to predate human arrival. Conversely, the survival into historic time of large, easily captured animals such as the dodo of Mauritius, the giant tortoises of Aldabra and the Galapagos, and Steller's sea cow on the Commander Islands of the Bering Sea, occurred only on remote islands undiscovered by or unoccupied by prehistoric people. The sequence of extinction during the last 100,000 years follows human dispersal, spreading out of Afro-Asia into other continents and finally reaching various oceanic islands.

8. *The archaeology of extinction* is *obscure.* In Eurasia and Africa, where comparatively few late Pleistocene extinctions occurred, the bones of large extinct mammals are commonly found in Paleolithic sites of various ages. In America where there are many late Pleistocene extinctions, very few archaeological sites have yielded bones of extinct.

Conclusions

Late Pleistocene extinctions can be studied in detail by judicious application of radiocarbon dating. In Africa large mammal extinctions were comparatively few, seven in the last 100,000 years; in Australia nineteen genera and more than fifty species disappeared within the last 40,000 years; in America more than seventy genera of large mammals disappeared, most, perhaps all, within the last 15,000 yr. B.P. Many terrestrial mammals, some birds, a few reptiles, and virtually no plants or invertebrates became extinct. All continents except Antarctica were affected, America and Australia to a much greater degree than Africa and Asia.

On the continents small- or medium-size animals survived the late Pleistocene with minor losses at worst. Genera of small mammals had suffered appreciable extinction in the early Pleistocene or Pliocene, at least in America and Europe. Small mammals and birds suffered severe losses in the last few thousand years only on oceanic islands. The subfossil faunas of certain oceanic islands disclose a much larger endemic fauna than has been anticipated by biogeographers.

Timing of late Pleistocene extinctions is diachronous. Australia was stricken before North and South America. The last mammoths of China and England died out before those of America. The animals which became extinct on the continents disappeared before the Holocene. Extinctions continued throughout the Holocene on oceanic islands, first on archipelagos close to the continents such as the West Indies and islands of the Mediterranean, and later on more remote islands such as Hawaii, New Zealand, and Madagascar. The lack of synchroneity between the extinctions on different continents and their variable intensity, for example, heavier in America than Africa, appears to eliminate as a cause any sudden extraterrestrial or cosmic catastrophe.

Regional models that correlate climatic change with the extinctions are often encountered; regional extinctions often coincide with changes in climate. Attempts at a synthetic treatment that incorporate the intensity, timing, and character of late Pleistocene extinction into a worldwide model related to climatic changes are more difficult. On a global scale the late Pleistocene extinction patterns appear to track the prehistoric movements or activities of *Homo sapiens* much more closely than any widely agreed-upon pattern of especially severe global climatic change in the late Pleistocene. The moderate loss of animals in Afro-Asia can be related to the gradual history of human spread which allowed an ecological equilibrium to develop. The loss of the great majority of large animals from North America, South America, and Australia can be related to sudden and severe human impact on these continents, which were previously unexposed to evolving hominids.

Oceanic island extinctions can be correlated closely with arrival of prehistoric seafarers. While they did not necessarily hunt all the small mammals or island birds and other animals to extinction, the side effects of the arrival of prehistoric colonizers were severe and involved fire, habitat destruction, and the introduction of an alien fauna. Most native endemic fauna succumbed rapidly at the hands of prehistoric people.

The model of human impact of "overkill" as a unique cause of late Pleistocene extinctions has attracted considerable interest and criticism. The model will not serve to explain earlier extinction events, such as those of the late Miocene (Hemphillian). Details within the chronology of large animal extinctions and human arrivals have been challenged. Disputes have arisen about quality of radiocarbon dates, and even about the size of a "large" mammal. A controversy exists in America regarding when hunters first arrived.

A conceptual difficulty has centered on the failure of the fossil record of many regions to disclose ample evidence of extinct faunas in kill sites in any other cultural context. Continents in which extinction of higher categories is severe, such as America and Australia, provide much less evidence of human contact with the lost fauna than continents with far fewer losses, such as Africa and Asia. The lack of kill sites can be accommodated by blitzkrieg, a special case of faunal overkill that maximizes speed and intensity of human impact and minimizes time of overlap between the first human invader and the disappearance of native fauna. Paleontologists have noted that an effective new invader need not overlap in the fossil record with the species it replaces. In the case of the Paleolithic invaders of America, Janzen proposes that the native carnivores acted to enhance the hunter's impact, a "fifth column" that increased the rate of extinction.

A rapid rate of change can be anticipated whenever a land mass was first exposed to *Homo sapiens*. The vulnerable species in a fauna first encountering prehistoric people will suddenly vanish from the fossil record. By virtue of the youth of its archaeological sites, New Zealand provides an exception. It yields abundant evidence of temporal overlap between prehistoric people and an extinct megafauna, the moas, whose extinction occurred in this millennium. Furthermore the North Island, environmentally more suitable for rapid prehistoric Polynesian settlement, has yielded much less evidence of "moa hunting" than the South Island.

How is the overkill model to be used? Outside Europe and Africa it means that prehistorians in search of kill sites should devote themselves to the first centuries following the arrival of early hunters and not be discouraged if the search proves difficult. In America this means increased attention to deposits of the ninth millenium B.C. The Australians should search for much older sites, as indeed they are. Success could be fatal to the model of sudden overkill if archaeological sites yielding extinct fauna prove to be variable in age, that is, scattered over thousands of years. Success at finding much evidence of killing or processing of the extinct fauna is not predicted by the blitzkrieg version of overkill.

Australia may offer the best opportunity for refutation of prehistoric blitzkrieg. Horton finds that some faunal extinctions occurred more than 10,000 years after the absolute minimum date for human arrival. This is especially intriguing since, although the fist Australians may not have been specialized big game hunters to the same degree as the first Americans, the extinct Australian large marsupials appear to have been especially ill-equipped to escape human predators. Oceanic islands, such as New Zealand and Hawaii, should provide an opportunity for the archaeological investigation of prehistoric extinctions under circumstances attending human colonization. Many oceanic islands, such as the Solomons, New Caledonia, and the Canaries, appear to be worthy of prospecting by those who wish to discover extinct prehistoric faunas yet unknown.

One striking advantage of the overkill model over others is its testability. The organism lost must be of a size or in a location potentially vulnerable to human impacts. The pattern and radiocarbon dating of late Pleistocene extinction can be studied on a global scale independent of the pattern and timing of prehistoric cultural changes. Extinctions and cultural changes can then be compared for degree of fit. Serious anomalies, if they emerge, will refute the model. To date those that have been advanced appear to be conceptual and not evidential discrepancies, such as the claims of a New World population long before 11,000 yr B.P. (no unequivocal evidence of an earlier invasion of big game hunters has been established) or the lack of kill sites in Australia and on some islands (archaeological associations will be rare or absent if extinction occurred rapidly, i.e. blitzkrieg). The late Pleistocene extinction pattern bears an intriguing relationship to prehistoric events. The model of overkill and its special case, blitzkrieg, offer a challenging alternative to other explanations of the last geological revolution in the history of life.

Critical Thinking

1. Why would human activities cause large animals to go extinct more than smaller animals?
2. What is the "blitzkrieg" version of the overkill hypothesis?
3. If humans are responsible for the Pleistocene extinctions, why should it be difficult to find explicit evidence?

The Historical Roots of Our Ecological Crisis

Lynn White, Jr.

Few people would deny that the world's religions have been a major influence in shaping human philosophical attitudes and social actions. No attempt to assess this influence has proven to be as controversial as the lecture delivered by Lynn White, Jr., (1907–1987) at the 1966 meeting of the American Association for the Advancement of Science entitled "The Historical Roots of Our Ecological Crisis," from which the following selection has been taken. In it, White, a distinguished professor of medieval and renaissance history at the University of California, Los Angeles, argues that Christianity, as an institution in the Western world, is to blame for the attitudes that have resulted in environmental degradation. In his view, Christian dogma has interpreted the dominion over nature that God granted to man to mean that "nature has no reason for existence save to serve man." According to White, this attitude, coupled with the power unleashed by the development of modern science and technology, has produced ecological crises.

Many of White's critics have based their arguments on biblical texts. They cite passages in Judeo-Christian scripture that suggest that dominion should be interpreted as requiring stewardship rather than wanton exploitation of nature. Such criticism fails to address the central thesis of White's analysis. White acknowledges that Biblical references to nature can be interpreted in many ways. His focus is on the actual preaching of the Christian church, which in his view emphasizes a reading of the Bible that condones the thoughtless conquest of nature but says little or nothing about the need for sensitive stewardship.

Key Concept: Judeo-Christian justification for the exploitation of nature

A conversation with Aldous Huxley not infrequently put one at the receiving end of an unforgettable monologue. About a year before his lamented death he was discoursing on a favorite topic: Man's unnatural treatment of nature and its sad results. To illustrate his point he told how, during the previous summer, he had returned to a little valley in England where he had spent many happy months as a child. Once it had been composed of delightful grassy glades; now it was becoming overgrown with unsightly brush because the rabbits that formerly kept such growth under control had largely succumbed to a disease, myxomatosis, that was deliberately introduced by the local farmers to reduce the rabbits' destruction of crops. Being something of a Philistine, I could be silent no longer, even in the interests of great rhetoric. I interrupted to point out that the rabbit itself had been brought as a domestic animal to England in 1176, presumably to improve the protein diet of the peasantry.

All forms of life modify their contexts. The most spectacular and benign instance is doubtless the coral polyp. By serving its own ends, it has created a vast undersea world favorable to thousands of other kinds of animals and plants. Ever since man became a numerous species he has affected his environment notably. The hypothesis that his fire-drive method of hunting created the world's great grasslands and helped to exterminate the monster mammals of the Pleistocene from much of the globe is plausible, if not proved. For 6 millennia at least, the banks of the lower Nile have been a human artifact rather than the swampy African jungle which nature, apart from man, would have made it. The Aswan Dam, flooding 5000 square miles, is only the latest stage in a long process. In many regions terracing or irrigation, overgrazing, the cutting of forests by Romans to build ships to fight Carthaginians or by Crusaders to solve the logistics problems of their expeditions, have profoundly changed some ecologies. Observation that the French landscape falls into two basic types, the open fields of the north and the *bocage* of the south and west, inspired Marc Bloch to undertake his classic study of medieval agricultural methods. Quite unintentionally, changes in human ways often affect nonhuman nature. It has been noted, for example, that the advent of the automobile eliminated huge flocks of sparrows that once fed on the horse manure littering every street.

The history of ecologic change is still so rudimentary that we know little about what really happened, or what the results were. The extinction of the European aurochs as late as 1627 would seem to have been a simple case of overenthusiastic hunting. On more intricate matters it often is impossible to find solid information. For a thousand years or more the Frisians and Hollanders have been pushing back the North Sea, and the process is culminating in our own time in the reclamation of the Zuider Zee. What, if any, species of animals, birds, fish, shore life, or plants have died out in the process? In their epic combat with Neptune have the Netherlanders overlooked ecological values in such a way that the quality of human life in the Netherlands has suffered? I cannot discover that the questions have ever been asked, much less answered.

People, then, have often been a dynamic element in their own environment, but in the present state of historical scholarship we usually do not know exactly when, where, or with what effects man-induced changes came. As we enter the last third of the 20th century, however, concern for the problem of ecologic backlash is mounting feverishly. Natural science, conceived as the effort to understand the nature of things, had flourished in several eras and among several peoples. Similarly there had been an age-old accumulation of technological skills, sometimes growing rapidly, sometimes slowly. But it was not until about four generations ago that Western Europe and North America arranged a marriage between science and technology, a union of the theoretical and the empirical approaches to our natural environment. The emergence in widespread practice of the Baconian creed that scientific knowledge means technological power over nature can scarcely be dated before about 1850, save in the chemical industries, where it is anticipated in the 18th century. Its acceptance as a normal pattern of action may mark the greatest event in human history since the invention of agriculture, and perhaps in nonhuman terrestrial history as well.

Almost at once the new situation forced the crystallization of the novel concept of ecology; indeed, the word *ecology* first appeared in the English language in 1873. Today, less than a century later, the impact of our race upon the environment has so increased in force that it has changed in essence. When the first cannons were fired, in the early 14th century, they affected ecology by sending workers scrambling to the forests and mountains for more potash, sulfur, iron ore, and charcoal, with some resulting erosion and deforestation. Hydrogen

bombs are of a different order: a war fought with them might alter the genetics of all life on this planet. By 1285 London had a smog problem arising from the burning of soft coal, but our present combustion of fossil fuels threatens to change the chemistry of the globe's atmosphere as a whole, with consequences which we are only beginning to guess. With the population explosion, the carcinoma of planless urbanism, the now geological deposits of sewage and garbage, surely no create other than man has ever managed to foul its nest in such short order.

There are many calls to action, but specific proposals, however worthy as individual items, seem too partial, palliative, negative: ban the bomb, tear down the billboards, give the Hindus contraceptives and tell them to eat their sacred cows. The simplest solution to any suspect change is, of course, to stop it, or, better yet, to revert to a romanticized past: make those ugly gasoline stations look like Anne Hathaway's cottage or (in the Far West) like ghost-town saloons. The "wilderness area" mentality invariably advocates deep-freezing an ecology, whether San Gimignano or the High Sierra, as it was before the first Kleenex was dropped. But neither atavism nor prettification will cope with the ecologic crisis of our time.

What shall we do? No one yet knows. Unless we think about fundamentals, our specific measures may produce new backlashes more serious than those they are designed to remedy.

As a beginning we should try to clarify our thinking by looking, in some historical depth, at the presuppositions that underlie modern technology and science. Science was traditionally aristocratic, speculative, intellectual in intent; technology was lower-class, empirical, action-oriented. The quite sudden fusion of these two, towards the middle of the 19th century, is surely related to the slightly prior and contemporary democratic revolutions which, by reducing social barriers, tended to assert a functional unity of brain and hand. Our ecologic crisis is the product of an emerging, entirely novel, democratic culture. The issue is whether a democratized world can survive its own implications. Presumably we cannot unless we rethink our axioms.

The Western Traditions of Technology and Science

One thing is so certain that it seems stupid to verbalize it: both modern technology and modern science are distinctively *Occidental*. Our technology has absorbed elements from all over the world, notably from China; yet everywhere today, whether in Japan or in Nigeria, successful technology is Western. Our science is the heir to all the sciences of the past, especially perhaps to the work of the great Islamic scientists of the Middle Ages, who so often outdid the ancient Greeks in skill and perspicacity: al Rāzi in medicine, for example; or ibn-al-Haytham in optics; or Omar Khayyám in mathematics. Indeed, not a few works of such geniuses seem to have vanished in the original Arabic and to survive only in medieval Latin translations that helped to lay foundations for later Western developments. Today, around the globe, all significant science is Western in style and method, whatever the pigmentation or language of the scientists.

A second pair of fact is less well recognized because they result from quite recent historical scholarship. The leadership of the West, both in technology and in science, is far older than the so-called Scientific Revolution of the 17th century or the so-called Industrial Revolution of the 18th century. These terms are in fact outmoded and obscure the true nature of what they try to describe—significant stages in two long and separate developments. By A.D. 1000 at the latest—and perhaps, feebly, as much as 200 years earlier—the West began to apply water power to industrial processes other than milling grain. This was followed in the late 12th century by the harnessing of wind power. From simple beginnings, but with remarkable consistency of style, the West rapidly expanded its skills in the development of power machinery, labor-saving devices, and automation. . . .

Since both our technological and our scientific movements got their start, acquired their character, and achieved world dominance in the Middle Ages, it would seem that we cannot understand their nature or their present impact upon ecology without examining fundamental medieval assumptions and developments.

Medieval View of Man and Nature

Until recently, agriculture has been the chief occupation even in "advanced" societies; hence, any change of methods of tillage has much importance. Early plows, drawn by two oxen, did not normally turn the sod but merely scratched it. Thus, cross-plowing was needed and fields tended to be squarish. In the fairly light soils and semiarid climates of the Near East and Mediterranean, this worked well. But such a plow was inappropriate to the wet climate and often sticky soils of northern Europe. By the latter part of the 7th century after Christ, however, following obscure beginnings, certain northern peasants were using an entirely new kind of plow, equipped with a vertical knife to cut the line of the furrow, a horizontal share to slice under the sod, and a moldboard to turn it over. The friction of this plow with the soil was so great that it normally required not two but eight oxen. It attacked the land with such violence that cross-plowing was not needed, and fields tended to be shaped in long strips.

In the days of the scratch-plow, fields were distributed generally in units capable of supporting a single family. Subsistence farming was the presupposition. But no peasant owned eight oxen: to use the new and more efficient plow, peasants pooled their oxen to form large plow-teams, originally receiving (it would appear) plowed strips in proportion to their contribution. Thus, distribution of land was based no longer on the needs of

a family but, rather, on the capacity of a power machine to till the earth. Man's relation to the soil was profoundly changed. Formerly man had been part of nature; now he was the exploiter of nature. Nowhere else in the world did farmers develop any analogous agricultural implement. Is it coincidence that modern technology, with its ruthlessness toward nature, has so largely been produced by descendants of these peasants of northern Europe?

This same exploitive attitude appears slightly before A.D. 830 in Western illustrated calendars. In older calendars the months were shown as passive personifications. The new Frankish calendars, which set the style for the Middle Ages, are very different: they show men coercing the world around them—plowing, harvesting, chopping trees, butchering pigs. Man and nature are two things, and man is master.

These novelties seem to be in harmony with larger intellectual patterns. What people do about their ecology depends on what they think about themselves in relation to things around them. Human ecology is deeply conditioned by beliefs about our nature and destiny—that is, by religion. To Western eyes this is very evident in, say, India or Ceylon. It is equally true of ourselves and of our medieval ancestors.

The victory of Christianity over paganism was the greatest psychic revolution in the history of our culture. It has become fashionable today to say that, for better or worse, we live in "the post-Christian age." Certainly the forms of our thinking and language have largely ceased to be Christian, but to my eye the substance often remains amazingly akin to that of the past. Our daily habits of action, for example, are dominated by an implicit faith in perpetual progress which was unknown either to Greco-Roman antiquity or to the Orient. It is rooted in, and is indefensible apart from, Judeo-Christian teleology. The fact that Communists share it merely helps to show what can be demonstrated on many other grounds: that Marxism, like Islam, is a Judeo-Christian heresy. We continue today to live, as we have lived for about 1700 years, very largely in a context of Christian axioms.

What did Christianity tell people about their relations with the environment?

While many of the world's mythologies provide stories of creation, Greco-Roman mythology was singularly incoherent in this respect. Like Aristotle, the intellectuals of the ancient West denied that the visible world had had a beginning. Indeed, the idea of a beginning was impossible in the framework of their cyclical notion of time. In sharp contrast, Christianity inherited from Judaism not only a concept of time as nonrepetitive and linear but also a striking story of creation. By gradual stages a loving and all-powerful God had created light and darkness, the heavenly bodies, the earth and all its plants, animals, birds, and fishes. Finally, God had created Adam and, as an afterthought, Eve to keep man from being lonely. Man named all the animals, thus establishing his dominance over them. God planned all of this explicitly for man's benefit and rule: no item in the physical creation had any purpose save to serve man's purposes. And, although man's body is made of clay, he is not simply part of nature: he is made in God's image.

Especially in its Western form, Christianity is the most anthropocentric religion the world has seen. As early as the 2nd century both Tertullian and Saint Irenaeus of Lyons were insisting that when God shaped Adam he was foreshadowing the image of the incarnate Christ, the Second Adam. Man shares, in great measure, God's transcendence of nature. Christianity, in absolute contrast to ancient paganism and Asia's religions (except, perhaps, Zoroastrianism), not only established a dualism of man and nature but also insisted that it is God's will that man exploit nature for his proper ends. . . .

The Christian dogma of creation, which is found in the first clause of all the Creeds, has another meaning for our comprehension of today's ecologic crisis. By revelation, God had given man the Bible, the Book of Scripture. But since God had made nature, nature also must reveal the divine mentality. The religious study of nature for the better understanding of God was known as natural theology. In the early Church, and always in the Greek East, nature was conceived primarily as a symbolic system through which God speaks to men: the ant is a sermon to sluggards; rising flames are the symbol of the soul's aspiration. This view of nature was essentially artistic rather than scientific. While Byzantium preserved and copied great numbers of ancient Greek scientific texts, science as we conceive it could scarcely flourish in such an ambience.

However, in the Latin West by the early 13th century natural theology was following a very different bent. It was ceasing to be the decoding of the physical symbols of God's communication with man and was becoming the effort to understand God's mind by discovering how his creation operates. The rainbow was no longer simply a symbol of hope first sent to Noah after the Deluge: Robert Grosseteste, Friar Roger Bacon, and Theodoric of Freiberg produced startlingly sophisticated work on the optics of the rainbow, but they did it as a venture in religious understanding. From the 13th century onward, up to and including Leibnitz and Newton, every major scientist, in effect, explained his motivations in religious terms. Indeed, if Galileo had not been so expert an amateur theologian he would have got into far less trouble: the professionals resented his intrusion. And Newton seems to have regarded himself more as a theologian than as a scientist. It was not until the late 18th century that the hypothesis of God became unnecessary to many scientists.

It is often hard for the historian to judge, when men explain why they are doing what they want to do, whether they are offering real reasons or merely culturally acceptable reasons. The consistency with which scientists during the long formative centuries of Western science said that the task and the reward of the scientist was "to think God's thoughts after him" leads one to believe that this was their real motivation. If so, then modern Western science was cast in a matrix of Christian theology. The dynamism of religious devotion, shaped by the Judeo-Christian dogma of creation, gave it impetus.

An Alternative Christian View

We would seem to be headed toward conclusions unpalatable to many Christians. Since both *science* and *technology* are blessed words in our contemporary vocabulary, some may be happy at the notions, first, that, viewed historically, modern science is an extrapolation of natural theology and, second, that modern technology is at least partly to be explained as an Occidental voluntarist realization of the Christian dogma of man's transcendence of, and rightful mastery over, nature. But, as we now recognize, somewhat over a century ago science and technology—hitherto quite separate activities—joined to give mankind powers which, to judge by many of the ecologic effects, are out of control. If so, Christianity bears a huge burden of guilt.

I personally doubt that disastrous ecologic backlash can be avoided simply by applying to our problems more science and more technology. Our science and technology have grown out of Christian attitudes toward man's relation to nature which are almost universally held not only by Christians and neo-Christians but also by those who fondly regard themselves as post-Christians. Despite Copernicus, all the cosmos rotates around our little globe. Despite Darwin, we are *not*, in our hearts, part of the natural process. We are superior to nature, contemptuous of it, willing to use it for our slightest whim. The newly elected Governor of California, like myself a churchman but less troubled than I, spoke for the Christian tradition when he said (as is alleged), "when you've seen one redwood tree, you've seen them all." To a Christian a tree can be no more than a physical fact. The whole concept of the sacred grove is alien to Christianity and to the ethos of the West. For nearly 2 millennia Christian missionaries have been chopping down sacred groves, which are idolatrous because they assume spirit in nature.

What we do about ecology depends on our ideas of the man-nature relationship. More science and more technology are not going to get us out of the present ecologic crisis until we find a new religion, or rethink our old one. The beatniks, who are the basic revolutionaries of our time, show a sound instinct in their affinity for Zen Buddhism, which conceives of the man-nature relationship as very nearly the mirror image of the Christian view. Zen, however, is as deeply conditioned by Asian history as Christianity is by the experience of the West, and I am dubious of its viability among us.

Possibly we should ponder the greatest radical in Christian history since Christ: Saint Francis of Assisi. The prime miracle of Saint Francis is the fact that he did not end at the stake, as many of his left-wing followers did. He was so clearly heretical that a General of the Franciscan Order, Saint Bonaventura, a great and perceptive Christian, tried to suppress the early account of Franciscanism. The key to an understanding of Francis is his belief in the virtue of humility—not merely for the individual but for man as a species. Francis tried to depose man from his monarchy over creation and set up a democracy of all God's creatures. With him the ant is no longer simply a homily for the lazy, flames a sign of the thrust of the soul toward union with God; now they are Brother Ant and Sister Fire, praising the Creator in their own ways as Brother Man does in his.

Later commentators have said that Francis preached to the birds as a rebuke to men who would not listen. The records do not read so: he urged the little birds to praise God, and in spiritual ecstasy they flapped their wings and chirped rejoicing. Legends of saints, especially the Irish saints, had long told of their dealings with animals but always, I believe, to show their human dominance over creatures. With Francis it is different. The land around Gubbio in the Apennines was being ravaged by a fierce wolf. Saint Francis, says the legend, talked to the wolf and persuaded him of the error of his ways. The wolf repented, died in the odor of sanctity, and was buried in consecrated ground.

What Sir Steven Ruciman calls "the Franciscan doctrine of the animal soul" was quickly stamped out. Quite possible it was in part inspired, consciously or unconsciously, by the belief in reincarnation held by the Cathar heretics who at that time teemed in Italy and southern France, and who presumably had got it originally from India. It is significant that at just the same moment, about 1200, traces of metempsychosis are found also in western Judaism, in the Provençal *Cabbala*. But Francis held neither to transmigration of souls nor to pantheism. His view of nature and of man rested on a unique sort of pan-psychism of all things animate and inanimate, designed for the glorification of their transcendent Creator, who, in the ultimate gesture of cosmic humility, assumed flesh, lay helpless in a manger, and hung dying on a scaffold.

I am not suggesting that many contemporary Americans who are concerned about our ecologic crisis will be either able or willing to counsel with wolves or exhort birds. However, the present increasing disruption of the global environment is the product of a dynamic technology and science which were originating in the Western medieval world against which Saint Francis was rebelling in so original a way. Their growth cannot be understood historically apart from distinctive attitudes toward nature which are deeply grounded in Christian dogma. The fact that most people do not think of these attitudes as Christian is irrelevant. No new set of basic values has been accepted in our society to displace those of Christianity. Hence we shall continue to have a worsening ecologic crisis until we reject the Christian axiom that nature has no reason for existence save to serve man.

The greatest spiritual revolutionary in Western history, Saint Francis, proposed what he thought was an alternative Christian view of nature and man's relation to it: he tried to substitute the idea of the equality of all creatures, including man, for the idea of man's limitless rule of creation. He failed. Both our present science and our present technology are so tinctured with orthodox Christian arrogance toward nature that no solution for our ecologic crisis can be expected from them alone. Since the roots of our trouble are so largely religious, the remedy must also be essentially religious, whether we call it that or not. We must rethink and refeel our nature and destiny. The profoundly religious, but heretical, sense of the primitive Franciscans for the spiritual autonomy of all parts of nature may point a direction. I propose Francis as a patron saint for ecologists.

Critical Thinking

1.   What is the "purpose" of nature?
2.   How might one apply the parable of the three stewards (Matthew 25: 14–30) to the human relationship with the environment?
3.   What makes Saint Francis an appropriate candidate for the position of patron saint of ecology?

The Tragedy of the Commons

Garrett Hardin

Since 1946 Garrett Hardin (1915–2003) was a professor of biology and human ecology at the University of California, Santa Barbara. The forthright, controversial views on population, evolution, and birth control he expressed in his numerous articles, books, and speeches earned him an international reputation as an influential, outspoken critic of the ecological and environmental community.

The best known and most often quoted of Hardin's writings on the social and ethical issues raised by a world of limited resources and increasing numbers of people is "The Tragedy of the Commons," which was printed in the December 1968 issue of *Science* and which is excerpted in the following selection. In this essay, Hardin relates a nineteenth-century tale about a common pasture becoming overgrazed and destroyed because each of the herdsmen whose animals grazed on it considered only the advantage to his own family of increasing his herd. From this parable, Hardin draws a general conclusion: that all resources, such as the oceans, which are held in common and are therefore not anyone's private property, will be overused and ultimately degraded. Among the policy implications he derives from this assessment is that programs that attempt to deal with hunger by providing free food to people are counterproductive. Indeed, Hardin decided to terminate his own research on the culture of algae as a potential major food source because he believes that more food simply encourages further increases in population, which will ultimately produce even greater starvation.

In subsequent writings, Hardin went on to develop his "lifeboat ethics" theory. He proposed a world model in which the developed, affluent nations that control and use most of the world's resources are in a lifeboat while the struggling developing nations are floundering in the surrounding ocean. He concluded that it is folly to try to rescue all the swimmers and suggested that the ethically appropriate strategy is one of triage, by which the "haves" permit the poorest and least developed of the "have nots" to drown in order to prevent the entire boat from sinking. While this harsh analysis has won praise from many environmentalists who share Hardin's predilection for "pragmatic" decisions based on a competitive "marketplace" model, it has been rejected by others who advocate a more egalitarian approach based on informed social planning.

Key Concept: commonly owned resources are doomed to destruction

An implicit and almost universal assumption of discussions published in professional and semipopular scientific journals is that the problem under discussion has a technical solution. A technical solution may be defined as one that requires a change only in the techniques of the natural sciences, demanding little or nothing in the way of change in human values or ideas of morality.

In our day (though not in earlier times) technical solutions are always welcome. Because of previous failures in prophecy, it takes courage to assert that a desired technical solution is not possible.... [T]he concern here is with the important concept of a class of human problems which can be called "no technical solution problems," and, more specifically, with the identification and discussion of one of these.

It is easy to show that the class is not a null class. Recall the game of tick-tack-toe. Consider the problem, "How can I win the game of tick-tack-toe?" It is well known that I cannot, if I assume (in keeping with the conventions of game theory) that my opponent understands the game perfectly. Put another way, there is no "technical solution" to the problem. I can win only by giving a radical meaning to the word "win." I can hit my opponent over the head; or I can drug him; or I can falsify the records. Every way in which I "win" involves, in some sense, an abandonment of the game, as we intuitively understand it. (I can also, of course, openly abandon the game—refuse to play it. This is what most adults do.)

The class of "No technical solution problems" has members. My thesis is that the "population problem," as conventionally conceived, is a member of this class. How it is conventionally conceived needs some comment. It is fair to say that most people who anguish over the population problem are trying to find a way to avoid the evils of overpopulation without relinquishing any of the privileges they now enjoy. They think that farming the seas or developing new strains of wheat will solve the problem—technologically. I try to show here that the solution they seek cannot be found. The population problem cannot be solved in a technical way, any more than can the problem of winning the game of tick-tack-toe. . . .

We can make little progress in working toward optimum population size until we explicitly exorcize the spirit of Adam Smith in the field of practical demography. In economic affairs, *The Wealth of Nations* (1776) popularized the "invisible hand," the idea that an individual who "intends only his own gain," is, as it were, "led by an invisible hand to promote. . . the public interest." Adam Smith did not assert that this was invariably true, and perhaps neither did any of his followers. But he contributed to a dominant tendency of thought that has ever since interfered with positive action based on rational analysis, namely, the tendency to assume that decisions reached individually will, in fact, be the best decisions for an entire society. If this assumption is correct it justifies the continuance of our present policy of laissez-faire in reproduction. If it is correct we can assume that men will control their individual fecundity so as to produce the optimum population. If the assumption is not correct, we need to reexamine our individual freedoms to see which ones are defensible.

Tragedy of Freedom in a Commons

The rebuttal to the invisible hand in population control is to be found in a scenario first sketched in a little-known pamphlet in 1833 by a mathematical amateur named William Forster Lloyd (1794–1852). We may well call it "the tragedy of the commons," using the word "tragedy" as the philosopher Whitehead used it: "The essence of dramatic tragedy is not unhappiness. It resides in the solemnity of the remorseless working of things." He then goes on to say, "This inevitableness of destiny can only be illustrated in terms of human life by incidents which in fact involve unhappiness. For it is only by them that the futility of escape can be made evident in the drama."

The tragedy of the commons develops in this way. Picture a pasture open to all. It is to be expected that each herdsman will try to keep as many cattle as possible on the commons. Such an arrangement may work reasonably satisfactorily for centuries because tribal wars, poaching, and disease keep the numbers of both man and beast well below the carrying capacity of the land. Finally, however, comes the day of reckoning, that is, the day when the long-desired goal of social stability becomes a reality. At this point, the inherent logic of the commons remorselessly generates tragedy.

As a rational being, each herdsman seeks to maximize his gain. Explicitly or implicitly, more or less consciously, he asks, "What is the utility *to me* of adding one more animal to my herd?" This utility has one negative and one positive component.

1. The positive component is a function of the increment of one animal. Since the herdsman receives all the proceeds from the sale of the additional animal, the positive utility is nearly +1.
2. The negative component is a function of the additional overgrazing created by one more animal. Since, however, the effects of overgrazing are shared by all the herdsmen, the negative utility for any particular decision-making herdsman is only a fraction of –1.

Adding together the component partial utilities, the rational herdsman concludes that the only sensible course for him to pursue is to add another animal to his herd. And another; and another. . . . But this is the conclusion reached by each and every rational herdsman sharing a commons. Therein is the tragedy. Each man is locked into a system that compels him to increase his herd without limit—in a world that is limited. Ruin is the destination toward which all men rush, each pursuing his own best interest in a society that believes in the freedom of the commons. Freedom in a commons brings ruin to all.

Some would say that this is a platitude. Would that it were! In a sense, it was learned thousands of years ago, but natural selection favors the forces of psychological denial. The individual benefits as an individual from his ability to deny the truth even though society as a whole, of which he is a part, suffers. Education can counteract the natural tendency to do the wrong thing, but the inexorable succession of generations requires that the basis for this knowledge be constantly refreshed.

A simple incident that occurred a few years ago in Leominster, Massachusetts, shows how perishable the knowledge is. During the Christmas shopping season the parking meters downtown were covered with plastic bags that bore tags reading: "Do not open until after Christmas. Free parking courtesy of the mayor and city council." In other words, facing the prospect of an increased demand for already scarce space, the city fathers reinstituted the system of the commons. (Cynically, we suspect that they gained more votes than they lost by this retrogressive act.)

In an approximate way, the logic of the commons has been understood for a long time, perhaps since the discovery of agriculture or the invention of private property in real estate. But it is understood mostly only in special cases which are not sufficiently generalized. Even at this late date, cattlemen leasing national land on the western ranges demonstrate no more than an ambivalent understanding, in constantly pressuring federal authorities to increase the head count to the point where overgrazing produces erosion and weed-dominance. Likewise, the oceans of the world continue to suffer from the survival of the philosophy of the commons. Maritime nations still respond automatically to the shibboleth of the "freedom of the seas." Professing to believe in the "inexhaustible resources of the oceans," they bring species after species of fish and whales closer to extinction.

The National Parks present another instance of the working out of the tragedy of the commons. At present, they are open to all, without limit. The parks themselves are limited in extent—there is only one Yosemite Valley—whereas population seems to grow without limit. The values that visitors seek in the parks are steadily eroded. Plainly, we must soon cease to treat the parks as commons or they will be of no value to anyone.

What shall we do? We have several options. We might sell them off as private property. We might keep them as public property, but allocate the right to enter them. The allocation might be on the basis of wealth, by the use of an auction system. It might be on the basis of merit, as defined by some agreed-upon standards. It might be by lottery. Or it might be on a first-come, first-served basis, administered to long queues. These, I think, are all the reasonable possibilities. They are all objectionable. But we must choose—or acquiesce in the destruction of the commons that we call our national parks.

Pollution

In a reverse way, the tragedy of the commons reappears in problems of pollution. Here it is not a question of taking something out of the commons, but of putting something in—sewage, or chemical, radioactive, and heat wastes into water; noxious and dangerous fumes into the air; and distracting and unpleasant advertising signs into the line of sight. The calculations of utility are much the same as before. The rational man finds that his share of the cost of the wastes he discharges into the commons is less than the cost of purifying his wastes before releasing them. Since this is true for everyone, we are locked into a system of "fouling our own nest," so long as we behave only as independent, rational, free-enterprisers.

The tragedy of the commons as a food basket is averted by private property, or something formally like it. But the air and waters surrounding us cannot readily be fenced, and so the tragedy of the commons as a cesspool must be prevented by different means, by coercive laws or taxing devices that make it cheaper for the polluter to treat his pollutants than to discharge them untreated. We have not progressed as far with the solution of this problem as we have with the first. Indeed, our particular concept of private property, which deters us from exhausting the positive resources of the earth, favors pollution. The owner of a factory on the bank of a stream—whose property extends to the middle of the stream—often has difficulty seeing why it is not his natural right to muddy the waters flowing past his door. The law, always behind the times, requires elaborate stitching and fitting to adapt it to this newly perceived aspect of the commons.

The pollution problem is a consequence of population. It did not much matter how a lonely American frontiersman disposed of his waste. "Flowing water purifies itself every 10 miles," my grandfather used to say, and the myth was near enough to the truth when he was a boy, for there were not too many people. But as population became denser, the natural chemical and biological recycling processes became overloaded, calling for a redefinition of property rights. . . .

Freedom to Breed is Intolerable

The tragedy of the commons is involved in population problems in another way. In a world governed solely by the principle of "dog eat dog"—if indeed there ever was such a world—how many children a family had would not be a matter of public concern. Parents who bred too exuberantly would leave fewer descendants, not more, because they would be unable to care adequately for their children. David Lack and others have found that such a negative feedback demonstrably controls the fecundity of birds. But men are not birds, and have not acted like them for millenniums, at least.

If each human family were dependent only on its own resources; *if* the children of improvident parents starved to death; *if*, thus, overbreeding brought its own "punishment" to the germ line—*then* there would be no public interest in controlling the breeding of families. But our society is deeply committed to the welfare state, and hence is confronted with another aspect of the tragedy of the commons.

In a welfare state, how shall we deal with the family, the religion, the race, or the class (or indeed any distinguishable and cohesive group) that adopts overbreeding as a policy to secure its own aggrandizement? To couple the concept of freedom to breed with the belief that everyone born has an equal right to the commons is to lock the world into a tragic course of action.

Unfortunately this is just the course of action that is being pursued by the United Nations. In late 1967, some 30 nations agreed to the following:

> The Universal Declaration of Human Rights describes the family as the natural and fundamental unit of society. It follows that any choice and decision with regard to the size of the family must irrevocably rest with the family itself, and cannot be made by anyone else.

It is painful to have to deny categorically the validity of this right; denying it, one feels as uncomfortable as a resident of Salem, Massachusetts, who denied the reality of witches in the 17th century. At the present time, in liberal quarters, something like a taboo acts to inhibit criticism of the United Nations. There is a feeling that the United Nations is "our last and best hope," that we shouldn't find fault with it; we shouldn't play into the hands of the archconservatives. However, let us not forget what Robert Louis Stevenson said: "The truth that is suppressed by friends is the readiest weapon of the enemy." If we love the truth we must openly deny the validity of the Universal Declaration of Human Rights, even though it is promoted by the United Nations. We should also join with Kingsley Davis in attempting to get Planned Parenthood-World Population to see the error of its ways in embracing the same tragic ideal.

Conscience is Self-Eliminating

It is a mistake to think that we can control the breeding of mankind in the long run by an appeal to conscience. Charles Galton Darwin made this point when he spoke on the centennial of the publication of his grandfather's great book. The argument is straightforward and Darwinian.

People vary. Confronted with appeals to limit breeding, some people will undoubtedly respond to the plea more than others. Those who have more children will produce a larger fraction of the next generation than those with more susceptible consciences. The difference will be accentuated, generation by generation.

In C. G. Darwin's words: "It may well be that it would take hundreds of generations for the progenitive instinct to develop in this way, but if it should do so, nature would have taken her revenge, and the variety *Homo contracipiens* would become extinct and would be replaced by the variety *Homo progenitivus.*"

The argument assumes that conscience or the desire for children (no matter which) is hereditary—but hereditary only in the most general formal sense. The result will be the same whether the attitude is transmitted through germ cells, or exosomatically, to use A. J. Lotka's term. (If one denies the latter possibility as well as the former, then what's the point of education?) The argument has here been stated in the context of the population problem, but it applies equally well to any instance in which society appeals to an individual exploiting a commons to restrain himself for the general good—by means of his conscience. To make such an appeal is to set up a selective system that works toward the elimination of conscience from the race. . . .

Mutual Coercion Mutually Agreed Upon

The social arrangements that produce responsibility are arrangements that create coercion, of some sort. Consider bank-robbing. The man who takes money from a bank acts as if the bank were a commons. How do we prevent such action? Certainly not by trying to control his behavior solely by a verbal appeal to his sense of responsibility. Rather than rely on propaganda we follow [Charles] Frankel's lead and insist that a bank is not a commons; we seek the definitive

social arrangements that will keep it from becoming a commons. That we thereby infringe on the freedom of would-be robbers we neither deny nor regret.

The morality of bank-robbing is particularly easy to understand because we accept complete prohibition of this activity. We are willing to say "Thou shalt not rob banks," without providing for exceptions. But temperance also can be created by coercion. Taxing is a good coercive device. To keep downtown shoppers temperate in their use of parking space we introduce parking meters for short periods, and traffic fines for longer ones. We need not actually forbid a citizen to park as long as he wants to; we need merely make it increasingly expensive for him to do so. Not prohibition, but carefully biased options are what we offer him. A Madison Avenue man might call this persuasion; I prefer the greater candor of the word coercion.

Coercion is a dirty word to most liberals now, but it need not forever be so. As with the four-letter words, its dirtiness can be cleansed away by exposure to the light, by saying it over and over without apology or embarrassment. To many, the word coercion implies arbitrary decisions of distant and irresponsible bureaucrats; but this is not a necessary part of its meaning. The only kind of coercion I recommend is mutual coercion, mutually agreed upon by the majority of the people affected.

To say that we mutually agree to coercion is not to say that we are required to enjoy it, or even to pretend we enjoy it. Who enjoys taxes? We all grumble about them. But we accept compulsory taxes because we recognize that voluntary taxes would favor the conscienceless. We institute and (grumblingly) support taxes and other coercive devices to escape the horror of the commons.

An alternative to the commons need not be perfectly just to be preferable. With real estate and other material goods, the alternative we have chosen is the institution of private property coupled with legal inheritance. Is this system perfectly just? As a genetically trained biologist I deny that it is. It seems to me that, if there are to be differences in individual inheritance, legal possession should be perfectly correlated with biological inheritance—that those who are biologically more fit to be the custodians of property and power should legally inherit more. But genetic recombination continually makes a mockery of the doctrine of "like father, like son" implicit in our laws of legal inheritance. An idiot can inherit millions, and a trust fund can keep his estate intact. We must admit that our legal system of private property plus inheritance is unjust—but we put up with it because we are not convinced, at the moment, that anyone has invented a better system. The alternative of the commons is too horrifying to contemplate. Injustice is preferable to total ruin.

It is one of the peculiarities of the warfare between reform and the status quo that it is thoughtlessly governed by a double standard. Whenever a reform measure is proposed it is often defeated when its opponents triumphantly discover a flaw in it. As Kingsley Davis has pointed out, worshippers of the status quo sometimes imply that no reform is possible without unanimous agreement, an implication contrary to historical fact. As nearly as I can make out, automatic rejection of proposed reforms is based on one of two unconscious assumptions: (i) that the status quo is perfect; or (ii) that the choice we face is between reform and no action; if the proposed reform is imperfect, we presumably should take no action at all, while we wait for a perfect proposal.

But we can never do nothing. That which we have done for thousands of years is also action. It also produces evils. Once we are aware that the status quo is action, we can then compare its discoverable advantages and disadvantages with the predicted advantages and disadvantages of the proposed reform, discounting as best we can for our lack of experience. On the basis of such a comparison, we can make a rational decision which will not involve the unworkable assumption that only perfect systems are tolerable.

Recognition of Necessity

Perhaps the simplest summary of this analysis of man's population problems is this: the commons, if justifiable at all, is justifiable only under conditions of low-population density. As the human population has increased, the commons has had to be abandoned in one aspect after another.

First we abandoned the commons in food gathering, enclosing farm land and restricting pastures and hunting and fishing areas. These restrictions are still not complete throughout the world.

Somewhat later we saw that the commons as a place for waste disposal would also have to be abandoned. Restrictions on the disposal of domestic sewage are widely accepted in the Western world; we are still struggling to close the commons to pollution by automobiles, factories, insecticide sprayers, fertilizing operations, and atomic energy installations.

In a still more embryonic state is our recognition of the evils of the commons in matters of pleasure. There is almost no restriction on the propagation of sound waves in the public medium. The shopping public is assaulted with mindless music, without its consent. Our government is paying out billions of dollars to create supersonic transport which will disturb 50,000 people for every one person who is whisked from coast to coast 3 hours faster. Advertisers muddy the airwaves of radio and television and pollute the view of travelers. We are a long way from outlawing the commons in matters of pleasure. Is this because our Puritan inheritance makes us view pleasure as something of a sin, and pain (that is, the pollution of advertising) as the sign of virtue?

Every new enclosure of the commons involves the infringement of somebody's personal liberty. Infringements made in the distant past are accepted because no contemporary complains of a loss. It is the newly proposed infringements that we vigorously oppose; cries of "rights" and "freedom" fill the air. But what does "freedom" mean? When men mutually agreed to pass laws against robbing, mankind became more free, not less so. Individuals locked into the logic of the commons are free only to bring on universal ruin; once they see the necessity of mutual coercion, they become free to pursue other goals. I believe it was Hegel who said, "Freedom is the recognition of necessity."

The most important aspect of necessity that we must now recognize, is the necessity of abandoning the commons in breeding. No technical solution can rescue us from the misery of overpopulation. Freedom to breed will bring ruin to all. At the moment, to avoid hard decisions many of us are tempted to propagandize for conscience and responsible parenthood. The temptation must be resisted, because an appeal to independently acting consciences selects for the disappearance of all conscience in the long run, and an increase in anxiety in the short.

The only way we can preserve and nurture other and more precious freedoms is by relinquishing the freedom to breed, and that very soon. "Freedom is the recognition of necessity"—and it is the role of education to reveal to all the necessity of abandoning the freedom to breed. Only so, can we put an end to this aspect of the tragedy of the commons.

Critical Thinking

1. Why should people not have as many children as possible?
2. List some examples of resources held in common by all the citizens of society that have suffered by overuse.
3. How might that overuse be prevented?

Ecosystems and Ecosystem Services

Selection 8

JOHN TEAL AND MILDRED TEAL, from *Life and Death of the Salt Marsh*
(Ballantine Books, 1969)

Selection 9

PETER M. VITOUSEK, HAROLD A. MOONEY, AND JERRY M. MELILLO, from
"Human Domination of Earth's Ecosystem," *Science* (July 25, 1997)

Selection 10

MILLENNIUM ECOSYSTEM ASSESSMENT, from *Ecosystems and Human Well-Being:
Synthesis* (Island Press, 2005)

Learning Outcomes

After reading this unit, you should be able to:

1. Explain how different uses of the land can conflict.
2. Explain why natural ecosystems can be more valuable to humans than ecosystems that have been "developed" for human use.
3. Describe the extent to which human activities impact the global environment.
4. Evaluate proposed methods of managing ecosystems for human benefit.

Life and Death of the Salt Marsh

John Teal and Mildred Teal

John and Mildred Teal began their intensive study of North American coastal wetlands at the Sapelo Island Marine Institute of the University of Georgia in 1955. They credit Dalhousie University in Halifax, Nova Scotia, Canada, with providing them the opportunity to explore the northern marshes of the Maritimes—the Canadian provinces of New Brunswick, Nova Scotia, and Prince Edward Island. The observations they made during these investigations were subsequently analyzed and systematized with the help of colleagues at the Woods Hole Oceanographic Institute in Falmouth, Massachusetts. This research led to a deep understanding of the flora and fauna that are supported by the salt marshes, their overall ecological importance, and their sensitivity to human intrusion.

    The essence of the Teals' findings is described in their classic book *Life and Death of the Salt Marsh* (Ballantine Books, 1969), from which the following selection is taken. It is one of the most frequently referenced works in environmental and ecological courses that include discussions of wetland ecosystems. Through their work, the Teals provided much of the foundation for the present efforts of environmental activists and grassroots organizations to convince elected officials of the urgent need for wetland protection.

Key Concept: the ecological significance of salt marshes

Introduction

Along the eastern coast of North America, from the north where ice packs grate upon the shore to the tropical mangrove swamps tenaciously holding the land together with a tangle of roots, lies a green ribbon of soft, salty, wet, low-lying land, the salt marshes.

The ribbon of green marshes, part solid land, part mobile water, has a definite but elusive border, now hidden, now exposed, as the tides of the Atlantic fluctuate. At one place and tide there is a line at which you can say, "Here begins the marsh." At another tide, the line, the "beginning of the marsh," is completely inundated and looks as though it had become part of the sea. The marsh reaches as far inland as the tides can creep and as far into the sea as marsh plants can find a roothold and live in saline waters.

The undisturbed salt marshes offer the inland visitor a series of unusual perceptions. At low tide, the wind blowing across *Spartina* grass sounds like wind on the prairie. When the tide is in, the gentle music of moving water is added to the prairie rustle. There are sounds of birds living in the marshes. The marsh wren advertises his presence with a reedy call, even at night, when most birds are still. The marsh hen, or clapper rail, calls in a loud, car-rying cackle. You can hear the tiny, high-pitched rustling thunder of the herds of crabs moving through the grass as they flee before advancing feet or the more leisurely sound of movement they make on their daily migrations in search of food. At night, when the air is still and other sounds are quieted, an attentive listener can hear the bubbling of air from the sandy soil as a high tide floods the marsh.

The wetlands are filled with smells. They smell of the sea and salt water and of the edge of the sea, the sea with a little iodine and trace of dead life. The marshes smell of *Spartina*, a fairly strong odor mixed from the elements of sea and the smells of grasses. These are clean, fresh smells, smells that are pleasing to one who lives by the sea but strange and not altogether pleasant to one who has always lived inland.

Unfortunately, in marshes which have been disturbed, dug up, suffocated with loads of trash and fill, poisoned and eroded with the wastes from large cities, there is another smell. Sick marshes smell of hydrogen sulfide, a rotten egg odor. This odor is very faint in a healthy marsh.

As the sound and smell of the salt marsh are its own, so is its feel. Some of the marshes can be walked on, especially the landward parts. In the north, the *Spartina patens* marsh is covered with dense grass that may be cut for salt hay. Its roots bind the wet mud into a firm surface. But the footing is spongy on an unused hay marsh as the mat of other years' grass, hidden under the green growth, resists the walker's weight and springs back as he moves along.

In the southern marshes, only one grass covers the entire marsh area, *Spartina alterniflora*. On the higher parts of the marsh, near the land, the roots have developed into a mass that provides firm footing although the plants are much more separated than in the northern hay marshes and you squish gently on mud rather than grass. It is like walking on a huge trampoline. The ground is stiff. It is squishy and wet, to be sure, but still solid as you walk about. However, jump and you can feel the ground give under the impact and waves spread out in all directions. The ground is a mat of plant roots and mud on top of a more liquid layer underneath which gives slightly by flowing to all sides when you jump down on it.

As you walk toward the edge of the marsh, the seaward edge, each step closer to open water brings a change in footing. The mud has less root material in it and is less firmly bound together. It begins to ooze around your shoes. On the edges of the creeks, especially the larger ones, there may be natural levees where the ground is higher. Here the rising tide meets its first real resistance as it spills over the creek banks and has to flow between the close-set plants. Here it is slowed and drops the mud it may be carrying. Here too, especially after a series of tides, lower than usual, the ground is firm and even dry and hard.

Down toward the creek, where the mud is watered at each tide, the soil is as muddy as you can find anywhere. When you try to walk across to the water at low tide, across the exposed mud where the marsh grass does not yet grow, hip boots are not high enough to keep you from getting muddy. The boots are pulled off on the first or second step when they have sunk deep into the clutching zone. There are no roots to give solidarity, nothing but the mud and water fighting a shifting battle to hold the area.

At low tide the salt marsh is a vast field of grasses with slightly higher grasses sticking up along the creeks and uniformly tall grass elsewhere. The effect is like that of a great flat meadow. At high tide, the look is the same, a wide flat sea of grass but with a great deal of water showing. The marsh is still marsh, but spears of grass are sticking up through water, a world of water where land was before, each blade of grass a little island, each island a refuge for the marsh animals which do not like or cannot stand submersion in salt water. . . .

Solutions and Suggestions

The dangers to salt marshes stem from human activities, not natural processes. We destroy wetlands and shallow water bottoms directly by dredging, filling, and building. Indirectly we destroy them by pollution. Much of this destruction is simply foolish. The marsh would often have been much more valuable as a marsh than it is in its subsequent desecrated form.

The increase in population pressure along the coast will inevitably destroy more and more of the frail marsh estuarine system. We do not propose the preservation of the marshes simply for the sake of their preservation. Instead, we regard them in light of their benefit to the growing population. The benefit of marshes will accrue to everyone, not only those who venture onto the surface of marshes but to fishermen along the coast and to consumers of fishery products who may live far inland.

Some destruction is inevitable. Even for those marshes preserved as wildlife areas, an access must be constructed so that people who want to enjoy these pieces of nature can do so. Roads must be built to the marshes, along the edges of marshes, and to impoundments that are designed for mosquito control and waterfowl hunting. Also, building roads to boat-launching ramps so that the network of creeks and rivers in the wetlands can be enjoyed is not only a convenience but a preservative: damage to the marshes will be less if adequate access is provided from the waterside.

But after having conceded that we cannot avoid destroying some marshes, how are we to decide which should be destroyed and which preserved? And by what means shall we preserve them? It is obvious that overall planning is necessary. The very minimum of planning could be approached on the state level, but a more rational approach demands planning on the national level, as it is the whole marsh system with its high productivity, rather than individual marshes, that needs preservation. Overall planning demands that we have a classification of the value and importance of every area of marsh along the coast. . . .

Whatever method is used to preserve marshes, it must include safeguards against the increased pressures to develop because of the ever increasing population. There have been too many cases in which the last land in the town, land reserved for park and playground, has been diverted to industrial use. The diversion occurred because the industry threatened to move to another town or even another state if it were not allowed to secure the land. This sort of corporate blackmail is hard to withstand and will inevitably bring pressures on organizations controlling the marshes.

Pressure even comes from the state officials who are trying to encourage industries to come to their area by offering filled marsh for building. The battle between the forces of development and conservation need be won only once by the developers but must be fought and won every year for conservation to triumph.

In the last hundred years, our nation has been called on to preserve certain unique natural resources such as those contained in many of our national parks: Grand Canyon, Yosemite Valley, the Yellowstone hot springs,

Mammoth Cave, the Everglades, and the Cape Cod seashore. Now we are confronted with the problem of preserving a different sort of natural resource. This resource is much more extensive—the ribbon of green marshes along the eastern coast of North America, which must be preserved almost in its entirety if its preservation is to have any real meaning.

Critical Thinking

1. What is "corporate blackmail"?
2. In what sense is it more valuable to leave salt marshes alone than to fill them in and construct office parks, condos, and malls?
3. Should landowners have the right to do whatever they want with their land?

Human Domination of Earth's Ecosystems

Peter M. Vitousek, Harold A. Mooney, and Jerry M. Melillo

Political decisions that take into account the present and future impacts of human activities on the environment are necessary to preserve the long-term viability of the human species. This was not always true. For most of human history, the by-products of civilization and the ecological impacts of hunting, agricultural, and industrial activities were relatively small-scale and local. The industrial revolution brought about the potential to cause greater environmental devastation. Humans began to clear-cut forests, pollute the air, and degrade the water of major rivers and lakes. However, at the beginning of the twentieth century these effects were still primarily local and regional, rather than truly global.

In "Human Domination of Earth's Ecosystems," *Science* (July 25, 1997), from which the following selection is taken, Stanford University biologists Peter M. Vitousek and Harold Mooney, Oregon State University zoologist Jane Lubchenko, and Jerry M. Mellilo of the U.S. Office of Science and Technology Policy present a well-documented overview that illustrates the extent to which human activity has exerted a global impact on the Earth's ecosystems during the past century. They consider the consequences of land transformation, alterations of marine ecosystems, modifications of major biogeochemical cycles, and biotic disruptions. They report that almost 50% of the Earth's land surface has been transformed by human endeavors. The concentration of carbon dioxide in the atmosphere has been increased by 30%, more than half of all available fresh water is presently being used, and approximately 25% of the planet's bird species have been driven to extinction. Furthermore, they point out that activities that have resulted in these impacts are continuing, and in most cases, accelerating.

In the face of this grim assessment Vitousek et al. point out that humans have considerable power to control and reduce the negative impacts of our expanding environmental influence. They end their selection with several recommendations about how humankind should ensure responsible management of the planet in the future.

Key Concept: the impacts of accelerating human dominance of global ecosystems

All organisms modify their environment, and humans are no exception. As the human population has grown and the power of technology has expanded, the scope and nature of this modification has changed drastically. Until recently, the term "human-dominated ecosystems" would have elicited images of agricultural fields, pastures, or urban landscapes; now it applies with greater or lesser force to all of Earth. Many ecosystems are dominated directly by humanity, and no ecosystem on Earth's surface is free of pervasive human influence.

This [selection] provides an overview of human effects on Earth's ecosystems. It is not intended as a litany of environmental disaster, though some disastrous situations are described; nor is it intended either to downplay or to celebrate environmental successes, of which there have been many. Rather, we explore how large humanity looms as a presence on the globe—how, even on the grandest scale, most aspects of the structure and functioning of Earth's ecosystems cannot be understood without accounting for the strong, often dominant influence of humanity.

We view human alterations to the Earth system as operating through the interacting processes summarized in Figure 1. The growth of the human population, and growth in the resource base used by humanity, is maintained by a suite of human enterprises such as agriculture, industry, fishing, and international commerce. These enterprises transform the land surface (through cropping, forestry, and urbanization), alter the major biogeochemical cycles, and add or remove species and genetically distinct populations in most of Earth's ecosystems. Many of these changes are substantial and reasonably well quantified; all are ongoing. These relatively well-documented changes in turn entrain further alterations to the functioning of the Earth system, most notably by

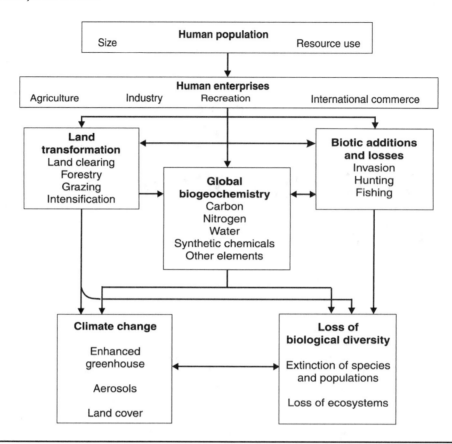

Figure 1 A conceptual model illustrating humanity's direct and indirect effects on the Earth system [modified from (*56*)].

driving global climatic change[1] and causing irreversible losses of biological diversity[2].

Land Transformation

The use of land to yield goods and services represents the most substantial human alteration of the Earth system. Human use of land alters the structure and functioning of ecosystems, and it alters how ecosystems interact with the atmosphere, with aquatic systems, and with surrounding land. Moreover, land transformation interacts strongly with most other components of global environmental change.

The measurement of land transformation on a global scale is challenging; changes can be measured more or less straightforwardly at a given site, but it is difficult to aggregate these changes regionally and globally. In contrast to analyses of human alteration of the global carbon cycle, we cannot install instruments on a tropical mountain to collect evidence of land transformation. Remote sensing is a most useful technique, but only recently has there been a serious scientific effort to use high-resolution civilian satellite imagery to evaluate even the more visible forms of land transformation, such as deforestation, on continental to global scales[3].

Land transformation encompasses a wide variety of activities that vary substantially in their intensity and consequences. At one extreme, 10 to 15% of Earth's land surface is occupied by row-crop agriculture or by urban-industrial areas, and another 6 to 8% has been converted to pastureland[4]; these systems are wholly changed by human activity. At the other extreme, every terrestrial ecosystem is affected by increased atmospheric carbon dioxide (CO_2), and most ecosystems have a history of hunting and other low-intensity resource extraction. Between these extremes lie grassland and semiarid ecosystems that are grazed (and sometimes degraded) by domestic animals, and forests and woodlands from which wood products have been harvested; together, these represent the majority of Earth's vegetated surface.

The variety of human effects on land makes any attempt to summarize land transformations globally a matter of semantics as well as substantial uncertainty. Estimates of the fraction of land transformed or degraded by humanity (or its corollary, the fraction of the land's biological production that is used or dominated) fall in the range of 39 to 50%[5] (Figure 2). These numbers have large uncertainties, but the fact that they are large is not at all uncertain. Moreover, if anything these estimates understate the global impact of land transformation, in that land that has not been transformed often has been divided into fragments by human alteration of the surrounding areas. This fragmentation affects the species

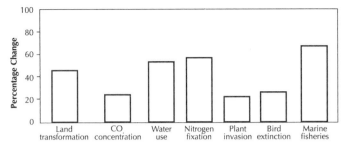

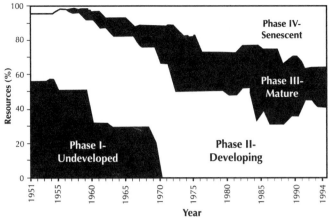

Figure 2 Human dominance or alteration of several major components of the Earth system, expressed as (from left to right) percentage of the land surface transformed (*5*); percentage of the current atmospheric CO_2 concentration that results from human action (*17*); percentage of accessible surface fresh water used (*20*); percentage of terrestrial N fixation that is human-caused (*28*); percentage of plant species in Canada that humanity has introduced from elsewhere (*48*); percentage of bird species on Earth that have become extinct in the past two millennia, almost all of them as a consequence of human activity (*42*); and percentage of major marine fisheries that are fully exploited, overexploited, or depleted (*14*).

Figure 3 Percentage of major world marine fish resources in different phases of development, 1951 to 1994 [from (*57*)]. Undeveloped = a low and relatively constant level of catches; developing = rapidly increasing catches; mature = a high and plateauing level of catches; senescent = catches declining from higher levels.

composition and functioning of otherwise little modified ecosystems[6].

Overall, land transformation represents the primary driving force in the loss of biological diversity worldwide. Moreover, the effects of land transformation extend far beyond the boundaries of transformed lands. Land transformation can affect climate directly at local and even regional scales. It contributes 20% to current anthropogenic CO_2 emissions, and more substantially to the increasing concentrations of the greenhouse gases methane and nitrous oxide; fires associated with it alter the reactive chemistry of the troposphere, bringing elevated carbon monoxide concentrations and episodes of urban-like photochemical air pollution to remote tropical areas of Africa and South America; and it causes runoff of sediment and nutrients that drive substantial changes in stream, lake, estuarine, and coral reef ecosystems[7-10].

The central importance of land transformation is well recognized within the community of researchers concerned with global environmental change. Several research programs are focused on aspects of it[9, 11]; recent and substantial progress toward understanding these aspects has been made[3], and much more progress can be anticipated. Understanding land transformation is a difficult challenge; it requires integrating the social, economic, and cultural causes of land transformation with evaluations of its biophysical nature and consequences. This interdisciplinary approach is essential to predicting the course, and to any hope of affecting the consequences, of human-caused land transformation.

Oceans

Human alterations of marine ecosystems are more difficult to quantify than those of terrestrial ecosystems, but several kinds of information suggest that they are substantial. The human population is concentrated near coasts—about 60% within 100 km—and the oceans' productive coastal margins have been affected strongly by humanity. Coastal wetlands that mediate interactions between land and sea have been altered over large areas; for example, approximately 50% of mangrove ecosystems globally have been transformed or destroyed by human activity[12]. Moreover, a recent analysis suggested that although humans use about 8% of the primary production of the oceans, that fraction grows to more than 25% for upwelling areas and to 35% for temperate continental shelf systems[13].

Many of the fisheries that capture marine productivity are focused on top predators, whose removal can alter marine ecosystems out of proportion to their abundance. Moreover, many such fisheries have proved to be unsustainable, at least at our present level of knowledge and control. As of 1995, 22% of recognized marine fisheries were overexploited or already depleted, and 44% more were at their limit of exploitation[14] (Figures 2 and 3). The consequences of fisheries are not restricted to their target organisms; commercial marine fisheries around the world discard 27 million tons of nontarget animals annually, a quantity nearly one-third as large as total landings[15]. Moreover, the dredges and trawls used in some fisheries damage habitats substantially as they are dragged along the sea floor.

A recent increase in the frequency, extent, and duration of harmful algal blooms in coastal areas[16] suggests that human activity has affected the base as well as the top of marine food chains. Harmful algal blooms are sudden increases in the abundance of marine phytoplankton that produce harmful structures or chemicals. Some but not all of these phytoplankton are strongly pigmented (red or brown tides). Algal blooms usually are correlated with changes in temperature, nutrients, or salinity; nutrients in coastal waters, in particular, are much modified by human activity. Algal blooms can cause extensive fish kills through toxins and by causing anoxia; they also lead to paralytic shellfish poisoning and amnesic shellfish poisoning in humans. Although the existence of harmful algal blooms has long been recognized, they have spread widely in the past two decades[16].

Alterations of the Biogeochemical Cycles

Carbon. Life on Earth is based on carbon, and the CO_2 in the atmosphere is the primary resource for photosynthesis. Humanity adds CO_2 to the atmosphere by mining and burning fossil fuels, the residue of life from the distant past, and by converting forests and grasslands to agricultural and other low-biomass ecosystems. The net result of both activities is that organic carbon from rocks, organisms, and soils is released into the atmosphere as CO_2.

The modern increase in CO_2 represents the clearest and best documented signal of human alteration of the Earth system. Thanks to the foresight of Roger Revelle, Charles Keeling, and others who initiated careful and systematic measurements of atmospheric CO_2 in 1957 and sustained them through budget crises and changes in scientific fashions, we have observed the concentration of CO_2 as it has increased steadily from 315 ppm to 362 ppm. Analysis of air bubbles extracted from the Antarctic and Greenland ice caps extends the record back much further; the CO_2 concentration was more or less stable near 280 ppm for thousands of years until about 1800, and has increased exponentially since then[17].

There is no doubt that this increase has been driven by human activity, today primarily by fossil fuel combustion. The sources of CO_2 can be traced isotopically; before the period of extensive nuclear testing in the atmosphere, carbon depleted in ^{14}C was a specific tracer of CO_2 derived from fossil fuel combustion, whereas carbon depleted in ^{13}C characterized CO_2 from both fossil fuels and land transformation. Direct measurements in the atmosphere, and analyses of carbon isotopes in tree rings, show that both ^{13}C and ^{14}C in CO_2 were diluted in the atmosphere relative to ^{12}C as the CO_2 concentration in the atmosphere increased.

Fossil fuel combustion now adds 5.5 ± 0.5 billion metric tons of CO_2-C to the atmosphere annually, mostly in economically developed regions of the temperate zone[18]. The annual accumulation of CO_2-C has averaged 3.2 ± 0.2 billion metric tons recently[17]. The other major terms in the atmospheric carbon balance are net ocean-atmosphere flux, net release of carbon during land transformation, and net storage in terrestrial biomass and soil organic matter. All of these terms are smaller and less certain than fossil fuel combustion or annual atmospheric accumulation; they represent rich areas of current research, analysis, and sometimes contention.

The human-caused increase in atmospheric CO_2 already represents nearly a 30% change relative to the pre-industrial era (Figure 2), and CO_2 will continue to increase for the foreseeable future. Increased CO_2 represents the most important human enhancement to the greenhouse effect; the consensus of the climate research community is that it probably already affects climate detectably and will drive substantial climate change in the next century[1]. The direct effects of increased CO_2 on plants and ecosystems may be even more important. The growth of most plants is enhanced by elevated CO_2, but to very different extents; the tissue chemistry of plants that respond to CO_2 is altered in ways that decrease food quality for animals and microbes; and the water use efficiency of plants and ecosystems generally is increased. The fact that increased CO_2 affects species differentially means that it is likely to drive substantial changes in the species composition and dynamics of all terrestrial ecosystems[19].

Water. Water is essential to all life. Its movement by gravity, and through evaporation and condensation, contributes to driving Earth's biogeochemical cycles and to controlling its climate. Very little of the water on Earth is directly usable by humans; most is either saline or frozen. Globally, humanity now uses more than half of the runoff water that is fresh and reasonably accessible, with about 70% of this use in agriculture[20] (Figure 2). To meet increasing demands for the limited supply of fresh water, humanity has extensively altered river systems through diversions and impoundments. In the United States only 2% of the rivers run unimpeded, and by the end of this century the flow of about two-thirds of all of Earth's rivers will be regulated[21]. At present, as much as 6% of Earth's river runoff is evaporated as a consequence of human manipulations[22]. Major rivers, including the Colorado, the Nile, and the Ganges, are used so extensively that little water reaches the sea. Massive inland water bodies, including the Aral Sea and Lake Chad, have been greatly reduced in extent by water diversions for agriculture. Reduction in the volume of the Aral Sea resulted in the demise of native fishes and the loss of other biota; the loss of a major fishery; exposure of the salt-laden sea bottom, thereby providing a major source of windblown dust; the production of a drier and more continental local climate and a decrease in water quality in the general region; and an increase in human diseases[23].

Impounding and impeding the flow of rivers provides reservoirs of water that can be used for energy generation as well as for agriculture. Waterways also are managed for transport, for flood control, and for the dilution of chemical wastes. Together, these activities have altered Earth's freshwater ecosystems profoundly, to a greater extent than terrestrial ecosystems have been altered. The construction of dams affects biotic habitats indirectly as well; the damming of the Danube River, for example, has altered the silica chemistry of the entire Black Sea. The large number of operational dams (36,000) in the world, in conjunction with the many that are planned, ensure that humanity's effects on aquatic biological systems will continue[24]. Where surface water is sparse or overexploited, humans use groundwater—and in many areas the groundwater that is drawn upon is nonrenewable, or fossil, water[25]. For example, three-quarters of the water supply of Saudi Arabia currently comes from fossil water[26].

Alterations to the hydrological cycle can affect regional climate. Irrigation increases atmospheric humidity in semiarid areas, often increasing precipitation and thunderstorm frequency[27]. In contrast, land transformation from forest to agriculture or pasture increases albedo and decreases surface roughness; simulations suggest that the net effect of this transformation is to increase temperature and decrease precipitation regionally[7, 26].

Conflicts arising from the global use of water will be exacerbated in the years ahead, with a growing human population and with the stresses that global changes will impose on water quality and availability. Of all of the environmental security issues facing nations, an adequate supply of clean water will be the most important.

Nitrogen. Nitrogen (N) is unique among the major elements required for life, in that its cycle includes a vast atmospheric reservoir (N_2) that must be fixed (combined with carbon, hydrogen, or oxygen) before it can be used by most organisms. The supply of this fixed N controls (at least in part) the productivity, carbon storage, and species composition of many ecosystems. Before the extensive human alteration of the N cycle, 90 to 130 million metric tons of N (Tg N) were fixed biologically on land each year; rates of biological fixation in marine systems are less certain, but perhaps as much was fixed there[28].

Human activity has altered the global cycle of N substantially by fixing N_2—deliberately for fertilizer and inadvertently during fossil fuel combustion. Industrial fixation of N fertilizer increased from <10 Tg/year in 1950 to 80 Tg/year in 1990; after a brief dip caused by economic dislocations in the former Soviet Union, it is expected to increase to >135 Tg/year by 2030[29]. Cultivation of soybeans, alfalfa, and other legume crops that fix N symbiotically enhances fixation by another ~40 Tg/year, and fossil fuel combustion puts >20 Tg/year of reactive N into the atmosphere globally—some by fixing N_2, more from the mobilization of N in the fuel. Overall, human activity adds at least as much fixed N to terrestrial ecosystems as do all natural sources combined (Figure 2), and it mobilizes >50 Tg/year more during land transformation[28, 30].

Alteration of the N cycle has multiple consequences. In the atmosphere, these include (i) an increasing concentration of the greenhouse gas nitrous oxide globally; (ii) substantial increases in fluxes of reactive N gases (two-thirds or more of both nitric oxide and ammonia emissions globally are human-caused); and (iii) a substantial contribution to acid rain and to the photochemical smog that afflicts urban and agricultural areas throughout the world[31]. Reactive N that is emitted to the atmosphere is deposited downwind, where it can influence the dynamics of recipient ecosystems. In regions where fixed N was in short supply, added N generally increases productivity and C storage within ecosystems, and ultimately increases losses of N and cations from soils, in a set of processes termed "N saturation"[32]. Where added N increases the productivity of ecosystems, usually it also decreases their biological diversity[33].

Human-fixed N also can move from agriculture, from sewage systems, and from N-saturated terrestrial systems to streams, rivers, groundwater, and ultimately the oceans. Fluxes of N through streams and rivers have increased markedly as human alteration of the N cycle has accelerated; river nitrate is highly correlated with the human population of river basins and with the sum of human-caused N inputs to those basins[8]. Increases in river N drive the eutrophication of most estuaries, causing blooms of nuisance and even toxic algae, and threatening the sustainability of marine fisheries[16, 34].

Other cycles. The cycles of carbon, water, and nitrogen are not alone in being altered by human activity. Humanity is also the largest source of oxidized sulfur gases in the atmosphere; these affect regional air quality, biogeochemistry, and climate. Moreover, mining and mobilization of phosphorus and of many metals exceed their natural fluxes; some of the metals that are concentrated and mobilized are highly toxic (including lead, cadmium, and mercury)[35]. Beyond any doubt, humanity is a major biogeochemical force on Earth.

Synthetic organic chemicals. Synthetic organic chemicals have brought humanity many beneficial services. However, many are toxic to humans and other species, and some are hazardous in concentrations as low as 1 part per billion. Many chemicals persist in the environment for decades; some are both toxic and persistent. Long-lived organochlorine compounds provide the clearest examples of environmental consequences of persistent compounds. Insecticides such as DDT and its relatives, and industrial compounds like polychlorinated biphenyls (PCBs), were used widely in North America in the 1950s and 1960s. They were transported globally, accumulated in organisms, and magnified in concentration through food chains; they devastated populations of some predators (notably falcons and eagles) and entered parts of the human food supply in concentrations higher than was prudent. Domestic use of these compounds was phased out in the 1970s in the United States and Canada, and their concentrations declined thereafter. However, PCBs in particular remain readily detectable in many organisms, sometimes approaching thresholds of public health concern[36]. They will continue to circulate through organisms for many decades.

Synthetic chemicals need not be toxic to cause environmental problems. The fact that the persistent and volatile chlorofluorocarbons (CFCs) are wholly nontoxic contributed to their widespread use as refrigerants and even aerosol propellants. The subsequent discovery that CFCs drive the breakdown of stratospheric ozone, and especially the later discovery of the Antarctic ozone hole and their role in it, represent great surprises in global environmental science[37]. Moreover, the response of the international political system to those discoveries is the best extant illustration that global environmental change can be dealt with effectively[38].

Particular compounds that pose serious health and environmental threats can be and often have been phased out (although PCB production is growing in Asia). Nonetheless, each year the chemical industry produces more than 100 million tons of organic chemicals representing some 70,000 different compounds, with about 1000 new ones being added annually[39]. Only a small fraction of the many chemicals produced and released into the environment are tested adequately for health hazards or environmental impact[40].

Biotic Changes

Human modification of Earth's biological resources—its species and genetically distinct populations—is substantial and growing. Extinction is a natural process, but the current rate of loss of genetic variability, of populations, and of species is far above background rates; it is ongoing; and it represents a wholly irreversible global change. At the same time, human transport of species around Earth is homogenizing Earth's biota, introducing many species into new areas where they can disrupt both natural and human systems.

Losses. Rates of extinction are difficult to determine globally, in part because the majority of species on Earth have not yet been identified. Nevertheless, recent calculations suggest that rates of species extinction are now on the order of 100 to 1000 times those before humanity's dominance of Earth[41]. For particular well-known groups, rates of loss are even greater; as many as one-quarter of Earth's bird species have been driven to extinction by human activities over the past two millennia, particularly on oceanic islands[42] (Figure 2). At present, 11% of the remaining birds, 18% of the mammals, 5% of fish, and 8% of plant species on Earth are threatened with extinction[43]. There has been a disproportionate loss of large mammal species because of hunting; these species played a dominant role in many ecosystems, and their loss has resulted in a fundamental change in the dynamics of those systems[44], one that could lead to further extinctions. The

largest organisms in marine systems have been affected similarly, by fishing and whaling. Land transformation is the single most important cause of extinction, and current rates of land transformation eventually will drive many more species to extinction, although with a time lag that masks the true dimensions of the crisis[45]. Moreover, the effects of other components of global environmental change—of altered carbon and nitrogen cycles, and of anthropogenic climate change—are just beginning.

As high as they are, these losses of species understate the magnitude of loss of genetic variation. The loss to land transformation of locally adapted populations within species, and of genetic material within populations, is a human-caused change that reduces the resilience of species and ecosystems while precluding human use of the library of natural products and genetic material that they represent[46].

Although conservation efforts focused on individual endangered species have yielded some successes, they are expensive—and the protection or restoration of whole ecosystems often represents the most effective way to sustain genetic, population, and species diversity. Moreover, ecosystems themselves may play important roles in both natural and human-dominated landscapes. For example, mangrove ecosystems protect coastal areas from erosion and provide nurseries for offshore fisheries, but they are threatened by transformation in many areas.

Invasions. In addition to extinction, humanity has caused a rearrangement of Earth's biotic systems, through the mixing of floras and faunas that had long been isolated geographically. The magnitude of transport of species, termed "biological invasion," is enormous[47]; invading species are present almost everywhere. On many islands, more than half of the plant species are non-indigenous, and in many continental areas the Figure is 20% or more[48] (Figure 2).

As with extinction, biological invasion occurs naturally—and as with extinction, human activity has accelerated its rate by orders of magnitude. Land transformation interacts strongly with biological invasion, in that human-altered ecosystems generally provide the primary foci for invasions, while in some cases land transformation itself is driven by biological invasions[49]. International commerce is also a primary cause of the breakdown of biogeographic barriers; trade in live organisms is massive and global, and many other organisms are inadvertently taken along for the ride. In freshwater systems, the combination of upstream land transformation, altered hydrology, and numerous deliberate and accidental species introductions has led to particularly widespread invasion, in continental as well as island ecosystems[50].

In some regions, invasions are becoming more frequent. For example, in the San Francisco Bay of California, an average of one new species has been established every 36 weeks since 1850, every 24 weeks since 1970, and every 12 weeks for the last decade[51]. Some introduced species quickly become invasive over large areas (for example, the Asian clam in the San Francisco Bay), whereas others become widespread only after a lag of decades, or even over a century[52].

Many biological invasions are effectively irreversible; once replicating biological material is released into the environment and becomes successful there, calling it back is difficult and expensive at best. Moreover, some species introductions have consequences. Some degrade human health and that of other species; after all, most infectious diseases are invaders over most of their range. Others have caused economic losses amounting to billions of dollars; the recent invasion of North America by the zebra mussel is a well-publicized example. Some disrupt ecosystem processes, altering the structure and functioning of whole ecosystems. Finally, some invasions drive losses in the biological diversity of native species and populations; after land transformation, they are the next most important cause of extinction[53].

Conclusions

The global consequences of human activity are not something to face in the future—as Figure 2 illustrates, they are with us now. All of these changes are ongoing, and in many cases accelerating; many of them were entrained long before their importance was recognized. Moreover, all of these seemingly disparate phenomena trace to a single cause—the growing scale of the human enterprise. The rates, scales, kinds, and combinations of changes occurring now are fundamentally different from those at any other time in history; we are changing Earth more rapidly than we are understanding it. We live on a human-dominated planet—and the momentum of human population growth, together with the imperative for further economic development in most of the world, ensures that our dominance will increase.

The [information in these pages] summarize our knowledge of and provide specific policy recommendations concerning major human-dominated ecosystems. In addition, we suggest that the rate and extent of human alteration of Earth should affect how we think about Earth. It is clear that we control much of Earth, and that our activities affect the rest. In a very real sense, the world is in our hands—and how we handle it will determine its composition and dynamics, and our fate.

Recognition of the global consequences of the human enterprise suggests three complementary directions. First, we can work to reduce the rate at which we alter the Earth system. Humans and human-dominated systems may be able to adapt to slower change, and ecosystems and the species they support may cope more effectively with the changes we impose, if those changes are slow. Our footprint on the planet[54] might then be stabilized at a point where enough space and resources remain to sustain most of the other species on Earth, for their sake and our own. Reducing the rate of growth in human effects on Earth involves slowing human population growth and using resources as efficiently as is practical. Often it is the waste products and by-products of human activity that drive global environmental change.

Second, we can accelerate our efforts to understand Earth's ecosystems and how they interact with the numerous components of human-caused global change. Ecological research is inherently complex and demanding: It requires measurement and monitoring of populations and ecosystems; experimental studies to elucidate the regulation of ecological processes; the development, testing, and validation of regional and global models; and integration with a broad range of biological, earth, atmospheric, and marine sciences. The challenge of understanding a human-dominated planet further requires that the human dimensions of global change—the social, economic, cultural, and other drivers of human actions—be included within our analyses.

Finally, humanity's dominance of Earth means that we cannot escape responsibility for managing the planet. Our activities are causing rapid, novel, and substantial changes to Earth's ecosystems. Maintaining populations, species, and ecosystems in the face of those changes, and maintaining the flow of goods and services they provide humanity[55], will require active management for the foreseeable future. There is no clearer illustration of the extent of human dominance of Earth than the fact that maintaining the diversity of "wild" species and the functioning of "wild" ecosystems will require increasing human involvement.

References

1. Intergovernmental Panel on Climate Change, *Climate Change 1995* (Cambridge Univ. Press, Cambridge, 1996), pp. 49.
2. United Nations Environment Program, *Global Biodiversity Assessment*, V. H. Heywood, Ed. (Cambridge Univ. Press, Cambridge, 1995).
3. D. Skole and C. J. Tucker, *Science* **260**, 1905 (1993).
4. J. S. Olson, J. A. Watts, L. J. Allison, *Carbon in Live Vegetation of Major World Ecosystems* (Office of Energy Research, U.S. Department of Energy, Washington, DC, 1983).
5. P. M. Vitousek, P. R. Ehrlich, A. H. Ehrlich, P. A. Matson, *Bioscience* **36**, 368 (1986); R. W. Kates, B. L. Turner, W. C. Clark, in (35), pp. 1–17; G. C. Daily, *Science* **269**, 350 (1995).
6. D. A. Saunders, R. J. Hobbs, C. R. Margules, *Conserv. Biol.* **5**, 18 (1991).
7. J. Shukla, C. Nobre, P. Sellers, *Science* **247**, 1322 (1990).
8. R. W. Howarth *et al.*, *Biogeochemistry* **35**, 75 (1996).
9. W. B. Meyer and B. L. Turner II, *Changes in Land Use and Land Cover: A Global Perspective* (Cambridge Univ. Press, Cambridge, 1994).
10. S. R. Carpenter, S. G. Fisher, N. B. Grimm, J. F. Kitchell, *Annu. Rev. Ecol. Syst.* **23**, 119 (1992); S. V. Smith and R. W. Buddemeier, *ibid.*, p. 89; J. M. Melillo, I. C. Prentice, G. D. Farquhar, E.-D. Schulze, O. E. Sala, in (1), pp. 449–481.
11. R. Leemans and G. Zuidema, *Trends Ecol. Evol.* **10**, 76 (1995).

12. World Resources Institute, *World Resources 1996–1997* (Oxford Univ. Press, New York, 1996).

13. D. Pauly and V. Christensen, *Nature* **374**, 257 (1995).

14. Food and Agricultural Organization (FAO), *FAO Fisheries Tech. Pap. 335* (1994).

15. D. L. Alverson, M. H. Freeberg, S. A. Murawski, J. G. Pope, *FAO Fisheries Tech. Pap. 339* (1994).

16. G. M. Hallegraeff, *Phycologia* **32**, 79 (1993).

17. D. S. Schimel *et al.*, in *Climate Change 1994: Radiative Forcing of Climate Change*, J. T. Houghton *et al.*, Eds. (Cambridge Univ. Press, Cambridge, 1995), pp. 39–71.

18. R. J. Andres, G. Marland, I. Y. Fung, E. Matthews, *Global Biogeochem. Cycles* **10**, 419 (1996).

19. G. W. Koch and H. A. Mooney, *Carbon Dioxide and Terrestrial Ecosystems* (Academic Press, San Diego, CA, 1996); C. Körner and F. A. Bazzaz, *Carbon Dioxide, Populations, and Communities* (Academic Press, San Diego, CA, 1996).

20. S. L. Postel, G. C. Daily, P. R. Ehrlich, *Science* **271**, 785 (1996).

21. J. N. Abramovitz, *Imperiled Waters, Impoverished Future: The Decline of Freshwater Ecosystems* (Worldwatch Institute, Washington. DC, 1996).

22. M. I. L'vovich and G. F. White, in (35), pp. 235–252; M. Dynesius and C. Nilsson, *Science* **266**, 753 (1994).

23. P. Micklin, *Science* **241**, 1170 (1988); V. Kotlyakov, *Environment* **33**, 4 (1991).

24. C. Humborg, V. Ittekkot, A. Cociasu, B. Bodungen, *Nature* **386**, 385 (1997).

25. P. H. Gleick, Ed., *Water in Crisis* (Oxford Univ. Press, New York, 1993).

26. V. Gornitz, C. Rosenzweig, D. Hillel, *Global Planet Change* **14**, 147 (1997).

27. P. C. Milly and K. A. Dunne, *J. Clim.* **7**, 506 (1994).

28. J. N. Galloway, W. H. Schlesinger, H. Levy II, A. Michaels, J. L. Schnoor, *Global Biogeochem. Cycles* **9**, 235 (1995).

29. J. N. Galloway, H. Levy II, P. S. Kasibhatla, *Ambio* **23**, 120 (1994).

30. V. Smil, in (35), pp. 423–436.

31. P. M. Vitousek *et al.*, *Ecol. Appl.*, in press.

32. J. D. Aber, J. M. Melillo, K. J. Nadelhoffer, J. Pastor, R. D. Boone, *ibid.* **1**, 303 (1991).

33. D. Tilman, *Ecol. Monogr.* **57**, 189 (1987).

34. S. W. Nixon *et al.*, *Biogeochemistry* **35**, 141 (1996).

35. B. L. Turner II *et al.*, Eds., *The Earth As Transformed by Human Action* (Cambridge Univ. Press, Cambridge, 1990).

36. C. A. Stow, S. R. Carpenter, C. P. Madenjian, L. A. Eby, L. J. Jackson, *Bioscience* **45**, 752 (1995).

37. F. S. Rowland, *Am. Sci.* **77**, 36 (1989): S. Solomon, *Nature* **347**, 347 (1990).

38. M. K. Tolba *et al.*, Eds., *The World Environment 1972–1992* (Chapman & Hall, London, 1992).

39. S. Postel, *Defusing the Toxics Threat: Controlling Pesticides and Industrial Waste* (World-watch Institute, Washington, DC, 1987).

40. United Nations Environment Program (UNEP), *Saving Our Planet—Challenges and Hopes* (UNEP, Nairobi, 1992).

41. J. H. Lawton and R. M. May, Eds., *Extinction Rates* (Oxford Univ. Press, Oxford, 1995); S. L. Pimm, G. J. Russell, J. L. Gittleman, T. Brooks, *Science* **269**, 347 (1995).

42. S. L. Olson, in *Conservation for the Twenty-First Century*, D. Western and M. C. Pearl, Eds. (Oxford Univ. Press, Oxford, 1989), p. 50; D. W. Steadman, *Science* **267**, 1123 (1995).

43. R. Barbault and S. Sastrapradja, in (2), pp. 193–274.

44. R. Dirzo and A. Miranda, in *Plant-Animal Interactions*, P. W. Price, T. M. Lewinsohn, W. Fernandes, W. W. Benson, Eds. (Wiley Interscience, New York, 1991), p. 273.

45. D. Tilman, R. M. May, C. Lehman, M. A. Nowak, *Nature* **371**, 65 (1994).

46. H. A. Mooney, J. Lubchenco, R. Dirzo, O. E. Sala, in (2), pp. 279–325.

47. C. Elton, *The Ecology of Invasions by Animals and Plants* (Methuen, London, 1958); J. A. Drake *et al.*, Eds., *Biological Invasions. A Global Perspective* (Wiley, Chichester, UK, 1989).

48. M. Rejmanek and J. Randall, *Madrono* **41**, 161 (1994).

49. C. M. D'Antonio and P. M. Vitousek, *Annu. Rev. Ecol. Syst.* **23**, 63 (1992).

50. D. M. Lodge, *Trends Ecol. Evol.* **8**, 133 (1993).

51. A. N. Cohen and J. T. Carlton, *Biological Study: Nonindigenous Aquatic Species in a United States Estuary: A Case Study of the Biological Invasions of the San Francisco Bay and Delta* (U.S. Fish and Wildlife Service, Washington, DC, 1995).

52. I. Kowarik, in *Plant Invasions—General Aspects and Special Problems*, P. Pysek, K. Prach, M. Rejmánek, M. Wade, Eds. (SPB Academic, Amsterdam, 1995), p. 15.

53. P. M. Vitousek, C. M. D'Antonio, L. L. Loope, R. Westbrooks, *Am. Sci.* **84**, 468 (1996).

54. W. E. Rees and M. Wackernagel, in *Investing in Natural Capital: The Ecological Economics Approach to Sustainability*, A. M. Jansson, M. Hammer, C. Folke, R. Costanza, Eds. (Island, Washington, DC, 1994).

55. G. C. Daily, Ed., *Nature's Services* (Island, Washington, DC, 1997).

56. Lubchenco *et al.*, *Ecology* **72**, 371 (1991); P. M. Vitousek, *ibid.* **75**, 1861 (1994).

57. S. M. Garcia and R. Grainger, *FAO Fisheries Tech. Pap. 359* (1996).

58. We thank G. C. Daily, C. B. Field, S. Hobbie, D. Gordon, P. A. Matson, and R. L. Naylor for constructive comments on this paper, A. S. Denning and S. M. Garcia for assistance with illustrations, and C. Nakashima and B. Lilley for preparing text and figures for publication.

Critical Thinking

1. Almost 50% of the land surface has been transformed by human endeavors and more than half of all available fresh water is presently being used by humans. We thus have a long way to go before we run out of land or water. What is wrong with that statement?

2. Would it help to reduce the human impact on the Earth if we could reduce the human population?

Ecosystems and Human Well-being

Millennium Ecosystem Assessment, 2005

In 2000, United Nations Secretary-General Kofi Annan called for the Millennium Ecosystem Assessment. The project was initiated in 2001, with its objective being to assess the consequences of ecosystem change for human well-being, the prospects that the Millennium Development goals[1] will be met, and the scientific basis for action needed to enhance the conservation and sustainable use of ecosystems. This meant pulling together a great deal of information on ecosystem changes and impacts on human well-being to date, on the conditions that lead to those changes and impacts, and on the prospects for further changes, given various patterns (or scenarios) of national and international economics, conflict, cooperation, and approaches to environmental management. The results, published in five technical volumes and six synthesis reports (see www.maweb.org/en/index.aspx), provide a state-of-the-art scientific appraisal of the condition of and trends in the world's ecosystems and the services they provide (such as clean water, food, flood control, and natural resources) and the prospects for a sustainable future.

Under some scenarios, the prospects are relatively rosy. But those scenarios call for using environmentally sound technology, managing and even engineering ecosystems to provide services, and anticipating problems and taking early steps to prevent them. The scenario that most resembles a continuation of the present—Order from Strength—is bleak, for it forecasts declining economic growth rates and rapidly growing population.

Key Concept: the need for substantial, basic changes in government, economics, technology, social behavior, and knowledge to achieve a sustainable future

Summary for Decision-Makers

Everyone in the world depends completely on Earth's ecosystems and the services they provide, such as food, water, disease management, climate regulation, spiritual fulfillment, and aesthetic enjoyment. Over the past 50 years, humans have changed these ecosystems more rapidly and extensively than in any comparable period of time in human history, largely to meet rapidly growing demands for food, fresh water, timber, fiber, and fuel. This transformation of the planet has contributed to substantial net gains in human well-being and economic development. But not all regions and groups of people have benefited from this process—in fact, many have been harmed. Moreover, the full costs associated with these gains are only now becoming apparent.

Three major problems associated with our management of the world's ecosystems are already causing significant harm to some people, particularly the poor, and unless addressed will substantially diminish the long-term benefits we obtain from ecosystems:

- First, approximately 60% (15 out of 24) of the ecosystem services examined during the Millennium Ecosystem Assessment are being degraded or used unsustainably, including fresh water, capture fisheries, air and water purification, and the regulation of regional and local climate, natural hazards, and

---

[1] The eight Millennium Development Goals are: eradicating extreme poverty and hunger, achieving universal primary education, promoting gender equality and empowering women, reducing child mortality, improving maternal health, combatting HIV/AIDS, malaria, and other diseases, ensuring environmental sustainability, and developing a global partnership for development. See www.un.org/millenniumgoals/.

pests. The full costs of the loss and degradation of these ecosystem services are difficult to measure, but the available evidence demonstrates that they are substantial and growing. Many ecosystem services have been degraded as a consequence of actions taken to increase the supply of other services, such as food. These trade-offs often shift the costs of degradation from one group of people to another or defer costs to future generations.

- Second, there is *established but incomplete* evidence that changes being made in ecosystems are increasing the likelihood of nonlinear changes in ecosystems (including accelerating, abrupt, and potentially irreversible changes) that have important consequences for human well-being. Examples of such changes include disease emergence, abrupt alterations in water quality, the creation of "dead zones" in coastal waters, the collapse of fisheries, and shifts in regional climate.

- Third, the harmful effects of the degradation of ecosystem services (the persistent decrease in the capacity of an ecosystem to deliver services) are being borne disproportionately by the poor, are contributing to growing inequities and disparities across groups of people, and are sometimes the principal factor causing poverty and social conflict. This is not to say that ecosystem changes such as increased food production have not also helped to lift many people out of poverty or hunger, but these changes have harmed other individuals and communities, and their plight has been largely overlooked. In all regions, and particularly in sub-Saharan Africa, the condition and management of ecosystem services is a dominant factor influencing prospects for reducing poverty.

The degradation of ecosystem services is already a significant barrier to achieving the Millennium Development Goals agreed to by the international community in September 2000 and the harmful consequences of this degradation could grow significantly worse in the next 50 years. The consumption of ecosystem services, which is unsustainable in many cases, will continue to grow as a consequence of a likely three- to sixfold increase in global GDP by 2050 even while global population growth is expected to slow and level off in mid-century. Most of the important direct drivers of ecosystem change are unlikely to diminish in the first half of the century and two drivers—climate change and excessive nutrient loading—will become more severe.

Already, many of the regions facing the greatest challenges in achieving the MDGs coincide with those facing significant problems of ecosystem degradation. Rural poor people, a primary target of the MDGs, tend to be most directly reliant on ecosystem services and most vulnerable to changes in those services. More generally, any progress achieved in addressing the MDGs of poverty and hunger eradication, improved health, and environmental sustainability is unlikely to be sustained if most of the ecosystem services on which humanity relies continue to be degraded. In contrast, the sound management of ecosystem services provides cost-effective opportunities for addressing multiple development goals in a synergistic manner.

There is no simple fix to these problems since they arise from the interaction of many recognized challenges, including climate change, biodiversity loss, and land degradation, each of which is complex to address in its own right. Past actions to slow or reverse the degradation of ecosystems have yielded significant benefits, but these improvements have generally not kept pace with growing pressures and demands. Nevertheless, there is tremendous scope for action to reduce the severity of these problems in the coming decades. Indeed, three of four detailed scenarios examined by the MA suggest that significant changes in policies, institutions, and practices can mitigate some but not all of the negative consequences of growing pressures on ecosystems. But the changes required are substantial and are not currently under way.

An effective set of responses to ensure the sustainable management of ecosystems requires substantial changes in institutions and governance, economic policies and incentives, social and behavior factors, technology, and knowledge. Actions such as the integration of ecosystem management goals in various sectors (such as agriculture, forestry, finance, trade, and health), increased transparency and accountability of government and private-sector performance in ecosystem management, elimination of perverse subsidies, greater use of economic instruments and market-based approaches, empowerment of groups dependent on ecosystem services or affected by their degradation, promotion of technologies enabling increased crop yields without harmful environmental impacts, ecosystem restoration, and the incorporation of nonmarket values of ecosystems and their services in management decisions all could substantially lessen the severity of these problems in the next several decades.

The remainder of this Summary for Decision-makers presents the four major findings of the Millennium Ecosystem Assessment on the problems to be addressed and the actions needed to enhance the conservation and sustainable use of ecosystems.

> Finding #1: Over the past 50 years, humans have changed ecosystems more rapidly and extensively than in any comparable period of time in human history, largely to meet rapidly growing demands for food, fresh water, timber, fiber, and fuel. This has resulted in a substantial and largely irreversible loss in the diversity of life on Earth.

The structure and functioning of the world's ecosystems changed more rapidly in the second half of the twentieth century than at any time in human history.

More land was converted to cropland in the 30 years after 1950 than in the 150 years between 1700 and 1850. Cultivated systems (areas where at least 30% of the landscape is in croplands, shifting cultivation, confined liveslock production, or freshwater aquaculture) now cover one quarter of Earth's terrestrial surface.

Approximately 20% of the world's coral reefs were lost and an additional 20% degraded in the last several decades of the twentieth century, and approximately 35% of mangrove area was lost during this time (in countries for which sufficient data exist, which encompass about half of the area of mangroves).

The amount of water impounded behind dams quadrupled since 1960, and three to six times as much water is held in reservoirs as in natural rivers. Water withdrawals from rivers and lakes doubled since 1960; most water use (70% worldwide) is for agriculture.

Since 1960, flows of reactive (biologically available) nitrogen in terrestrial ecosystems have doubled, and flows of phosphorus have tripled. More than half of all the synthetic nitrogen fertilizer, which was first manufactured in 1913, ever used on the planet has been used since 1985.

Since 1750, the atmospheric concentration of carbon dioxide has increased by about 32% (from about 280 to 376 parts per million in 2003), primarily due to the combustion of fossil fuels and land use changes. Approximately 60% of that increase (60 parts per million) has taken place since 1959.

Humans are fundamentally, and to a significant extent irreversibly, changing the diversity of life on Earth, and most of these changes represent a loss of biodiversity.

More than two thirds of the area of 2 of the world's 14 major terrestrial biomes and more than half of the area of 4 other biomes had been converted by 1990, primarily to agriculture.

Across a range of taxonomic groups, either the population size or range or both of the majority of species is currently declining.

The distribution of species on Earth is becoming more homogenous; in other words, the set of species in any one region of the world is becoming more similar to the set in other regions primarily as a result of introductions of species, both intentionally and inadvertently in association with increased travel and shipping.

The number of species on the planet is declining. Over the past few hundred years, humans have increased the species extinction rate by as much as 1,000 times over background rates typical over the planet's history (*medium certainty*). Some 10–30% of mammal, bird, and amphibian species are currently threatened with extinction (*medium to high certainty*). Freshwater ecosystems tend to have the highest proportion of species threatened with extinction.

Genetic diversity has declined globally, particularly among cultivated species.

Most changes to ecosystems have been made to meet a dramatic growth in the demand for food, water, timber, fiber, and fuel. Some ecosystem changes have been the inadvertent result of activities unrelated to the use of ecosystem services, such as the construction of roads, ports, and cities and the discharge of pollutants. But most ecosystem changes were the direct or indirect result of changes made to meet growing demands for ecosystem services, and in particular growing demands for food, water, timber, fiber, and fuel (fuelwood and hydropower). Between 1960 and 2000, the demand for ecosystem services grew significantly as world population doubled to 6 billion people and the global economy increased more than sixfold. To meet this demand, food production increased by roughly two-and-a-half times, water use doubled, wood harvests for pulp and paper production tripled, installed hydropower capacity doubled, and timber production increased by more than half.

The growing demand for these ecosystem services was met both by consuming an increasing fraction of the available supply (for example, diverting more water for irrigation or capturing more fish from the sea) and by raising the production of some services, such as crops and livestock. The latter has been accomplished through the use of new technologies (such as new crop varieties, fertilization, and irrigation) as well as through increasing the area managed for the services in the case of crop and livestock production and aquaculture.

Finding #2: The changes that have been made to ecosystems have contributed to substantial net gains in human well-being and economic development, but these gains have been achieved at growing costs in the form of the degradation of many ecosystem services, increased risks of nonlinear changes, and the exacerbation of poverty for some groups of people. These problems, unless addressed, will substantially diminish the benefits that future generations obtain from ecosystems.

In the aggregate, and for most countries, changes made to the world's ecosystems in recent decades have provided substantial benefits for human well-being and national development. Many of the most significant changes to ecosystems have been essential to meet growing needs for food and water; these changes have helped reduce the proportion of malnourished people and improved human health. Agriculture, including fisheries and forestry, has been the mainstay of strategies for the development of countries for centuries, providing revenues that have enabled investments in industrialization and poverty alleviation. Although the value of food production in 2000 was only about 3% of gross world product, the agricultural labor force accounts for approximately 22% of the world's population, half the world's total labor force, and 24% of GDP in countries with per capita incomes of less than $765 (the low-income developing countries, as defined by the World Bank).

These gains have been achieved, however, at growing costs in the form of the degradation of many ecosystem services, increased risks of nonlinear changes in ecosystems, the exacerbation of poverty for some people, and growing inequities and disparities across groups of people. . . .

Finding #3: The degradation of ecosystem services could grow significantly worse during the first half of this century and is a barrier to achieving the Millennium Development Goals.

The MA developed four scenarios to explore plausible futures for ecosystems and human well-being.

(See Box 1.) The scenarios explored two global development paths, one in which the world becomes increasingly globalized and the other in which it becomes increasingly regionalized, as well as two different approaches to ecosystem management, one in which actions are reactive and most problems are addressed only after they become obvious and the other in which ecosystem management is proactive and policies deliberately seek to maintain ecosystem services for the long term.

Most of the direct drivers of change in ecosystems currently remain constant or are growing in intensity in most ecosystems. In all four MA scenarios, the pressures on ecosystems are projected to continue to grow during the first half of this century. The most important direct drivers of change in ecosystems are habitat change (land

---

BOX 1. MA SCENARIOS

The MA developed four scenarios to explore plausible futures for ecosystems and human well-being based on different assumptions about driving forces of change and their possible interactions:

Global Orchestration – This scenario depicts a globally connected society that focuses on global trade and economic liberalization and takes a reactive approach to ecosystem problems but that also takes strong steps to reduce poverty and inequality and to invest in public goods such as infrastructure and education. Economic growth in this scenario is the highest of the four scenarios, while it is assumed to have the lowest population in 2050.

Order from Strength – This scenario represents a regionalized and fragmented world, concerned with security and protection, emphasizing primarily regional markets, paying little attention to public goods, and taking a reactive approach to ecosystem problems. Economic growth rates are the lowest of the scenarios (particularly low in developing countries) and decrease with time, while population growth is the highest.

Adapting Mosaic – In this scenario, regional watershed-scale ecosystems are the focus of political and economic activity. Local institutions are strengthened and local ecosystem management strategies are common; societies develop a strongly proactive approach to the management of ecosystems. Economic growth rates are somewhat low initially but increase with time, and population in 2050 is nearly as high as in *Order from Strength*.

TechnoGarden – This scenario depicts a globally connected world relying strongly on environmentally sound technology, using highly managed, often engineered, ecosystems to deliver ecosystem services, and taking a proactive approach to the management of ecosystems in an effort to avoid problems. Economic growth is relatively high and accelerates, while population in 2050 is in the midrange of the scenarios.

The scenarios are not predictions; instead they were developed to explore the unpredictable features of change in drivers and ecosystem services. No scenario represents business as usual, although all begin from current conditions and trends.

Both quantitative models and qualitative analyses were used to develop the scenarios. For some drivers (such as land use change and carbon emissions) and ecosystem services (water withdrawals, food production), quantitative projections were calculated using established, peer-reviewed global models. Other drivers (such as rates of technological change and economic growth), ecosystem services (particularly supporting and cultural services, such as soil formation and recreational opportunities), and human well-being indicators (such as human health and social relations) were estimated qualitatively. In general, the qualitative models used for these scenarios addressed incremental changes but failed to address thresholds, risk of extreme events, or impacts of large, extremely costly, or irreversible changes in ecosystem services. These phenomena were addressed qualitatively by considering the risks and impacts of large but unpredictable ecosystem changes in each scenario.

Three of the scenarios – *Global Orchestration, Adapting Mosaic,* and *TechnoGarden* incorporate significant changes in policies aimed at addressing sustainable development challenges. In *Global Orchestration* trade barriers are eliminated, distorting subsidies are removed, and a major emphasis is placed on eliminating poverty and hunger. In *Adapting Mosaic,* by 2010, most countries are spending close to 13% of their GDP on education (as compared to an average of 3.5% in 2000), and institutional arrangements to promote transfer of skills and knowledge among regional groups proliferate. In *TechnoGarden* policies are put in place to provide payment to individuals and companies that provide or maintain the provision of ecosystem services. For example, in this scenario, by 2015, roughly 50% of European agriculture, and 10% of North American agriculture is aimed at balancing the production of food with the production of other ecosystem services. Under this scenario, significant advances occur in the development of environmental technologies to increase production of services, create substitutes, and reduce harmful trade-offs.

use change and physical modification of rivers or water withdrawal from rivers), overexploitation, invasive alien species, pollution, and climate change. These direct drivers are often synergistic. For example, in some locations land use change can result in greater nutrient loading (if the land is converted to high-intensity agriculture), increased emissions of greenhouse gases (if forest is cleared), and increased number of invasive species (due to the disturbed habitat). . . .

The degradation of ecosystem services poses a significant barrier to the achievement of the Millennium Development Goals and the MDG targets for 2015. The eight Millennium Development Goals adopted by the United Nations in 2000 aim to improve human well-being by reducing poverty, hunger, child and maternal mortality, by ensuring education for all, by controlling and managing diseases, by tackling gender disparity, by ensuring environmental sustainability, and by pursuing global partnerships. Under each of the MDGs, countries have agreed to targets to be achieved by 2015. Many of the regions facing the greatest challenges in achieving these targets coincide with regions facing the greatest problems of ecosystem degradation.

Although socioeconomic policy changes will play a primary role in achieving most of the MDGs, many of the targets (and goals) are unlikely to be achieved without significant improvement in management of ecosystems. The role of ecosystem changes in exacerbating poverty (Goal 1, Target 1) for some groups of people has been described already, and the goal of environmental sustainability, including access to safe drinking water (Goal 7, Targets 9, 10, and 11), cannot be achieved as long as most ecosystem services are being degraded. Progress toward three other MDGs is particularly dependent on sound ecosystem management:

- Hunger (Goal 1, Target 2): All four MA scenarios project progress in the elimination of hunger but at rates far slower than needed to attain the internationally agreed target of halving, between 1990 and 2015, the share of people suffering from hunger. Moreover, the improvements are slowest in the regions in which the problems are greatest: South Asia and sub-Saharan Africa. Ecosystem condition, in particular climate, soil degradation, and water availability, influences process toward this goal through its effect on crop yields as well as through impacts on the availability of wild sources of food.
- Child mortality (Goal 4): Undernutrition is the underlying cause of a substantial proportion of all child deaths. Three of the MA scenarios project reductions in child undernourishment by 2050 of between 10% and 60% but undernourishment increases by 10% in *Order from Strength* (low certainty). Child mortality is also strongly influenced by diseases associated with water quality. Diarrhea is one of the predominant causes of infant deaths worldwide. In sub-Saharan

Africa, malaria additionally plays an important part in child mortality in many countries of the region.
- Disease (Goal 6): In the more promising MA scenarios, progress toward Goal 6 is achieved, but under *Order from Strength* it is plausible that health and social conditions for the North and South could further diverge, exacerbating health problems in many low-income regions. Changes in ecosystems influence the abundance of human pathogens such as malaria and cholera as well as the risk of emergence of new diseases. Malaria is responsible for 11% of the disease burden in Africa, and it is estimated that Africa's GDP could have been $100 billion larger in 2000 (roughly a 25% increase) if malaria had been eliminated 35 years ago. The prevalence of the following infectious diseases is particularly strongly influenced by ecosystem change: malaria, schistosomiasis, lymphatic filariasis, Japanese encephalitis, dengue fever, leishmaniasis, Chagas disease, meningitis, cholera, West Nile virus, and Lyme disease.

Finding #4: The challenge of reversing the degradation of ecosystems while meeting increasing demands for their services can be partially met under some scenarios that the MA considered, but these involve significant changes in policies, institutions, and practices that are not currently under way. Many options exist to conserve or enhance specific ecosystem services in ways that reduce negative trade-offs or that provide positive synergies with other ecosystem services.

Three of the four MA scenarios show that significant changes in policies, institutions, and practices can mitigate many of the negative consequences of growing pressures on ecosystems, although the changes required are large and not currently under way. All provisioning, regulating, and cultural ecosystem services are projected to be in worse condition in 2050 than they are today in only one of the four MA scenarios (*Order from Strength*). At lease one of the three categories of services is in better condition in 2050 than in 2000 in the other three scenarios. The scale of interventions that result in these positive outcomes are substantial and include significant investments in environmentally sound technology, active adaptive management, proactive action to address environmental problems before their full consequences are experienced, major investments in public goods (such as education and health), strong action to reduce socioeconomic disparities and eliminate poverty, and expanded capacity of people to manage ecosystems adaptively. However, even in scenarios where one or more categories of ecosystem services improve, biodiversity continues to be lost and thus the long-term sustainability of actions to mitigate degradation of ecosystem services is uncertain.

BOX 2. EXAMPLES OF PROMISING
AND EFFECTIVE RESPONSES FOR SPECIFIC SECTORS

Illustrative examples of response options specific to particular sectors judged to be promising or effective are listed below. A response is considered effective when it enhances the target ecosystem services and contributes to human well-being without significant harm to other services or harmful impacts on other groups of people. A response is considered promising if it does not have a long track record to assess but appears likely to succeed or if there are known ways of modifying the response so that it can become effective.

Agriculture

Removal of production subsidies that have adverse economic, social, and environmental effects.

Investment in, and diffusion of, agricultural science and technology that can sustain the necessary increase of food supply without harmful tradeoffs involving excessive use of water, nutrients, or pesticides.

Use of response polices that recognize the role of women in the production and use of food and that are designed to empower women and ensure access to and control of resources necessary for food security.

Application of a mix of regulatory and incentive- and market-based mechanisms to reduce overuse of nutrients.

Fisheries and Aquaculture

Reduction of marine fishing capacity.

Strict regulation of marine fisheries both regarding the establishment and implementation of quotas and steps to address unreported and unregulated harvest. Individual transferable quotas may be appropriate in some cases, particularly for cold water, single species fisheries.

Establishment of appropriate regulatory systems to reduce the detrimental environmental impacts of aquaculture.

Establishment of marine protected areas including flexible no-take zones.

Water

Payments for ecosystem services provided by watersheds.

Improved allocation of rights to freshwater resources to align incentives with conservation needs.

Increased transparency of information regarding water management and improved representation of marginalized stakeholders.

Development of water markets.

Increased emphasis on the use of the natural environment and measures other than dams and levees for flood control.

Investment is science and technology to increase the efficiency of water use in agriculture.

Forestry

Integration of agreed sustainable forest management practices in financial institutions, trade rules, global environment programs, and global security decision-making.

Empowerment of local communities in support of initiatives for sustainable use of forest products; these initiatives are collectively more significant than efforts led by governments or international processes but require their support to spread.

Reform of forest governance and development of country-led, strategically focused national forest programs negotiated by stakeholders.

Past actions to slow or reverse the degradation of ecosystems have yielded significant benefits, but these improvements have generally not kept pace with growing pressures and demands. Although most ecosystem services assessed in the MA are being degraded, the extent of that degradation would have been much greater without responses implemented in past decades. For example, more than 100,000 protected areas (including strictly protected areas such as national parks as well as areas managed for the sustainable use of natural ecosystems, including timber or wildlife harvest) covering about 11.7% of the terrestrial surface have now been established, and these play an important role in the conservation of biodiversity and ecosystem services (although important gaps in the distribution of protected areas remain, particularly in marine and freshwater systems). Technological advances have also helped lessen the increase in pressure on ecosystems caused per unit increase in demand for ecosystem services.

Substitutes can be developed for some but not all ecosystem services, but the cost of substitutes is generally high, and substitutes may also have other negative environmental consequences. For example, the substitution of vinyl, plastics, and metal for wood has contributed to relatively slow growth in global timber consumption in recent years. But while the availability of substitutes can reduce pressure on specific ecosystem services, they may not always have positive net benefits on the environment. Substitution of fuelwood by fossil fuels, for example, reduces pressure on forests and lowers indoor air pollution but it also increases net greenhouse gas emissions. Substitutes are also often costlier to provide than the original ecosystem services.

Ecosystem degradation can rarely be reversed without actions that address the negative effects or enhance the positive effects of one or more of the five indirect drivers of change: population change (including growth and migration), change in economic activity (including economic growth, disparities in wealth, and trade patterns), sociopolitical factors (including factors ranging from the presence of conflict to public participation in decision-making), cultural factors, and technological change. Collectively these factors influence the level of production and consumption of ecosystem services and the sustainability of the production. Both economic growth and population growth lead to increased consumption of ecosystem services, although the harmful environmental impacts of any particular level of consumption depend on the efficiency of the technologies used to produce the service. Too often, actions to slow ecosystem degradation do not address these indirect drivers. For example, forest management is influenced more strongly by actions outside the forest sector, such as trade policies and institutions, macroeconomic policies, and policies in other sectors such as agriculture, infrastructure, energy, and mining, than by those within it.

An effective set of responses to ensure the sustainable management of ecosystems must address the indirect and drivers just described and must overcome barriers related to:

- Inappropriate institutional and governance arrangements, including the presence of corruption and weak systems of regulation and accountability.
- Market failures and the misalignment of economic incentives.
- Social and behavioral factors, including the lack of political and economic power of some groups (such as poor people, women, and indigenous peoples) that are particularly dependent on ecosystem services or harmed by their degradation.
- Underinvestment in the development and diffusion of technologies that could increase the efficiency of use of ecosystem services and could reduce the harmful impacts of various drivers of ecosystem change.
- Insufficient knowledge (as well as the poor use of existing knowledge) concerning ecosystem services and management, policy, technological, behavioral, and institutional responses that could enhance benefits from these services while conserving resources.

All these barriers are further compounded by weak human and institutional capacity related to the assessment and management of ecosystem services, underinvestment in the regulation and management of their use, lack of public awareness, and lack of awareness among decision-makers of both the threats posed by the degradation of ecosystem services and the opportunities that more sustainable management of ecosystems could provide. . . .

Critical Thinking

1. Why is it difficult to manage ecosystems sustainably?
2. In what ways does damage to ecosystems affect human well-being?

Part 2: Energy

Energy and Ecosystems

Selection 11

CHANCEY JUDAY, from "The Annual Energy Budget of an Inland Lake,"
Ecology (October 1940)

Selection 12

JOHN M. FOWLER, from *Energy and the Environment* (McGraw-Hill, 1975)

Learning Outcomes

After reading this unit, you should be able to:

1. Describe how energy enters ecosystems and is converted from one form to another, including forms of interest to humans.
2. Explain why energy cannot be recycled endlessly.
3. Explain how human energy use damages the environment.

The Annual Energy Budget of an Inland Lake

Chancey Juday

The study of the role played by energy in an ecosystem and the energy interchanges that occur among the biotic (living) and nonbiotic components of the system is a key ecological organizing principle used in understanding the dynamics of nature's biogeochemical cycles. Chancey Juday (1871–1944) was a pioneer in the study of freshwater biotic communities. He was responsible for developing a program that made the University of Wisconsin the center for such studies in the United States.

Juday's study "The Annual Energy Budget of an Inland Lake," *Ecology* (October 1940), from which the following selection has been taken, was a groundbreaking piece of work. It was the first time a scientist attempted to examine the energy input and subsequent energy uses and transformations that occur within a complex biotic community. Juday's work represented an important step toward transforming ecology into a more quantitative discipline, one in which sophisticated analyses of entire ecosystems were possible.

Key Concept: the roles of energy in ecosystem dynamics

Introduction

The variation in the quantity of solar radiation delivered to the surface of an inland lake during the course of the year is the principal factor in determining the physical, chemical and biological cycle of changes that take place within the water. This is true especially of lakes which are situated in temperate latitudes where there are considerable differences between summer and winter temperatures of the air and of the water.

In the deeper lakes which become covered with ice for a few to several weeks during the winter, the annual cycle consists of four phases which correspond roughly to the four seasons of the year. (1) There is a winter stagnation period in which the water is inversely stratified while the lake is covered with ice; that is, the warmest water is at the bottom and the coldest at the surface. (2) There is an overturning and circulation of the entire body of water following the disappearance of the ice in spring.

(3) As the temperature rises above the point of maximum density (4°C.) in the spring, the free circulation of the water is hindered as a result of the difference in density between the warm upper layer and the cold lower stratum; this is due to the fact that most of the warming takes place in a comparatively thin upper stratum. From 65 to 90 percent or more of the sun and sky radiation is cut off by the upper meter of water, depending upon the color and the transparency. This large reduction is due chiefly to reflection at the surface, to the rapid absorption of the radiation by the water itself and by the stains and suspensoids that may be present in the water.

With the further rise in the temperature of the upper water as the season advances, a summer stratification is established. This is a direct stratification in which there is a warm upper stratum, a cold lower stratum and a transition zone between them in which the temperature changes rapidly from that of the warm upper water to that of the cold stratum below; naming them in order from surface to bottom, they are known as the epilimnion, the thermocline or mesolimnion, and the hypolimnion. In the deeper lakes these three strata persist throughout the summer.

(4) The fourth stage is represented by the autumnal cooling and overturning, which is followed by a complete circulation of the water until the lake becomes covered with ice.

Table 1 Quantity of Solar and Sky Radiation Used by Lake Mendota in Various Physical and Biological Processes

Melting of ice in spring	3,500
Annual heat budget of water	24,200
Annual heat budget of bottom	2,000
Energy lost by evaporation	29,300
Annual surface loss	28,500
Loss by conduction, convection and radiation	30,324
Biological energy budget (maximum)	1,048

The results are indicated in gram calories per square centimeter of surface.

The circulation of the water in spring and autumn distributes the dissolved substances uniformly from surface to bottom, but a more or less marked difference in the chemical character of the upper and lower strata develops during the two stratification periods, especially in summer.

The seasonal changes in the physical and chemical characteristics of the water have an important effect upon aquatic life. The spring rise in temperature speeds up life processes and the increase in the amount of solar radiation at this time makes conditions more favorable for photosynthesis. Likewise the circulation of the water in spring and autumn brings dissolved substances into the upper stratum, which is the zone of photosynthesis, where they can be readily obtained by the aquatic plants. The phytoplankton responds promptly to these favorable growing conditions and, as a result, the standing crop of plankton is usually much larger during the two circulation periods than it is at the time of summer and winter stratification.

The annual cycle of changes that is induced by the march of the seasons naturally raises the question of the amount of energy actually involved in the phenomenon as a whole and also in the various phases of the cycle. This problem is a very complex one and a large amount of data is necessary to evaluate each of the several items.

Limnological studies have been in progress on Lake Mendota at Madison, Wisconsin, for many years and the data accumulated in these investigations make it possible to give approximations of the quantity of energy involved in a number of the items. Quantitative studies of some of the factors have not been made up to the present time and a more complete assessment of the energy budget of the lake will have to await such investigations.

The annual energy budget of a lake may be regarded as comprising the energy received from sun and sky each year and the expenditures or uses which the lake makes of this annual income of radiation. In general the annual income and outgo substantially balance each other. This is true more particularly of the physical energy budget. Considerable biological material produced in one energy year lives over into the next, but this overlapping crop of organisms is much the same in quantity from year to year so that it plays approximately the same annual role. For this reason it does not require any special consideration.

There is a certain amount of organic material contributed to the bottom deposits in the deeper water and to peat formation in the shallow water which lasts for long periods of time, but the annual energy value of these materials is so small in most cases that they may be neglected. . . .

The respective amounts of energy included in the four items of the physical energy budget. . . are indicated in Table 1. The sum of these four items is 71,000 calories, which is approximately 60 percent of the mean quantity of energy delivered to the surface of Lake Mendota annually by sun and sky. . . .

Biological Energy Budget

A certain amount of the solar radiation that passes into the water of Lake Mendota is utilized by aquatic plants in the process of photosynthesis. The products of this assimilation, namely, proteins, fats and carbohydrates, thus constitute the primary accumulation or storage of the energy derived from the sub-surface illumination. Since this organic material manufactured by the plants serves, either directly or indirectly, as a source of food for all of the nonchlorophyllaceous organisms that inhabit the lake, these latter forms, therefore, constitute a secondary stage in the storage of the energy accumulated by the aquatic plants. The original amount of energy represented by these secondary organisms varies with the different forms, depending upon the number of links in their respective food chains; in general they represent a comparatively small proportion of the primary organic material manufactured by the plants.

Chemical analyses of the various aquatic organisms have now progressed far enough to enable one to compute their energy values from the standards that have been established by food chemists. The standard values are 5,650 calories per gram of protein, 9,450 calories per gram of fat and 4,100 calories per gram of carbohydrate, on a dry weight basis. These values do not represent the total quantity of energy utilized by the aquatic organisms, however, because a part of the synthesized material is oxidized in the metabolic processes of the living organisms. These metabolic oxidations result in the production of heat which is transmitted to the water, but the quantity of heat derived from this source

Table 2 Annual Production of Plankton, Bottom Flora, Bottom Fauna and Fish, as Well as Crude Protein, Ether Extract (Fat), and Carbohydrate Constituents of the Organic Matter

	Dry Organic Matter	**Crude Protein**	**Ether Extract**	**Carbohydrate**
Total plankton	6,240	2,704	431	3,105
Phytoplankton	5,850	2,501	383	2,966
Zooplankton	390	203	48	139
Bottom flora	512	64	6	442
Bottom fauna	45	33	4	8
Fish	5	3.4	1	0.6
Dissolved organic matter	1,523	334	68	1,121
Total organic matter	8,325	3,138.4	510	4,676.6

The results are stated in kilograms per hectare on a dry, ash-free basis. The plankton yield is based on a turnover every two weeks during the year. The average quantity of dissolved organic matter is included also.

is extremely small in comparison with that which comes from direct insolation.

The amount of organic matter consumed in the metabolism of plants is much smaller than that in animals because several grams of plant material may be consumed in the production of one gram of animal tissue even in animals that feed directly on plants; the predaceous animals represent a still larger quantity of the original photosynthesized material.

. . . [I]t may be estimated that the average turnover in the organic matter of the mean standing crop of plankton takes place about every two weeks throughout the year. It would be more frequent than this in spring and summer, and less frequent in winter. A turnover of 26 times per year would give an annual yield of 6,240 kilograms of dry organic matter per hectare of surface as indicated in Table 2. This material would consist of 2,704 kilograms of protein, 431 kilograms of fat and 3,105 kilograms of carbohydrate. Approximately 94 percent of the organic matter comes from the phytoplankton and 6 percent from the zooplankton. . . .

Schuette and some of his students (1922–29) made chemical analyses of certain species of. . . plants and several others have been analyzed in more recent years, so that the energy values of all of the more common forms can now be computed from the data in hand. The results of such computations are given in Table 2. The 512 kilograms of dry organic matter per hectare consisted of 64 kilograms of protein, 6 kilograms of ether extract or fat and 442 kilograms of carbohydrate.

The bottom deposits, especially in the deeper water, contain a rather large population of bacteria. . . . While these organisms are present in considerable numbers, they are so small in size that they add very little to the crop of organic matter in the lake; so they have been disregarded. Likewise fungi are fairly abundant in the bottom deposits, but no quantitative study of them has yet been made; it seems probable that their contribution to the organic content of the lake is negligible from an energy standpoint. . . .

The macroscopic bottom fauna yielded 45 kilograms of dry organic matter per hectare; of this amount 33 kilograms consisted of protein, 4 kilograms of ether extract or fat and 8 kilograms of carbohydrate. While some of these organisms live more than one year, others pass through two or three generations in a year; the two groups of organisms are generally considered as balancing each other, so that the above quantities may be taken as the annual crop of this material as shown in Table 2.

Considerable numbers of protozoa and other microscopic animals have been found in the bottom deposits, but no quantitative study of them has been made. It seems probable, however, that these minute forms would not add an appreciable amount of organic matter to the total weight of the bottom fauna.

Fish. No accurate census of the fish caught by anglers in Lake Mendota each year has ever been made so that the assessment of this part of the biological crop can be estimated only roughly. . . .

The average yield of carp between 1933 and 1936, inclusive, was 16 kilograms per hectare; adding the carp crop to that of the game and pan fish gives an annual fish yield of 22 kilograms per hectare, live weight. On a dry, ash-free basis, the total yield amounts to a little more than 5 kilograms per hectare as indicated in Table 2. By far the greater part of this material consists of protein and fat.

Energy value of annual crop. Table 2 shows that the total quantity of stored and accumulated energy in the form of dry organic matter in the annual crop of plants and animals amounts to 6,802 kilograms per hectare; of this quantity protein constitutes a little more than 2,804 kilograms, ether extract or fat 442 kilograms and carbohydrates 3,556 kilograms. On the basis of the energy equivalents of these three classes of organic matter, . . . the total energy value of the annual crop amounts to 346 gram calories per square centimeter of lake surface (Table 3).

In addition to the organic material in the plants and animals, the water contains a certain amount of organic matter which cannot be recovered with a high speed

Table 3   Energy Values of the Organic Matter in the Organisms, Together With the Estimated Amounts of Energy Represented in Their Metabolism, and in the Dissolved Organic Matter

Phytoplankton	299
Metabolism	100
Zooplankton	22
Metabolism	110
Bottom flora	22
Metabolism	7
Bottom fauna and fish	3
Metabolism	15
Dissolved organic matter	71
Total	649

The values are stated in gram calories per square centimeter of lake surface. The results for phytoplankton and zooplankton are based on a turnover every two weeks during the year.

centrifuge. It is either in true solution or is in such a finely divided state that it cannot be obtained with a centrifuge; for lack of a better term it has been called "dissolved organic matter" as compared with the "particulate organic matter" which can be recovered from the water with a centrifuge. The water of Lake Mendota contains 10 to 14 milligrams per liter, dry weight, of this dissolved organic matter; the mean of some 60 determinations is 12 milligrams per liter. When computed to an area basis, the average weight of this material is 1,523 kilograms per hectare, dry weight, of which 334 kilograms are protein, 68 kilograms fat and 1,121 kilograms carbohydrate. The energy value of this dissolved organic matter is about 71 gram calories per square centimeter.

Utilization of solar energy. The chlorophyll-bearing aquatic plants are responsible for the utilization of the sub-surface radiation; that is, the sun furnishes the power and the chlorophyll and associated pigments of the plants serve as the machines for the manufacture of the fundamental organic matter of the lake. Table 2 shows that the phytoplankton and the large aquatic plants constitute the major item in the annual yield of biological material. Together they contribute 6,362 kilograms of dry organic matter per hectare as compared with 440 kilograms of zooplankton, bottom fauna and fish; that is, the plant contribution is 93 percent and the animal part is 7 percent of the total organic matter.

Table 3 gives the energy value of the various constituents of the annual biological crop. The two groups of plants, namely phytoplankton and large aquatics, have an energy value of 321 gram calories as compared with 25 gram calories per square centimeter in the animals. The 321 gram calories represented in the organic matter of the plants is only 0.27 of one percent of the mean annual radiation delivered to the surface of the lake, namely 118,872 calories. Two corrections need to be made in this result, however. (1) As already indicated

some 28,500 calories of solar energy are lost at the surface of the water and thus do not reach the aquatic vegetation. Deducting this amount leaves 90,372 calories which pass into the water and thus become available for the plants. On this basis the percentage of utilization is increased to a little more than 0.35 of one percent of the available radiation. (2) A certain amount of the organic matter synthesized by the plants is used in their metabolism and this does not appear in the percentage of utilization given above. Experiments show that some of the algae utilize in their metabolic processes about one-third of the organic matter that they synthesize. No data are available for the large aquatics, but assuming that they also utilize a similar proportion in their metabolism, the two groups of plants would represent a utilization of 428 calories which is equivalent to 0.47 of one percent of the annual quantity of solar energy that actually enters the water.

This percentage is based on an average turnover in the phytoplankton every two weeks throughout the year, but there is some evidence that the turnover takes place more frequently, especially from April to October. With an average turnover once a week in the organic matter of the phytoplanton during the year, the energy value of this crop would be 798 calories, including metabolism; adding to this amount the 29 calories in the annual crop of bottom flora gives a total of 827 calories which is utilized by the plants. This is 0.91 of one percent of the 90,372 calories of energy that penetrate the water and become available to the plants; in round numbers this may be regarded as a utilization of one percent.

This small percentage of utilization of solar energy by aquatic plants shows that Lake Mendota is not a very efficient manufacturer of biological products in so far as utilizing the annual supply of solar and sky radiation is concerned; on the other hand it belongs to the group of highly productive lakes.

While the aquatic plant crop appears to be inefficient in its utilization of solar energy, it compares very favorably with some of the more important land crops in this respect. Transeau ('26) states that only 1.6 percent of the total available energy is used by the corn plant in photosynthesis during a growing period of 100 days, or from June 1 to September 8. Spoehr ('26) gives a table or Pütter's calculations for various crop plants in which the general average of the utilization of solar energy is about 3.0 percent; summer wheat is given as 3.2 percent, potatoes 3.0 percent and beets 2.1 percent. These computations for cultivated crops, however, take into account only the quantity of solar radiation available during comparatively brief growing periods and thus do not cover the entire year as indicated for the aquatic plants.

Energy value of animals. The organic content of the animal population of the lake represents a conversion and further storage of the material manufactured by the plants, but no direct utilization of solar energy is involved in the transformation. It may be regarded as an expensive method of prolonging the existence of a certain portion of the original plant material. As previously indicated, it may take five grams of plant food to produce one gram of animal tissue, so that the plant equivalent of the animal crop may be reckoned as five times as large as the organic content of the animals; in the predatory animals, however, it would be much larger.

Table 3 shows that the energy value of the bottom and fish population is 25 gram calories per square centimeter; on the five fold basis, this would represent the conversion of at least 125 gram calories of original plant organic matter. This utilization is approximately 40 percent of the potential energy stored in the annual plant crop of 321 gram calories which is based on a turnover in the phytoplankton every two weeks during the year. A turnover in the phytoplankton every week would give a plant crop of 620 calories and an animal utilization of a little more than 20 percent.

Dissolved organic matter.   The energy value of the dissolved organic matter is indicated as 71 gram calories per square centimeter in Table 3. This material is constantly being supplied to the water by the various organisms and the standing crop of it remains fairly uniform in quantity during the different seasons of the year as well as in different years. While there is a regular turnover in this organic matter, it needs to be taken into account only once in computing the organic crop of the lake because it has its source in the plants and animals for which an annual yield has already been computed.

Summary

1. The mean annual quantity of sun and sky radiation delivered to the surface of Lake Mendota over a period of 28 years was 118,872 gram calories per square centimeter of surface.

2. In the physical energy budget, the melting of the ice utilized 3,500 calories; the annual heat budget of the water was 24,200 calories and of the bottom 2,000 calories; the loss of energy by evaporation amounted to 29,300 calories; the surface loss by reflection, upward scattering and absorption was 28,500 calories and about 31,000 calories were lost by conduction, convection and radiation.

3. On the basis of a turnover in the organic content of the plankton every two weeks during the year, the energy value of the annual crop of plants and animals was 346 gram calories per square centimeter; of this amount, 321 calories were contributed by the plants and 25 calories by the animals. Adding to this the organic matter utilized by the plants and animals in their metabolic processes (232 calories) gives the annual crop an energy value of 578 gram calories. In addition the dissolved organic matter had a value of 71 gram calories.

4. Assuming an average turnover of once a week in the organic matter of the phytoplankton, instead of every two weeks, would raise the energy budget of the annual crop of plants and animals to 977 gram calories, including the metabolized material. Adding the 71 calories in the dissolved organic matter gives a total of 1,048 calories in the biological energy budget.

5. An average turnover of once a week in the phytoplankton would give an annual utilization by these organisms and by the large aquatic plants of about one percent of the subsurface solar energy.

Critical Thinking

1. In what forms does energy enter an ecosystem such as an inland lake? In what forms does energy leave the ecosystem?
2. Describe the transfers of energy within the inland lake ecosystem.

Energy and the Environment

John M. Fowler

Physicist John M. Fowler has devoted most of his professional career to the promotion of effective science education for both scientists and the general public. He served for several years as the executive officer of the American Physical Society's Commission on College Physics and more recently as director of special projects for the National Science Teachers Association. He was also an active participant in the Scientist's Institute for Public Information, an organization devoted to providing the public with the scientific knowledge and understanding required to make intelligent decisions on issues of science and public policy. One of Fowler's specific educational preoccupations has been with promoting the public's understanding of the key role played by the production and use of energy in our increasingly technological society.

Most of the world's environmental problems, including air, water, and land pollution, are directly affected by the particular energy sources that have been developed and the efficiency with which they are used. Fowler wrote *Energy and the Environment* (McGraw-Hill, 1975), from which the following selection has been taken, when the United States was experiencing its first "energy crisis" as a result of the 1973–1974 Arab oil embargo. There was suddenly serious debate about the future of America's growing dependence on petroleum. At the same time, a movement was gaining momentum and swelling the ranks of grassroots organizations concerned about the growing signs of environmental deterioration. Through his book, Fowler was one of the first to successfully explain the links between these two issues in a manner that was accessible to the nonscientific, literate public.

Key Concept: limits on the efficiency of energy conversion

Energy for Life

Let's . . . take up the trail of energy through the eons. We left it 5 billion years ago with the formation of the sun. We will skip over the 4.5 billion years during which the earth was formed, mountains were raised and washed away by the great seas, and life began. We will take up the trail, again, 300 million years ago, in the middle part of the Late Paleozoic Era. Plants have emerged from the sea and the tropical marshlands are covered by ferns, horsetails, and mosses grown to enormous size. Man is still far in the future, his genes are beginning their evolution in some of the creatures of these jungles.

A small fraction of the sun's energy which beat down on these strange jungles was used by the huge fern trees and other vegetation in their life processes. They converted the radiant kinetic energy of sunlight into *chemical potential energy*, the third great form of potential energy.

The energy-converting process of photosynthesis can be summarized in chemical shorthand as

$$CO_2 + H_2O + energy \rightarrow C_x(H_2O)_y + O_2$$

which, in words, states that in a plant, carbon dioxide (CO_2) and water (H_2O) are combined, with the addition of energy from sunlight, to form the carbohydrate group $C_x(H_2O)_y$ and oxygen (O_2). $C_x(H_2O)_y$ is a general formula for the carbohydrate group which is an important part of the molecule of sugars and starches. Ordinary sugar, for instance, is $C_{12}(H_2O)_{11}$. This is an oversimplification of a step-wise process which proceeds more correctly as

$$CO_2 + H_2O + energy \rightarrow intermediate\ products$$
$$intermediate\ products + energy \rightarrow C_x(H_2O)_y + O_2$$

The radiant energy from the sun, kinetic energy (energy on the move), is caught by plants and used to break up the molecules of CO_2 and H_2O and rearrange their atoms. To form carbohydrates from CO_2 and H_2O, forces must be operating so that molecules are pulled apart and atoms moved around. Since the forces which hold atoms and molecules together are electrical forces, the chemical potential energy which we have just introduced is a form of *electrical potential energy*.

It is certainly potential energy. The carbohydrate formed with the help of the sun's energy is either food or fuel, depending on the use we make of it. Eating or burning turns the photosynthesis reactions around; carbohydrates in the plant sugars or starches combine with oxygen to re-form CO_2 and H_2O, releasing energy in the process.

$$C_x(H_2O)_y + O_2 \rightarrow CO_2 + H_2O + \text{energy}$$

The fern and moss jungle of 300 million years ago grew in a rich environment. Much of the world at that time had an almost tropical climate, wet and warm. In addition, the atmosphere was very rich in CO_2, released from the depths of the earth by many active volcanoes and hot springs.

Let us now focus down on one spot on Paleozoic earth, a swampy river delta in southern Illinois only a few feet above the level of the ocean which covered Missouri to the west and parts of Kentucky to the south. This swampy delta was formed of sediment washed down from the mountain ranges to the east and north. It was covered by a gloomy jungle where, for thousands of years, fern trees and huge mosses grew, died, and fell into the shallow water from which they emerged. Some of them decayed. Bacteria consumed their carbohydrates and recombined the carbon C with O_2 to form CO_2 which was returned to the atmosphere to feed future plants in the familiar cycle of life and death. Some of the plants, however, were covered with water, and the oxygen-requiring bacteria could not work on them. Over tens of thousands of years, huge amounts of this undecayed plant material built up into the spongy mass we call *peat*. (A modern peat bog, 6 feet deep, can be found in regions of the Dismal Swamp of Virginia and North Carolina.)

In peat, the *anaerobic* bacteria (those which do not require oxygen), plus the pressure and accompanying heat caused by the overbearing sediment, drove off some of the water, oxygen, nitrogen, and miscellaneous plant products so that the percentage of energy-rich carbon was increased. Peat, when burned as fuel, as it is in parts of the world, releases about 6,000 Btu per pound.

In our Illinois river delta the thick layer of peat was finally buried under tons of sea-born sediment and further changes took place. The pressure increased greatly and, along with it, the temperature. The peat was greatly compacted, perhaps by as much as a factor of 16 (in other words, a thickness of 16 feet became compressed to a thickness of one foot). More and more of the water, nitrogen, oxygen, and other materials were driven out so that more and more pure carbon was released from its earlier molecular combination. The percentage of carbon increased from an original 50 percent in the living material to 75 or 80 percent. The peat was now compacted into a hard, black mineral; the southern Illinois coal basin was formed.

What had happened to the energy? The radiant energy from the sun was converted to chemical potential energy in the plants. Through the compaction and reactions under pressure and heat, a great concentration of this energy had taken place. Thus, in the high-grade coals,

there are 11,000 to 12,000 Btu of energy per pound compared with, for instance, the 5,000 or so Btu per pound of the original wood.

Man Enters the Scene

The Illinois coal deposits, with their tremendous energy locked in the carbon atoms, lay buried for millions of years. As the earliest settlers came to Illinois, they found scattered about black outcroppings of this coal thrust to the surface by mammoth foldings of the earth's crust. Those early settlers knew what it was; coal had been used in Europe as early as the twelfth and thirteenth centuries. But they were not very interested in it as fuel, for all about them was the abundance of wood—wood which needed to be cleared from the land.

Not until the settlers had chopped their way across the land and wood no longer lay at their doorstep did they turn to coal. As the country became more and more industrialized, coal consumption grew, from perhaps 8 million tons in 1850 to 56 million tons in 1875 and 270 million tons in 1900. It reached a peak of 633 million tons in 1945 but production has since decreased, amounting to about 450 million tons in 1970.

Let us go back to our saga of energy and follow a ton of coal mined from that Illinois deposit in the 1970s. Had it been pulled from the ground thirty years earlier, odds are it would either have been made into coke for the smelting of iron ore or would have ended up in the firebox of a steam-driven locomotive. In 1970, however, its likely destination was an electric power plant, perhaps one on the outskirts of Chicago; the one whose plume of smoke you might have noticed while flying over that city.

In that ton of coal were some 24 million Btu of energy. Dumped into the huge furnace of the power plant the coal burned efficiently; most of the heat energy released (85 to 90 percent) went into the boiler which turned water into steam. The other 10 to 15 percent of the released heat, along with the sulphur impurities in the coal, went up the stack. This wasted heat which warmed the air above the power plant, is of interest in itself. . . . Inside the boiler, the super-heated steam with a temperature of 1000°F, was sent against the giant turbines to turn them, thus turning the electric generators which they drove to produce electricity.

Our story is almost over. We have followed this infinitesimal bit of the energy from that "Cosmic Egg" down through the ages from sun to plants to coal and now to electricity. We must keep careful track of what is left. Ten percent of the 24 million Btu (that is, 2.4 million Btu) were lost as heat to the atmosphere. We lost a lot more in the turbine; the steam went in at 1000°F, but it left still hot; it was considerably above the 212°F boiling point of water when it was exhausted from the turbine. Some 60 percent of the heat energy was not used to turn the turbine but instead was carried to a nearby river by the power plant's cooling system. Thirteen million Btu were lost that way.

The generator converted the kinetic energy of the turbine into that most important form of kinetic energy, electricity. This was done very efficiently, so that practically all of the remaining 8.6 million Btu went out into the high voltage transmission lines as electrical energy to be distributed throughout Chicago.

Who can say where it ended? Electricity has many varied uses. We can estimate with some assurance that 10 percent of it (about 1 million Btu of energy) was lost in the transmission and distributing lines; it was converted to heat and warmed the surrounding atmosphere. The remaining 7.7 million Btu was used throughout the city, in air conditioners, cooking stoves, water heaters, motors, industrial plants, and the like. In the end, all the kinetic energy of this electricity ended as heat energy, directly converted in ranges and heaters or converted by friction in motors.

Let's look back now with the help of Table 1. We began with 24 million Btu of chemical potential energy in our ton of coal, lost about 2.5 million Btu up the stack as heat when we burned it, lost 13 million Btu as heat to the river due to the inefficiency of the turbine, approximately another million Btu went as heat into the atmosphere from losses in the transmission lines, and finally the remaining nearly 8 million Btu did its job and ended up again as heat in the air above Chicago.

We need but one more night to finish the story, a clear night with no clouds between Chicago and deep space. The heated air and the warm water of the river now finally give back to the universe the loan made so long ago. The heat is radiated away from the earth and travels with the speed of light out into those seemingly empty infinite depths; but space is not empty. Here and there in those depths are atoms and molecules of gas and dust, the raw material of the stars, thinner now in the aging galaxy. These widely scattered particles absorb the heat energy traveling out from earth and increase their motion by the amount of that energy. The universe becomes a bit warmer. . . .

Energy cannot be created or destroyed; that was the statement of the First Law of Thermodynamics. None of the energy we have followed was destroyed, or created; it was only converted from one form to another. But at each conversion there was a *heat tax* imposed; some of the energy went from a "useful" form (either kinetic or potential) into heat energy, in which its usefulness has

Table 1   Where the Coal's Heat Went

Distribution	Amount (M Btu)*	
Original total heat	24	
Lost with stack gases (10%)		2.4
Lost to cooling water (60% of remainder)		13.0
Electrical output	8.6	
Lost in transmission (10%)		0.9
Total losses		16.3
Useful work	7.7	

*M Btu = million Btu

been decreased. The conversion to heat energy is a one-way street. While it is possible to convert any form of energy to heat with 100 percent efficiency, to reverse the process is difficult. As we saw in the case of the steam turbine, in the conversion to mechanical energy of the heat released by burning coal, we lost 60 percent of it.

The one-way nature of energy conversion is also a law of nature. For our study of the "energy crisis," more important than the First Law we have just reviewed, is the *Second Law of Thermodynamics*, which states that:

> No device can be constructed which, operating in a cycle (like an engine), accomplishes *only* the extraction of heat energy from a reservoir and its complete conversion to mechanical energy (work).

There are many other ways to state this important law. . . . Its consequence, however, should already be clear. Heat energy cannot be completely converted to mechanical energy. In any conversion some of it is irrevocably lost; it remains in the form of heat and cannot be reclaimed for useful purposes.

Critical Thinking

1. Why is it impossible to convert energy from one form to another with 100 percent efficiency?
2. If modern energy use damages the environment, then using less energy must damage the environment less. What are the difficulties involved in cutting energy use back to a pre-industrial level?

Renewable and Nonrenewable Energy

Selection 13

MARK Z. JACOBSON AND MARK A. DELUCCHI, from "A Path to Sustainable Energy by 2030," *Scientific American* (November 2009)

Learning Outcomes

After reading this unit, you should be able to:

1. Explain why a shift from fossil-fuel based energy to other forms of energy is desirable.
2. Explain the difficulties in shifting to a new energy system.
3. Describe the best way to shift to a new energy system.
4. Describe the link between using fossil fuels and national security.

A Path to Sustainable Energy by 2030

Mark Z. Jacobson and Mark A. Delucchi

Unlike Al Gore, Mark Z. Jacobson and Mark A. Delucchi do not think the U.S. energy system can be replaced by wind and solar in just 10 years. Even with "extremely aggressive policies" it will take two or three times that long. More likely, it will take half a century. They do, however, agree on the need, and they have mapped out a feasible path to meeting that need. Their proposal, excerpted here, was published in *Scientific American* in November 2009.

Mark Z. Jacobson is Professor of Civil and Environmental Engineering at Stanford University. Mark A. Delucchi is a research scientist at the Institute of Transportation Studies at the University of California, Davis.

Key Concept: the feasibility of redesigning the world's energy systems to rely on alternative energy sources instead of fossil fuels

In December leaders from around the world will meet in Copenhagen to try to agree on cutting back greenhouse gas emissions for decades to come. The most effective step to implement that goal would be a massive shift away from fossil fuels to clean, renewable energy sources. If leaders can have confidence that such a transformation is possible, they might commit to an historic agreement. We think they can.

A year ago former vice president Al Gore threw down a gauntlet: to repower America with 100 percent carbon-free electricity within 10 years. As the two of us started to evaluate the feasibility of such a change, we took on an even larger challenge: to determine how 100 percent of the world's energy, for *all* purposes, could be supplied by wind, water and solar resources, by as early as 2030. Our plan is presented here.

Scientists have been building to this moment for at least a decade, analyzing various pieces of the challenge. Most recently, a 2009 Stanford University study ranked energy systems according to their impacts on global warming, pollution, water supply, land use, wildlife and other concerns. The very best options were wind, solar, geothermal, tidal and hydroelectric power—all of which are driven by wind, water or sunlight (referred to as WWS). Nuclear power, coal with carbon capture, and ethanol were all poorer options, as were oil and natural gas. The study also found that battery-electric vehicles and hydrogen fuel-cell vehicles recharged by WWS options would largely eliminate pollution from the transportation sector.

Our plan calls for millions of wind turbines, water machines and solar installations. The numbers are large, but the scale is not an insurmountable hurdle; society has achieved massive transformations before. During World War II, the U.S. retooled automobile factories to produce 300,000 aircraft, and other countries produced 486,000 more. In 1956 the U.S. began building the Interstate Highway System, which after 35 years extended for 47,000 miles, changing commerce and society.

Is it feasible to transform the world's energy systems? Could it be accomplished in two decades? The answers depend on the technologies chosen, the availability of critical materials, and economic and political factors.

Clean Technologies Only

Renewable energy comes from enticing sources: wind, which also produces waves; water, which includes hydroelectric, tidal and geothermal energy (water heated by hot underground rock); and sun, which includes photovoltaics and solar power plants that focus sunlight to heat a fluid that drives a turbine to generate electricity. Our plan includes only technologies that work or are close to working today on a large scale, rather than those that may exist 20 or 30 years from now.

To ensure that our system remains clean, we consider only technologies that have near-zero emissions of greenhouse gases and air pollutants over their entire life cycle, including construction, operation and decommissioning.

For example, when burned in vehicles, even the most ecologically acceptable sources of ethanol create air pollution that will cause the same mortality level as when gasoline is burned. Nuclear power results in up to 25 times more carbon emissions than wind energy, when reactor construction and uranium refining and transport are considered. Carbon capture and sequestration technology can reduce carbon dioxide emissions from coal-fired power plants but will *increase* air pollutants and will extend all the other deleterious effects of coal mining, transport and processing, because more coal must be burned to power the capture and storage steps. Similarly, we consider only technologies that do not present significant waste disposal or terrorism risks.

In our plan, WWS will supply electric power for heating and transportation—industries that will have to revamp if the world has any hope of slowing climate change. We have assumed that most fossil-fuel heating (as well as ovens and stoves) can be replaced by electric systems and that most fossil-fuel transportation can be replaced by battery and fuel-cell vehicles. Hydrogen, produced by using WWS electricity to split water (electrolysis), would power fuel cells and be burned in airplanes and by industry.

Plenty of Supply

Today the maximum power consumed worldwide at any given moment is about 12.5 trillion watts (terawatts, or TW), according to the U.S. Energy Information Administration. The agency projects that in 2030 the world will require 16.9 TW of power as global population and living standards rise, with about 2.8 TW in the U.S. The mix of sources is similar to today's, heavily dependent on fossil fuels. If, however, the planet were powered entirely by WWS, with no fossil-fuel or biomass combustion, an intriguing savings would occur. Global power demand would be only 11.5 TW, and U.S. demand would be 1.8 TW. That decline occurs because, in most cases, electrification is a more efficient way to use energy. For example, only 17 to 20 percent of the energy in gasoline is used to move a vehicle (the rest is wasted as heat), whereas 75 to 86 percent of the electricity delivered to an electric vehicle goes into motion.

Even if demand did rise to 16.9 TW, WWS sources could provide far more power. Detailed studies by us and others indicate that energy from the wind, worldwide, is about 1,700 TW. Solar, alone, offers 6,500 TW. Of course, wind and sun out in the open seas, over high mountains and across protected regions would not be available. If we subtract these and low-wind areas not likely to be developed, we are still left with 40 to 85 TW for wind and 580 TW for solar, each far beyond future human demand. Yet currently we generate only 0.02 TW of wind power and 0.008 TW of solar. These sources hold an incredible amount of untapped potential.

The other WWS technologies will help create a flexible range of options. Although all the sources can expand greatly, for practical reasons, wave power can be extracted only near coastal areas. Many geothermal sources are too deep to be tapped economically. And even though hydroelectric power now exceeds all other WWS sources, most of the suitable large reservoirs are already in use.

The Plan: Power Plants Required

Clearly, enough renewable energy exists. How, then, would we transition to a new infrastructure to provide the world with 11.5 TW? We have chosen a mix of technologies emphasizing wind and solar, with about 9 percent of demand met by mature water-related methods. (Other combinations of wind and solar could be as successful.)

Wind supplies 51 percent of the demand, provided by 3.8 million large wind turbines (each rated at five megawatts) worldwide. Although that quantity may sound enormous, it is interesting to note that the world manufactures 73 million cars and light trucks *every year.* Another 40 percent of the power comes from photovoltaics and concentrated solar plants, with about 30 percent of the photovoltaic output from rooftop panels on homes and commercial buildings. About 89,000 photovoltaic and concentrated solar power plants, averaging 300 megawatts apiece, would be needed. Our mix also includes 900 hydroelectric stations worldwide, 70 percent of which are already in place.

Only about 0.8 percent of the wind base is installed today. The worldwide footprint of the 3.8 million turbines would be less than 50 square kilometers (smaller than Manhattan). When the needed spacing between them is figured, they would occupy about 1 percent of the earth's land, but the empty space among turbines could be used for agriculture or ranching or as open land or ocean. The nonrooftop photovoltaics and concentrated solar plants would occupy about 0.33 percent of the planet's land. Building such an extensive infrastructure will take time. But so did the current power plant network. And remember that if we stick with fossil fuels, demand by 2030 will rise to 16.9 TW, requiring about 13,000 large new coal plants, which themselves would occupy a lot more land, as would the mining to supply them.

The Materials Hurdle

The scale of the WWS infrastructure is not a barrier. But a few materials needed to build it could be scarce or subject to price manipulation.

Enough concrete and steel exist for the millions of wind turbines, and both those commodities are fully recyclable. The most problematic materials may be rare-earth metals such as neodymium used in turbine gearboxes. Although the metals are not in short supply, the low-cost sources are concentrated in China, so countries such as the U.S. could be trading dependence on Middle Eastern oil for dependence on Far Eastern metals. Manufacturers are moving toward gear-less turbines, however, so that limitation may become moot.

Photovoltaic cells rely on amorphous or crystalline silicon, cadmium telluride, or copper iridium selenide and sulfide. Limited supplies of tellurium and indium could reduce the prospects for some types of thin-film solar cells, though not for all; the other types might be able to take up the slack. Large-scale production could be restricted by the silver that cells require, but finding ways to reduce the silver content could tackle that hurdle. Recycling parts from old cells could ameliorate material difficulties as well.

Three components could pose challenges for building millions of electric vehicles: rare-earth metals for electric motors, lithium for lithium-ion batteries and platinum for fuel cells. More than half the world's lithium reserves lie in Bolivia and Chile. That concentration, combined with rapidly growing demand, could raise prices significantly. More problematic is the claim by Meridian International Research that not enough economically recoverable lithium exists to build anywhere near the number of batteries needed in a global electric-vehicle economy. Recycling could change the equation, but the economics of recycling depend in part on whether batteries are made with easy recyclability in mind, an issue the industry is aware of. The long-term use of platinum also depends on recycling; current available reserves would sustain annual production of 20 million fuel-cell vehicles, along with existing industrial uses, for fewer than 100 years.

Smart Mix for Reliability

A new infrastructure must provide energy on demand at least as reliably as the existing infrastructure. WWS technologies generally suffer less downtime than traditional sources. The average U.S. coal plant is offline 12.5 percent of the year for scheduled and unscheduled maintenance. Modern wind turbines have a down time of less than 2 percent on land and less than 5 percent at sea. Photovoltaic systems are also at less than 2 percent. Moreover, when an individual wind, solar or wave device is down, only a small fraction of production is affected; when a coal, nuclear or natural gas plant goes offline, a large chunk of generation is lost.

The main WWS challenge is that the wind does not always blow and the sun does not always shine in a given location. Intermittency problems can be mitigated by a smart balance of sources, such as generating a base supply from steady geothermal or tidal power, relying on wind at night when it is often plentiful, using solar by day and turning to a reliable source such as hydroelectric that can be turned on and off quickly to smooth out supply or meet peak demand. For example, interconnecting wind farms that are only 100 to 200 miles apart can compensate for hours of zero power at any one farm should the wind not be blowing there. Also helpful is interconnecting geographically dispersed sources so they can back up one another, installing smart electric meters in homes that automatically recharge electric vehicles when demand is low and building facilities that store power for later use.

Because the wind often blows during stormy conditions when the sun does not shine and the sun often shines on calm days with little wind, combining wind and solar can go a long way toward meeting demand, especially when geothermal provides a steady base and hydroelectric can be called on to fill in the gaps.

As Cheap as Coal

The mix of WWS sources in our plan can reliably supply the residential, commercial, industrial and transportation sectors. The logical next question is whether the power would be affordable. For each technology, we calculated how much it would cost a producer to generate power and transmit it across the grid. We included the annualized cost of capital, land, operations, maintenance, energy storage to help offset intermittent supply, and transmission. Today the cost of wind, geothermal and hydroelectric are all less than seven cents a kilowatt-hour (¢/kWh); wave and solar are higher. But by 2020 and beyond wind, wave and hydro are expected to be 4¢/kWh or less.

For comparison, the average cost in the U.S. in 2007 of conventional power generation and transmission was about 7¢/kWh, and it is projected to be 8¢/kWh in 2020. Power from wind turbines, for example, already costs about the same or less than it does from a new coal or natural gas plant, and in the future wind power is expected to be the least costly of all options. The competitive cost of wind has made it the second-largest source of new electric power generation in the U.S. for the past three years, behind natural gas and ahead of coal.

Solar power is relatively expensive now but should be competitive as early as 2020. A careful analysis by Vasilis Fthenakis of Brookhaven National Laboratory indicates that within 10 years, photovoltaic system costs could drop to about 10¢/kWh, including long-distance transmission and the cost of compressed-air storage of power for use at night. The same analysis estimates that concentrated solar power systems with enough thermal storage to generate electricity 24 hours a day in spring, summer and fall could deliver electricity at 10¢/kWh or less.

Transportation in a WWS world will be driven by batteries or fuel cells, so we should compare the economics of these electric vehicles with that of internal-combustion-engine vehicles. Detailed analyses by one of us (Delucchi) and Tim Lipman of the University of California, Berkeley, have indicated that mass-produced electric vehicles with advanced lithium-ion or nickel metal-hydride batteries could have a full lifetime cost per mile (including battery replacements) that is comparable with that of a gasoline vehicle, when gasoline sells for more than $2 a gallon.

When the so-called externality costs (the monetary value of damages to human health, the environment and climate) of fossil-fuel generation are taken into account, WWS technologies become even more cost-competitive.

Overall construction cost for a WWS system might be on the order of $100 trillion worldwide, over 20 years, not including transmission. But this is not money handed out by governments or consumers. It is investment that is paid back through the sale of electricity and energy. And again, relying on traditional sources would raise output from 12.5 to 16.9 TW, requiring thousands more of those plants, costing roughly $10 trillion, not to mention tens of trillions of dollars more in health, environmental and security costs. The WWS plan gives the world a new, clean, efficient energy system rather than an old, dirty, inefficient one.

Political Will

Our analyses strongly suggest that the costs of WWS will become competitive with traditional sources. In the interim, however, certain forms of WWS power will be significantly more costly than fossil power. Some combination of WWS subsidies and carbon taxes would thus be needed for a time. A feed-in tariff (FIT) program to cover the difference between generation cost and wholesale electricity prices is especially effective at scaling-up new technologies. Combining FITs with a so-called declining clock auction, in which the right to sell power to the grid goes to the lowest bidders, provides continuing incentive for WWS developers to lower costs. As that happens, FITs can be phased out. FITs have been implemented in a number of European countries and a few U.S. states and have been quite successful in stimulating solar power in Germany.

Taxing fossil fuels or their use to reflect their environmental damages also makes sense. But at a minimum, existing subsidies for fossil energy, such as tax benefits for exploration and extraction, should be eliminated to level the playing field. Misguided promotion of alternatives that are less desirable than WWS power, such as farm and production subsidies for biofuels, should also be ended, because it delays deployment of cleaner systems. For their part, legislators crafting policy must find ways to resist lobbying by the entrenched energy industries.

Finally, each nation needs to be willing to invest in a robust, long-distance transmission system that can carry large quantities of WWS power from remote regions where it is often greatest—such as the Great Plains for wind and the desert Southwest for solar in the U.S.—to centers of consumption, typically cities. Reducing consumer demand during peak usage periods also requires a smart grid that gives generators and consumers much more control over electricity usage hour by hour.

A large-scale wind, water and solar energy system can reliably supply the world's needs, significantly benefiting climate, air quality, water quality, ecology and energy security. As we have shown, the obstacles are primarily political, not technical. A combination of feed-in tariffs plus incentives for providers to reduce costs, elimination of fossil subsidies and an intelligently expanded grid could be enough to ensure rapid deployment. Of course, changes in the real-world power and transportation industries will have to overcome sunk investments in existing infrastructure. But with sensible policies, nations could set a goal of generating 25 percent of their new energy supply with WWS sources in 10 to 15 years and almost 100 percent of new supply in 20 to 30 years. With extremely aggressive policies, all existing fossil-fuel capacity could theoretically be retired and replaced in the same period, but with more modest and likely policies full replacement may take 40 to 50 years. Either way, clear leadership is needed, or else nations will keep trying technologies promoted by industries rather than vetted by scientists.

A decade ago it was not clear that a global WWS system would be technically or economically feasible. Having shown that it is, we hope global leaders can figure out how to make WWS power politically feasible as well. They can start by committing to meaningful climate and renewable energy goals now.

Critical Thinking

1. Over the years, society has spent enormous amounts of money to build the current energy system. Why does this make it difficult to change to a new energy system?

2. What is the most effective way to cut back on greenhouse gas emissions?

Part 3: Environmental Degradation

Forests, Wilderness, and Wildlife

Selection 14

WILLIAM O. DOUGLAS, from "Sierra Club v. Morton," *Sierra Club v. Morton* (405 U.S. 727, 1972)

Selection 15

WILLIAM CRONON, from *Uncommon Ground: Toward Reinventing Nature* (W. W. Norton, 1995)

Learning Outcomes

After reading this unit, you should be able to:

1. Explain the difficulty of preventing harm to the environment when the harmful act violates no state or federal law.
2. Explain the need to give some form of legal standing to components of the environment.
3. Explain who, if inanimate objects are given legal standing, should be able to sue on their behalf.
4. Discuss the nature and functions (in human terms) of wilderness.

Sierra Club v. Morton

William O. Douglas

The courts have played a very significant role in the cause of environmental protection. In cases where conservation or antipollution legislation exists but is being ignored or violated, individuals or grassroots organizations have frequently been successful in bringing suit against the offending party or against the agency that is failing to enforce the law. More difficult is the task of preventing an act of environmental desecration that may do great ecological harm but that violates no specific state or federal law. In such cases it is generally necessary for the individuals or organizations seeking judicial restraint to demonstrate that they will suffer unreasonable personal injury or loss as a result of the project or activity in question.

This important problem is probably best illustrated by the 1972 U.S. Supreme Court case *Sierra Club v. Morton,* specifically, the dissenting opinion of Justice William O. Douglas (1898–1980), from which the following selection has been taken. The issue in question was the attempt by the Sierra Club to prevent Walt Disney Enterprises, Inc., from degrading the Mineral King Valley, which is adjacent to Sequoia National Park, by building a ski resort there. The Court denied the Sierra Club's petition because the club had not proven that it would be harmed by the Disney project and therefore did not have legal standing to sue.

Douglas's landmark dissent on this ruling was based in large part on arguments made by law professor Christopher Stone in his treatise entitled "Should Trees Have Standing?" Douglas also gives proper credit to U.S. Forest Service officer Aldo Leopold, who had earlier argued for a land ethic that extends the boundaries of the community to include animate and inanimate components of the environment as well as human beings. That Justice Douglas took this position is no surprise in view of his reputation as an ardent conservationist and advocate of environmental protection. Indeed, Douglas had foreshadowed his position in this case by arguments he had made in several of the numerous books he authored, including *A Wilderness Bill of Rights,* published in 1964.

Key Concept: the legal defense of the environment

Mr. Justice DOUGLAS, dissenting.

I share the views of my Brother BLACKMUN and would reverse the judgment below.

The critical question of "standing" would be simplified and also put neatly in focus if we fashioned a federal rule that allowed environmental issues to be litigated before federal agencies or federal courts in the name of the inanimate object about to be despoiled, defaced, or invaded by roads and bulldozers and where injury is the subject of public outrage. Contemporary public concern for protecting nature's ecological equilibrium should lead to the conferral of standing upon environmental objects to sue for their own preservation. See Stone, Should Trees Have Standing?—Toward Legal Rights for Natural Objects, 45 S.Cal.L.Rev. 450 (1972). This suit would therefore be more properly labeled as Mineral King v. Morton.

Inanimate objects are sometimes parties in litigation. A ship has a legal personality, a fiction found useful for maritime purposes. The corporation sole—a creature of ecclessiastical law—is an acceptable adversary and large fortunes ride on its cases. The ordinary corporation is a "person" for purposes of the adjudicatory processes, whether it represents proprietary, spiritual, aesthetic, or charitable causes.

So it should be as respects valleys, alpine meadows, rivers, lakes, estuaries, beaches, ridges, groves of trees, swampland, or even air that feels the destructive pressures of modern technology and modern life. The river, for example, is the living symbol of all the life it sustains or nourishes—fish, aquatic insects, water ouzels, otter, fisher, deer, elk, bear, and all other animals, including man, who are dependent on it or who enjoy it for its sight, its sound, or its life. The river as plaintiff speaks for the ecological unit of life that is part of it. Those people who have a meaningful relation to that body of water—whether it be a fisherman, a canoeist, a zoologist, or a logger—must be able to speak for the values which the river represents and which are threatened with destruction.

I do not know Mineral King. I have never seen it nor traveled it, though I have seen articles describing its proposed "development" notably Hano, Protectionists vs. recreationists—The Battle of Mineral King, N.Y. Times Mag., Aug. 17, 1969, P. 25; and Browning, Mickey Mouse in the Mountains, Harper's, March 1972, p. 65. The Sierra Club in its complaint alleges that "[o]ne of the principal purposes of the Sierra Club is to protect and conserve the national resources of the Sierra Nevada Mountains." The District Court held that this uncontested allegation made the Sierra Club "sufficiently aggrieved" to have "standing" to sue on behalf of Mineral King.

Mineral King is doubtless like other wonders of the Sierra Nevada such as Tuolumne Meadows and the John Muir Trail. Those who hike it, fish it, hunt it, camp in it, frequent it, or visit it merely to sit in solitude and wonderment are legitimate spokesmen for it, whether they may be few or many. Those who have that intimate relation with the inanimate object about to be injured, polluted, or otherwise despoiled are its legitimate spokesmen.

The Solicitor General . . . takes a wholly different approach. He considers the problem in terms of "government by the Judiciary." With all respect, the problem is to make certain that the inanimate objects, which are the very core of America's beauty, have spokesmen before they are destroyed. It is, of course, true that most of them are under the control of a federal or state agency. The standards given those agencies are usually expressed in terms of the "public interest." Yet "public interest" has so many differing shades of meaning as to be quite meaningless on the environmental front. Congress accordingly has adopted ecological standards in the National Environmental Policy Act of 1969, and guidelines for agency action have been provided by the Council on Environmental Quality of which Russell E. Train is Chairman.

Yet the pressures on agencies for favorable action one way or the other are enormous. The suggestion that Congress can stop action which is undesirable is true in theory; yet even Congress is too remote to give meaningful direction and its machinery is too ponderous to use very often. The federal agencies of which I speak are not venal or corrupt. But they are notoriously under the control of powerful interests who manipulate them through advisory committees, or friendly working relations, or who have that natural affinity with the agency which in time develops between the regulator and the regulated.

As early as 1894, Attorney General Olney predicted that regulatory agencies might become "industry-minded," as illustrated by his forecast concerning the Interstate Commerce Commission:

"The Commission . . . is, or can be made, of great use to the railroads. It satisfies the popular clamor for a government supervision of railroads, at the same time that that supervision is almost entirely nominal. Further, the older such a commission gets to be, the more inclined it will be found to take the business and railroad view of things." M. Josephson, The Politicos 526 (1938).

Years later a court of appeals observed, "the recurring question which has plagued public regulation of industry [is] whether the regulatory agency is unduly oriented toward the interests of the industry it is designed to regulate, rather than the public interest it is designed to protect." Moss v. CAB, 139, U.S.App.D.C. 150, 152, 430 F.2d 891, 893. . . .

The Forest Service—one of the federal agencies behind the scheme to despoil Mineral King—has been notorious for its alignment with lumber companies, although its mandate from Congress directs it to consider the various aspects of multiple use in its supervision of the national forests. . . .

The voice of the inanimate object, therefore, should not be stilled. That does not mean that the judiciary takes over the managerial functions from the federal agency. It merely means that before these priceless bits of Americana (such as a valley, an alpine meadow, a river, or a lake) are forever lost or are so transformed as to be reduced to the eventual rubble of our urban environment, the voice of the existing beneficiaries of these environmental wonders should be heard.

Perhaps they will not win. Perhaps the bulldozers of "progress" will plow under all the aesthetic wonders of this beautiful land. That is not the present question. The sole question is, who has standing to be heard?

Those who hike the Appalachian Trail into Sunfish Pond, New Jersey, and camp or sleep there, or run the Allagash in Maine, or climb the Guadalupes in West Texas, or who canoe and portage the Quetico Superior in Minnesota, certainly should have standing to defend those natural wonders before courts or agencies, though they live 3,000 miles away. Those who merely are caught up in environmental news or propaganda and flock to defend these waters or areas may be treated differently. That is why these environmental issues should be tendered by the inanimate object itself. Then there will be assurances that all of the forms of life which it represents will stand before the court—the pileated woodpecker as well as the coyote and bear, the lemmings as well as the trout in the streams. Those inarticulate members of the ecological group cannot speak. But those people who have so frequented the place as to know its values and wonders will be able to speak for the entire ecological community.

Ecology reflects the land ethic; and Aldo Leopold wrote in A Sand County Almanac 204 (1949), "The land ethic simply enlarges the boundaries of the community to include soils, waters, plants, and animals, or collectively: the land."

That, as I see it, is the issue of "standing" in the present case and controversy.

Critical Thinking

1. Should legal standing be granted to inanimate objects such as trees and mountains?
2. Clearly, if legal standing is granted to inanimate objects, those objects cannot sue to prevent damage by construction projects. Who should be able to sue on their behalf?

The Trouble With Wilderness; or, Getting Back to the Wrong Nature

William Cronon

Most environmentalists support the preservation of wilderness as an unquestioned ecological necessity. Thus, when respected University of Wisconsin environmental historian, William Cronon, wrote a short article, "The Trouble With Wilderness; or, Getting Back to the Wrong Nature," *The New York Times Magazine* (August 13, 1995), sharply attacking that perspective, he immediately became the focus of a heated controversy. Cronon further stoked the fires of that controversy by editing a book entitled *Uncommon Ground: Toward Reinventing Nature* (W.W. Norton, 1995), which contained essays by a select group of scholars from a variety of academic disciplines. These essays examine various aspects of the interface between human beings and nature. Most contain provocative, unconventional critiques of commonly held attitudes toward nature. Included in that book is a greatly expanded version of Cronon's *New York Times Magazine* article. It is from this version of the article that the following selection is taken.

Cronon attacks the very concept of wilderness. He views wilderness as being a human creation, reflecting particular cultural perspectives, rather than a pristine "natural" ecological state unaffected by human influence. He argues that the movement to preserve large tracts of land that are designated as wilderness is an elitist enterprise that is not only unnecessary, but ecologically counterproductive. Yet, Cronon considers himself to be an environmentalist and is even on the governing board of the Wilderness Society. In the face of his uncompromising denigration of the commonly held concept of wilderness, Cronon has confounded both his supporters and detractors by declaring that "wilderness is my religion." Such assertions have done little to reduce the animosity that Cronon's writings have engendered among many environmentalists who have accused him of misinterpreting the conservation movement, ignoring such serious issues as the need to preserve biodiversity, and paying too little attention to the political consequences of his work. With regard to this latter concern, it is a fact that whether wittingly or not, Cronon has become something of a hero to the antienvironmental lobby.

Key Concept: the concept of wilderness as a human construct

The time has come to rethink wilderness.

This will seem a heretical claim to many environmentalists, since the idea of wilderness has for decades been a fundamental tenet—indeed, a passion—of the environmental movement, especially in the United States. For many Americans wilderness stands as the last remaining place where civilization, that all too human disease, has not fully infected the earth. It is an island in the polluted sea of urban-industrial modernity, the one place we can turn for escape from our own too-muchness. Seen in this way, wilderness presents itself as the best antidote to our human selves, a refuge we must somehow recover if we hope to save the planet. As Henry David Thoreau, [a nineteenth-century writer and naturalist,] once famously declared, "In Wildness is the preservation of the World."

But is it? The more one knows of its peculiar history, the more one realizes that wilderness is not quite what it seems. Far from being the one place on earth that stands apart from humanity, it is quite profoundly a human creation—indeed, the creation of very particular human cultures at very particular moments in human history. It is not a pristine sanctuary where the last remnant of an untouched, endangered, but still transcendent nature can for at least a little while longer be encountered without the contaminating taint of civilization. Instead, it is a product of that civilization, and could hardly be contaminated by the very stuff of which it is made. Wilderness hides its unnaturalness behind a mask that is all the more beguiling because it seems so natural. As we gaze into the mirror it holds up for us, we too easily imagine that what we behold is Nature when in fact we see the reflection of our own unexamined longings and desires. For this reason, we mistake ourselves when we suppose that wilderness can be the solution to our culture's problematic relationships with the nonhuman world, for wilderness is itself no small part of the problem.

To assert the unnaturalness of so natural a place will no doubt seem absurd or even perverse to many readers, so let me hasten to add that the nonhuman world we encounter in wilderness is far from being merely our own invention. I celebrate with others who love wilderness the beauty and power of the things it contains. Each of us who has spent time there can conjure images and sensations that seem all the more hauntingly real for having engraved themselves so indelibly on our memories. Such memories may be uniquely our own, but they are also familiar enough to be instantly recognizable to others. Remember this? The torrents of mist shoot out from the base of a great waterfall in the depths of a Sierra canyon, the tiny droplets cooling your face as you listen to the roar of the water and gaze up toward the sky through a rainbow that hovers just out of reach. Remember this too: looking out across a desert canyon in the evening air, the only sound a lone raven calling in the distance, the rock walls dropping away into a chasm so deep that its bottom all but vanishes as you squint into the amber light of the setting sun. And this: the moment beside the trail as you sit on a sandstone ledge, your boots damp with the morning dew while you take in the rich smell of the pines, and the small red fox—or maybe for you it was a raccoon or a coyote or a deer—that suddenly ambles across your path, stopping for a long moment to gaze in your direction with cautious indifference before continuing on its way. Remember the feelings of such moments, and you will know as well as I do that you were in the presence of something irreducibly nonhuman, something profoundly Other than yourself. Wilderness is made of that too.

And yet: what brought each of us to the places where such memories became possible is entirely a cultural invention. . . .

Wilderness had once been the antithesis of all that was orderly and good—it had been the darkness, one might say, on the far side of the garden wall—and yet now it was frequently likened to Eden itself. When John Muir, [a nineteenth-century writer and explorer,] arrived in the Sierra Nevada in 1869, he would declare, "No description of Heaven that I have ever heard or read of seems half so fine." He was hardly alone in expressing such emotions. One by one, various corners of the American map came to be designated as sites whose wild beauty was so spectacular that a growing number of citizens had to visit and see them for themselves. Niagara Falls was the first to undergo this transformation, but it was soon followed by the Catskills, the Adirondacks, Yosemite, Yellowstone, and others. Yosemite was deeded by the U.S. government to the state of California in 1864 as the nation's first wildland park, and Yellowstone became the first true national park in 1872.

By the first decade of the twentieth century, in the single most famous episode in American conservation history, a national debate had exploded over whether the city of San Francisco should be permitted to augment its water supply by damming the Tuolumne River in Hetch Hetchy valley, well within the boundaries of Yosemite National Park. The dam was eventually built, but what today seems no less significant is that so many people fought to prevent its completion. Even as the fight was being lost, Hetch Hetchy became the battle cry of an emerging movement to preserve wilderness. Fifty years earlier, such opposition would have been unthinkable. Few would have questioned the merits of "reclaiming" a wasteland like this in order to put it to human use. Now the defenders of Hetch Hetchy attracted widespread national attention by portraying such an act not as improvement or progress but as desecration and vandalism. Lest one doubt that the old biblical metaphors had been turned completely on their heads, listen to John Muir attack the dam's defenders. "Their arguments," he wrote, "are curiously like those of the devil, devised for the destruction of the first garden—so much of the very best Eden fruit going to waste; so much of the best Tuolumne water and Tuolumne scenery going to waste." For Muir and the growing number of Americans who shared his views, Satan's home had become God's own temple.

The sources of this rather astonishing transformation were many, but for the purposes of this [selection] they can be gathered under two broad headings: the sublime and the frontier. Of the two, the sublime is the older and more pervasive cultural construct, being one of the most important expressions of that broad transatlantic movement we today label as romanticism; the frontier is more peculiarly American, though it too had its European antecedents and parallels. The two converged to remake wilderness in their own image, freighting it with moral values and cultural symbols that it carries to this day. Indeed, it is not too much to say that the modern environmental movement is itself a grandchild of romanticism and post-frontier ideology, which is why it is no accident that so much environmentalist discourse takes its bearings from the wilderness these intellectual movements helped create. Although wilderness may today seem to be just one environmental concern among many, it in fact serves as the foundation for a long list of other such concerns that on their face seem quite remote from it. That is why its influence is so pervasive and, potentially, so insidious. . . .

The emotions [John] Muir describes in [his writings about] Yosemite could hardly be more different from [Henry David] Thoreau's on Katahdin or [William] Wordsworth's, [a nineteenth-century poet,] on the Simplon Pass. Yet all three men are participating in the same cultural tradition and contributing to the same myth: the mountain as cathedral. The three may differ in the way they choose to express their piety—Wordsworth favoring an awe-filled bewilderment, Thoreau a stern loneliness, Muir a welcome ecstasy—but they agree completely about the church in which they prefer to worship. Muir's closing words on North Dome diverge from his older contemporaries only in mood, not in their ultimate content:

> Perched like a fly on this Yosemite dome, I gaze and sketch and bask, oftentimes settling down into dumb admiration without definite hope of ever learning much, yet with the longing, unresting effort that lies at the door of hope, humbly prostrate before the vast display of God's power, and eager to offer self-denial and renunciation with eternal toil to learn any lesson in the divine manuscript.

... But the romantic sublime was not the only cultural movement that helped transform wilderness into a sacred American icon during the nineteenth century. No less important was the powerful romantic attraction of primitivism, dating back at least to [Jean Jacques] Rousseau, [a French philosopher (1712–1778),]—the belief that the best antidote to the ills of an overly refined and civilized modern world was a return to simpler, more primitive living. In the United States, this was embodied most strikingly in the national myth of the frontier. The historian Frederick Jackson Turner wrote in 1893 the classic academic statement of this myth, but it had been part of American cultural traditions for well over a century. As Turner described the process, easterners and European immigrants, in moving to the wild unsettled lands of the frontier, shed the trappings of civilization, rediscovered their primitive racial energies, reinvented direct democratic institutions, and thereby reinfused themselves with a vigor, an independence, and a creativity that were the source of American democracy and national character. Seen in this way, wild country became a place not just of religious redemption but of national renewal, the quintessential location for experiencing what it meant to be an American.

One of Turner's most provocative claims was that by the 1890s the frontier was passing away. Never again would "such gifts of free land offer themselves" to the American people. "The frontier has gone," he declared, "and with its going has closed the first period of American history." Built into the frontier myth from its very beginning was the notion that this crucible of American identity was temporary and would pass away. . . .

This nostalgia for a passing frontier way of life inevitably implied ambivalence, if not downright hostility, toward modernity and all that it represented. If one saw the wild lands of the frontier as freer, truer, and more natural than other, more modern places, then one was also inclined to see the cities and factories of urban-industrial civilization as confining, false, and artificial. Owen Wister, [an American writer of cowboy fiction (1860–1938),] looked at the post-frontier "transition" that had followed "the horseman of the plains," and did not like what he saw: "a shapeless state, a condition of men and manners as unlovely as is that moment in the year when winter is gone and spring not come, and the face of Nature is ugly." In the eyes of writers who shared Wister's distaste for modernity, civilization contaminated its inhabitants and absorbed them into the faceless, collective, contemptible life of the crowd. For all of its troubles and dangers, and despite the fact that it must pass away, the frontier had been a better place. If civilization was to be redeemed, it would be by men like the Virginian who could retain their frontier virtues even as they made the transition to post-frontier life.

The mythic frontier individualist was almost always masculine in gender: here, in the wilderness, a man could be a real man, the rugged individual he was meant to be before civilization sapped his energy and threatened his masculinity. Wister's contemptuous remarks about Wall Street and Newport suggest what he and many others of his generation believed—that the comforts and seductions of civilized life were especially insidious for men, who all too easily became emasculated by the feminizing tendencies of civilization. More often than not, men who felt this way came, like Wister and [Theodore] Roosevelt, from elite class backgrounds. The curious result was that frontier nostalgia became an important vehicle for expressing a peculiarly bourgeois form of antimodernism. The very men who most benefited from urban-industrial capitalism were among those who believed they must escape its debilitating effects. If the frontier was passing, then men who had the means to do so should preserve for themselves some remnant of its wild landscape so that they might enjoy the regeneration and renewal that came from sleeping under the stars, participating in blood sports, and living off the land. The frontier might be gone, but the frontier experience could still be had if only wilderness were preserved.

Thus the decades following the Civil War saw more and more of the nation's wealthiest citizens seeking out wilderness for themselves. The elite passion for wild land took many forms: enormous estates in the Adirondacks and elsewhere (disingenuously called "camps" despite their many servants and amenities), cattle ranches for would-be rough riders on the Great Plains, guided big-game hunting trips in the Rockies, and luxurious resort hotels wherever railroads pushed their way into sublime landscapes. Wilderness suddenly emerged as the landscape of choice for elite tourists, who brought with them strikingly urban ideas of the countryside through which they traveled. For them, wild land was not a site for productive labor and not a permanent home; rather, it was a place of recreation. One went to the wilderness not as a producer but as a consumer, hiring guides and other backcountry residents who could serve as romantic surrogates for the rough riders and hunters of the frontier if one was willing to overlook their new status as employees and servants of the rich.

In just this way, wilderness came to embody the national frontier myth, standing for the wild freedom of America's past and seeming to represent a highly attractive natural alternative to the ugly artificiality of modern civilization. The irony, of course, was that in the process wilderness came to reflect the very civilization its devotees sought to escape. Ever since the nineteenth century, celebrating wilderness has been an activity mainly for well-to-do city folks. Country people generally know far too much about working the land to regard *unworked* land as their ideal. In contrast, elite urban tourists and wealthy sportsmen projected their leisure-time frontier fantasies onto the American landscape and so created wilderness in their own image.

There were other ironies as well. The movement to set aside national parks and wilderness areas followed hard on the heels of the final Indian wars, in which the prior human inhabitants of these areas were rounded up and moved onto reservations. The myth of the wilderness as "virgin," uninhabited land had always been especially cruel when seen from the perspective of the Indians who had once called that land home. Now they were forced to move elsewhere, with the result that tourists could safely enjoy the illusion that they were seeing their nation in its pristine, original state, in the new morning of God's own creation. Among the things that most marked the new national parks as reflecting a post-frontier consciousness was the relative absence of human violence within their boundaries. The actual frontier had often been a place of conflict, in which invaders and invaded fought for control of land and resources. Once set aside within the fixed and carefully policed boundaries of the modern bureaucratic state, the wilderness lost its savage image and became safe: a place more of reverie than of revulsion or fear. Meanwhile, its original inhabitants were kept out by dint of force, their earlier uses of the land redefined as inappropriate or even illegal. To this day, for instance, the Blackfeet continue to be accused of "poaching" on the lands of Glacier National Park that originally belonged to them and that were ceded by treaty only with the proviso that they be permitted to hunt there.

The removal of Indians to create an "uninhabited wilderness"—uninhabited as never before in the human history of the place—reminds us just how invented, just how constructed, the American wilderness really is. To return to my opening argument: there is nothing natural about the concept of wilderness. It is entirely a creation of the culture that holds it dear, a product of the very history it seeks to deny. Indeed, one of the most striking proofs of the cultural invention of wilderness is its thoroughgoing erasure of the history from which it sprang. In virtually all of its manifestations, wilderness represents a flight from history. Seen as the original garden, it is a place outside of time, from which human beings had to be ejected before the fallen world of history could properly begin. Seen as the frontier, it is a savage world at the dawn of civilization, whose transformation represents the very

beginning of the national historical epic. Seen as the bold landscape of frontier heroism, it is the place of youth and childhood, into which men escape by abandoning their pasts and entering a world of freedom where the constraints of civilization fade into memory. Seen as the sacred sublime, it is the home of a God who transcends history by standing as the One who remains untouched and unchanged by time's arrow. No matter what the angle from which we regard it, wilderness offers us the illusion that we can escape the cares and troubles of the world in which our past has ensnared us.

. . . Wilderness is the natural, unfallen antithesis of an unnatural civilization that has lost its soul. It is a place of freedom in which we can recover the true selves we have lost to the corrupting influences of our artificial lives. Most of all, it is the ultimate landscape of authenticity. Combining the sacred grandeur of the sublime with the primitive simplicity of the frontier, it is the place where we can see the world as it really is, and so know ourselves as we really are—or ought to be.

But the trouble with wilderness is that it quietly expresses and reproduces the very values its devotees seek to reject. The flight from history that is very nearly the core of wilderness represents the false hope of an escape from responsibility, the illusion that we can somehow wipe clean the slate of our past and return to the tabula rasa that supposedly existed before we began to leave our marks on the world. The dream of an unworked natural landscape is very much the fantasy of people who have never themselves had to work the land to make a living—urban folk for whom food comes from a supermarket or a restaurant instead of a field, and for whom the wooden houses in which they live and work apparently have no meaningful connection to the forests in which trees grow and die. Only people whose relation to the land was already alienated could hold up wilderness as a model for human life in nature, for the romantic ideology of wilderness leaves precisely nowhere for human beings actually to make their living from the land.

This, then, is the central paradox: wilderness embodies a dualistic vision in which the human is entirely outside the natural. If we allow ourselves to believe that nature, to be true, must also be wild, then our very presence in nature represents its fall. The place where we are is the place where nature is not. If this is so—if by definition wilderness leaves no place for human beings, save perhaps as contemplative sojourners enjoying their leisurely reverie in God's natural cathedral—then also by definition it can offer no solution to the environmental and other problems that confront us. To the extent that we celebrate wilderness as the measure with which we judge civilization, we reproduce the dualism that sets humanity and nature at opposite poles. We thereby leave ourselves little hope of discovering what an ethical, sustainable, *honorable* human place in nature might actually look like.

Worse: to the extent that we live in an urban-industrial civilization but at the same time pretend to ourselves that our *real* home is in the wilderness, to just that extent we give ourselves permission to evade responsibility for the lives we actually lead. We inhabit civilization while holding some part of ourselves—what we imagine to be the most precious part—aloof from its entanglements. We work our nine-to-five jobs in its institutions, we eat its food, we drive its cars (not least to reach the wilderness), we benefit from the intricate and all too invisible networks with which it shelters us, all the while pretending that these things are not an essential part of who we are. By imagining that our true home is in the wilderness, we forgive ourselves the homes we actually inhabit. In its flight from history, in its siren song of escape, in its reproduction of the dangerous dualism that sets human beings outside of nature—in all of these ways, wilderness poses a serious threat to responsible environmentalism at the end of the twentieth century.

By now I hope it is clear that my criticism in this [selection] is not directed at wild nature per se, or even at efforts to set aside large tracts of wild land, but rather at the specific habits of thinking that flow from this complex cultural construction called wilderness. It is not the things we label as wilderness that are the problem—for nonhuman nature and large tracts of the natural world *do* deserve protection—but rather what we ourselves mean when we use that label. Lest one doubt how pervasive these habits of thought actually are in contemporary environmentalism, let me list some of the places where wilderness serves as the ideological underpinning for environmental concerns that might otherwise seem quite remote from it. Defenders of biological diversity, for instance, although sometimes appealing to more utilitarian concerns, often point to "untouched" ecosystems as the best and richest repositories of the undiscovered species we must certainly try to protect. Although at first blush an apparently more "scientific" concept than wilderness, biological diversity in fact invokes many of the same sacred values, which is why organizations like the Nature Conservancy have been so quick to employ it as an alternative to the seemingly fuzzier and more problematic concept of wilderness. There is a paradox here, of course. To the extent that biological diversity (indeed, even wilderness itself) is likely to survive in the future only by the most vigilant and self-conscious management of the ecosystems that sustain it, the ideology of wilderness is potentially in direct conflict with the very thing it encourages us to protect.

The most striking instances of this have revolved around "endangered species," which serve as vulnerable symbols of biological diversity while at the same time standing as surrogates for wilderness itself. The terms of the Endangered Species Act in the United States have often meant that those hoping to defend pristine wilderness have had to rely on a single endangered species like the spotted owl to gain legal standing for their case—thereby making the full power of sacred land inhere in a single numinous organism whose habitat then becomes the object of intense debate about appropriate management and use. The ease with which anti-environmental forces like the wise-use movement have attacked such single-species preservation efforts suggests the vulnerability of strategies like these. . . .

Wilderness gets us into trouble only if we imagine that this experience of wonder and otherness is limited to the remote corners of the planet, or that it somehow depends on pristine landscapes we ourselves do not inhabit. Nothing could be more misleading. The tree in the garden is in reality no less other, no less worthy of our wonder and respect, than the tree in an ancient forest that has never known an ax or a saw—even though the tree in the forest reflects a more intricate web of ecological relationships. The tree in the garden could easily have sprung from the same seed as the tree in the forest, and we can claim only its location and perhaps its form as our own. Both trees stand apart from us; both share our common world. The special power of the tree in the wilderness is to remind us of this fact. It can teach us to recognize the wildness we did not see in the tree we planted in our own backyard. By seeing the otherness in that which is most unfamiliar, we can learn to see it too in that which at first seemed merely ordinary. If wilderness can do this—if it can help us perceive and respect a nature we had forgotten to recognize as natural—then it will become part of the solution to our environmental dilemmas rather than part of the problem.

This will only happen, however, if we abandon the dualism that sees the tree in the garden as artificial—completely fallen and unnatural—and the tree in the wilderness as natural—completely pristine and wild. Both trees in some ultimate sense are wild; both in a practical sense now depend on our management and care. We are responsible for both, even though we can claim credit for neither. Our challenge is to stop thinking of such things according to a set of bipolar moral scales in which the human and the nonhuman, the unnatural and the natural, the fallen and the unfallen, serve as our conceptual map for understanding and valuing the world. Instead, we need to embrace the full continuum of a natural landscape that is also cultural, in which the city, the suburb, the pastoral, and the wild each has its proper place, which we permit ourselves to celebrate without needlessly denigrating the others. We need to honor the Other within and the Other next door as much as we do the exotic Other that lives far away—a lesson that applies as much to people as it does to (other) natural things. In particular, we need to discover a common middle ground in which all of these things, from the city to the wilderness, can somehow be encompassed in the word "home." Home, after all, is the place where finally we make our living. It is the place for which we take responsibility, the place we try to sustain so we can pass on what is best in it (and in ourselves) to our children. . . .

Learning to honor the wild—learning to remember and acknowledge the autonomy of the other—means striving for critical self-consciousness in all of our actions. It means that deep reflection and respect must accompany each act of use, and means too that we must always consider the possibility of non-use. It means looking at the part of nature we intend to turn toward our own ends and asking whether we can use it again and again and again—sustainably—without its being diminished in the process. It means never imagining that we can flee into a mythical wilderness to escape history and the obligation to take responsibility for our own actions that history inescapably entails. Most of all, it means practicing remembrance and gratitude, for thanksgiving is the simplest and most basic of ways for us to recollect the nature, the culture, and the history that have come together to make the world as we know it. If wildness can stop being (just) out there and start being (also) in here, if it can start being as humane as it is natural, then perhaps we can get on with the unending task of struggling to live rightly in the world—not just in the garden, not just in the wilderness, but in the home that encompasses them both.

Critical Thinking

1. What is "wilderness"?
2. Is it more important to provide "wilderness" as public playgrounds or as "safe zones" for wildlife?

Biodiversity

Selection 16
Secretariat of the Convention on Biological Diversity, from *Global Biodiversity Outlook 3* (2010); http://gbo3.cbd.int

Selection 17
BORIS WORM ET AL., from "Impacts of Biodiversity Loss on Ocean Ecosystem Services," *Science* (November 3, 2006)

Selection 18
JOHN VANDERMEER AND IVETTE PERFECTO, from "Rethinking Rain Forests: Biodiversity and Social Justice," *Food First Backgrounder* (Summer 1995)

Learning Outcomes
After reading this unit, you should be able to:

1. Describe the benefits of intact ecosystems.
2. Explain how loss of biodiversity affects the world's poorest peoples.
3. Describe the factors that cause loss of biodiversity.
4. Describe how loss of marine biodiversity threatens commercial fisheries.
5. Explain why it is to difficult to say that environmental damage has a single cause.

Executive Summary from Secretariat of the Convention on Biological Diversity

Global Biodiversity Outlook 3

It may properly be said that the extinction of species of plants and animals is a natural phenomenon that has been occurring since the beginning of life on Earth. Why then is there growing concern among biologists and environmentalists about endangered species to the point where many consider it a crisis worthy of immediate international attention? The answer is that the ability of species to evolve and adapt to changing conditions requires the continued existence of a diverse pool of genetic material, which dwindles as species disappear. The loss of species also threatens the long-term maintenance of ecosystem stability, which is essential to human well-being. And although we lack sufficient information to deduce the precise rate of current species extinction, there is broad agreement among experts that the rate has been accelerating rapidly as the direct result of human activities.

At the World Summit on Sustainable Development held in Johannesburg, South Africa, in 2002, the world's governments agreed "to achieve by 2010 a significant reduction of the current rate of biodiversity loss at the global, regional and national level as a contribution to poverty alleviation and to the benefit of all life on Earth." By 2010, it was clear that this goal had not been met, and the Secretariat of the Convention on Biological Diversity reported in *Global Biodiversity Outlook 3* that what the world does in the near future "will determine whether the relatively stable environmental conditions on which human civilization has depended for the past 10,000 years will continue beyond this century. If we fail to use this opportunity, many ecosystems on the planet will move into new, unprecedented states in which the capacity to provide for the needs of present and future generations is highly uncertain."

The United Nation's Secretariat of the Convention on Biological Diversity was established to support the goals of the Convention on Biological Diversity (CBD), signed at the Earth Summit in Rio de Janeiro, Brazil, in 1992. The CBD was the first global agreement to cover all aspects of biological diversity: the conservation of biological diversity, the sustainable use of its components and the fair and equitable sharing of benefits arising from the use of genetic resources.

Key Concept: the urgency of maintaining biodiversity

*T*he target agreed by the world's Governments in 2002, "to achieve by 2010 a significant reduction of the current rate of biodiversity loss at the global, regional and national level as a contribution to poverty alleviation and to the benefit of all life on Earth," has not been met.

There are multiple indications of continuing decline in biodiversity in all three of its main components—genes, species and ecosystems—including:

- Species which have been assessed for extinction risk are on average moving closer to extinction. Amphibians face the greatest risk and coral species are deteriorating most rapidly in status. Nearly a quarter of plant species are estimated to be threatened with extinction.

- The abundance of vertebrate species, based on assessed populations, fell by nearly a third on average between 1970 and 2006, and continues to fall globally, with especially severe declines in the tropics and among freshwater species.

- Natural habitats in most parts of the world continue to decline in extent and integrity, although there has been significant progress in slowing the rate of loss for tropical forests and mangroves, in some regions. Freshwater wetlands, sea ice habitats, salt marshes, coral reefs, seagrass beds and shellfish reefs are all showing serious declines.

- Extensive fragmentation and degradation of forests, rivers and other ecosystems have also led to loss of biodiversity and ecosystem services.
- Crop and livestock genetic diversity continues to decline in agricultural systems.
- The five principal pressures directly driving biodiversity loss (habitat change, overexploitation, pollution, invasive alien species and climate change) are either constant or increasing in intensity.
- The ecological footprint of humanity exceeds the biological capacity of the Earth by a wider margin than at the time the 2010 target was agreed.

The loss of biodiversity is an issue of profound concern for its own sake. Biodiversity also underpins the functioning of ecosystems which provide a wide range of services to human societies. Its continued loss, therefore, has major implications for current and future human well-being. The provision of food, fibre, medicines and fresh water, pollination of crops, filtration of pollutants, and protection from natural disasters are among those ecosystem services potentially threatened by declines and changes in biodiversity. Cultural services such as spiritual and religious values, opportunities for knowledge and education, as well as recreational and aesthetic values, are also declining.

The existence of the 2010 biodiversity target has helped to stimulate important action to safeguard biodiversity, such as creating more protected areas (both on land and in coastal waters), the conservation of particular species, and initiatives to tackle some of the direct causes of ecosystem damage, such as pollution and alien species invasions. Some 170 countries now have national biodiversity strategies and action plans. At the international level, financial resources have been mobilized and progress has been made in developing mechanisms for research, monitoring and scientific assessment of biodiversity.

Many actions in support of biodiversity have had significant and measurable results in particular areas and amongst targeted species and ecosystems. This suggests that with adequate resources and political will, the tools exist for loss of biodiversity to be reduced at wider scales. For example, recent government policies to curb deforestation have been followed by declining rates of forest loss in some tropical countries. Measures to control alien invasive species have helped a number of species to move to a lower extinction risk category. It has been estimated that at least 31 bird species (out of 9,800) would have become extinct in the past century, in the absence of conservation measures.

However, action to implement the Convention on Biological Diversity has not been taken on a sufficient scale to address the pressures on biodiversity in most places. There has been insufficient integration of biodiversity issues into broader policies, strategies and programmes, and the underlying drivers of biodiversity loss have not been addressed significantly. Actions to promote the conservation and sustainable use of biodiversity receive a tiny fraction of funding compared to activities aimed at promoting infrastructure and industrial developments. Moreover, biodiversity considerations are often ignored when such developments are designed, and opportunities to plan in ways that minimize unnecessary negative impacts on biodiversity are missed. Actions to address the underlying drivers of biodiversity loss, including demographic, economic, technological, socio-political and cultural pressures, in meaningful ways, have also been limited.

Most future scenarios project continuing high levels of extinctions and loss of habitats throughout this century, with associated decline of some ecosystem services important to human well-being. For example:

- Tropical forests would continue to be cleared in favour of crops and pastures, and potentially for biofuel production.
- Climate change, the introduction of invasive alien species, pollution and dam construction would put further pressure on freshwater biodiversity and the services it underpins.
- Overfishing would continue to damage marine ecosystems and cause the collapse of fish populations, leading to the failure of fisheries.

Changes in the abundance and distribution of species may have serious consequences for human societies. The geographical distribution of species and vegetation types is projected to shift radically due to climate change, with ranges moving from hundreds to thousands of kilometres towards the poles by the end of the 21st century. Migration of marine species to cooler waters could make tropical oceans less diverse, while both boreal and temperate forests face widespread dieback at the southern end of their existing ranges, with impacts on fisheries, wood harvests, recreation opportunities and other services.

There is a high risk of dramatic biodiversity loss and accompanying degradation of a broad range of ecosystem services if ecosystems are pushed beyond certain thresholds or tipping points. The poor would face the earliest and most severe impacts of such changes, but ultimately all societies and communities would suffer. Examples include:

- The Amazon forest, due to the interaction of deforestation, fire and climate change, could undergo a widespread dieback, with parts of the forest moving into a self-perpetuating cycle of more frequent fires and intense droughts leading to a shift to savanna-like vegetation. While there are large uncertainties associated with these scenarios, it is known that such dieback becomes much more likely to occur if deforestation exceeds 20–30% (it is currently above 17% in the Brazilian Amazon). It would lead to regional rainfall reductions, compromising agricultural production. There would also be global impacts through increased carbon emissions, and massive loss of biodiversity.

- The build-up of phosphates and nitrates from agricultural fertilizers and sewage effluent can shift freshwater lakes and other inland water ecosystems into a long-term, algae dominated (eutrophic) state. This could lead to declining fish availability with implications for food security in many developing countries. There will also be loss of recreation opportunities and tourism income, and in some cases health risks for people and livestock from toxic algal blooms. Similar, nitrogen-induced eutrophication phenomena in coastal environments lead to more oxygen-starved dead zones, with major economic losses resulting from reduced productivity of fisheries and decreased tourism revenues.

- The combined impacts of ocean acidification, warmer sea temperatures and other human induced stresses make tropical coral reef ecosystems vulnerable to collapse. More acidic water—brought about by higher carbon dioxide concentrations in the atmosphere—decreases the availability of the carbonate ions required to build coral skeletons. Together with the bleaching impact of warmer water, elevated nutrient levels from pollution, overfishing, sediment deposition arising from inland deforestation, and other pressures, reefs worldwide increasingly become algae-dominated with catastrophic loss of biodiversity and ecosystem functioning, threatening the livelihoods and food security of hundreds of millions of people.

There are greater opportunities than previously recognized to address the biodiversity crisis while contributing to other social objectives. For example, analyses conducted for this Outlook identified scenarios in which climate change is mitigated while maintaining and even expanding the current extent of forests and other natural ecosystems (avoiding additional habitat loss from the widespread deployment of biofuels). Other opportunities include "rewilding" abandoned farmland in some regions, and the restoration of river basins and other wetland ecosystems to enhance water supply, flood control and the removal of pollutants.

Well-targeted policies focusing on critical areas, species and ecosystem services are essential to avoid the most dangerous impacts on people and societies. Preventing further human-induced biodiversity loss for the near term future will be extremely challenging, but biodiversity loss may be halted and in some aspects reversed in the longer term, if urgent, concerted and effective action is initiated now in support of an agreed long-term vision. Such action to conserve biodiversity and use its components sustainably will reap rich rewards—through better health, greater food security,

less poverty and a greater capacity to cope with, and adapt to, environmental change.

Placing greater priority on biodiversity is central to the success of development and poverty-alleviation measures. It is clear that continuing with "business as usual" will jeopardize the future of all human societies, and none more so than the poorest who depend directly on biodiversity for a particularly high proportion of their basic needs. The loss of biodiversity is frequently linked to the loss of cultural diversity, and has an especially high negative impact on indigenous communities.

The linked challenges of biodiversity loss and climate change must be addressed by policymakers with equal priority and in close co-ordination, if the most severe impacts of each are to be avoided. Reducing the further loss of carbon storing ecosystems such as tropical forests, salt marshes and peatlands will be a crucial step in limiting the build-up of greenhouse gases in the atmosphere. At the same time, reducing other pressures on ecosystems can increase their resilience, make them less vulnerable to those impacts of climate change which are already unavoidable, and allow them to continue to provide services to support people's livelihoods and help them adapt to climate change.

Better protection of biodiversity should be seen as a prudent and cost-effective investment in risk-avoidance for the global community. The consequences of abrupt ecosystem changes on a large scale affect human security to such an extent, that it is rational to minimize the risk of triggering them—even if we are not clear about the precise probability that they will occur. Ecosystem degradation, and the consequent loss of ecosystem services, has been identified as one of the main sources of disaster risk. Investment in resilient and diverse ecosystems, able to withstand the multiple pressures they are subjected to, may be the best-value insurance policy yet devised.

Scientific uncertainty surrounding the precise connections between biodiversity and human well-being, and the functioning of ecosystems, should not be used as an excuse for inaction. No one can predict with accuracy how close we are to ecosystem tipping points, and how much additional pressure might bring them about. What is known from past examples, however, is that once an ecosystem shifts to another state, it can be difficult or impossible to return it to the former conditions on which economies and patterns of settlement have been built for generations.

Effective action to address biodiversity loss depends on addressing the underlying causes or indirect drivers of that decline.

This will mean:

- Much greater efficiency in the use of land, energy, fresh water and materials to meet growing demand.
- Use of market incentives, and avoidance of perverse subsidies to minimize unsustainable resource use and wasteful consumption.

- Strategic planning in the use of land, inland waters and marine resources to reconcile development with conservation of biodiversity and the maintenance of multiple ecosystem services. While some actions may entail moderate costs or tradeoffs, the gains for biodiversity can be large in comparison.
- Ensuring that the benefits arising from use of and access to genetic resources and associated traditional knowledge, for example through the development of drugs and cosmetics, are equitably shared with the countries and cultures from which they are obtained.
- Communication, education and awareness raising to ensure that as far as possible, everyone understands the value of biodiversity and what steps they can take to protect it, including through changes in personal consumption and behaviour.

The real benefits of biodiversity, and the costs of its loss, need to be reflected within economic systems and markets. Perverse subsidies and the lack of economic value attached to the huge benefits provided by ecosystems have contributed to the loss of biodiversity. Through regulation and other measures, markets can and must be harnessed to create incentives to safeguard and strengthen, rather than to deplete, our natural infrastructure. The re-structuring of economies and financial systems following the global recession provides an opportunity for such changes to be made. Early action will be both more effective and less costly than inaction or delayed action.

Urgent action is needed to reduce the direct drivers of biodiversity loss. The application of best practices in agriculture, sustainable forest management and sustainable fisheries should become standard practice, and approaches aimed at optimizing multiple ecosystem services instead of maximizing a single one should be promoted. In many cases, multiple drivers are combining to cause biodiversity loss and degradation of ecosystems. Sometimes, it may be more effective to concentrate urgent action on reducing those drivers most responsive to policy changes. This will reduce the pressures on biodiversity and protect its value for human societies in the short to medium-term, while the more intractable drivers are addressed over a longer time-scale. For example the resilience of coral reefs—and their ability to withstand and adapt to coral bleaching and ocean acidification—can be enhanced by reducing overfishing, land-based pollution and physical damage.

Direct action to conserve biodiversity must be continued, targeting vulnerable as well as culturally-valued species and ecosystems, combined with steps to safeguard key ecosystem services, particularly those of importance to the poor. Activities could focus on the conservation of species threatened with extinction, those harvested for commercial purposes, or species of cultural significance. They should also ensure the protection of functional ecological groups—that is, groups of species that collectively perform particular, essential roles within

ecosystems, such as pollination, control of herbivore numbers by top predators, cycling of nutrients and soil formation.

Increasingly, restoration of terrestrial, inland water and marine ecosystems will be needed to re-establish ecosystem functioning and the provision of valuable services. Economic analysis shows that ecosystem restoration can give good economic rates of return. However the biodiversity and associated services of restored ecosystems usually remain below the levels of natural ecosystems. This reinforces the argument that, where possible, avoiding degradation through conservation is preferable (and even more cost-effective) than restoration after the event.

Better decisions for biodiversity must be made at all levels and in all sectors, in particular the major economic sectors, and government has a key enabling role to play. National programmes or legislation can be crucial in creating a favourable environment to support effective "bottom-up" initiatives led by communities, local authorities, or businesses. This also includes empowering indigenous peoples and local communities to take responsibility for biodiversity management and decision-making; and developing systems to ensure that the benefits arising from access to genetic resources are equitably shared.

We can no longer see the continued loss of and changes to biodiversity as an issue separate from the core concerns of society: to tackle poverty, to improve the health, prosperity and security of our populations, and to deal with climate change. Each of those objectives is undermined by current trends in the state of our ecosystems, and each will be greatly strengthened if we correctly value the role of biodiversity in supporting the shared priorities of the international community. Achieving this will involve placing biodiversity in the mainstream of decision-making in government, the private sector, and other institutions from the local to international scales.

The action taken over the next decade or two, and the direction charted under the Convention on Biological Diversity, will determine whether the relatively stable environmental conditions on which human civilization has depended for the past 10,000 years will continue beyond this century. If we fail to use this opportunity, many ecosystems on the planet will move into new, unprecedented states in which the capacity to provide for the needs of present and future generations is highly uncertain.

Critical Thinking

1. What services do ecosystems provide? Which of these services benefit human beings?
2. In what sense does the loss of biodiversity threaten the lives of the world's poorest peoples?
3. What are the chief factors that cause loss of biodiversity?
4. In what ways does this reading echo the concerns of Selection 11 (on the Millennium Ecosystem Assessment)?

Impacts of Biodiversity Loss on Ocean Ecosystem Services

Boris Worm et al.

Ecologists think the loss of biodiversity is a problem because it threatens the stability of ecosystems and the potential ability of life to adapt to changes in climate and other conditions. Yet there is a more human way to view the issue as well. A current concern for farmers in the United States and elsewhere is the decline in numbers of wild bees and other pollinators. Together with the increasing difficulty of keeping domestic bees free of parasites and disease, this means that the ability of farmers to produce crops (and of many wild plants to produce seed) is under threat. See, e.g., J. C. Biesmeijer, et al., "Parallel Declines in Pollinators and Insect-Pollinated Plants in Britain and the Netherlands," *Science* (July 21, 2006).

Pollination is, of course, only one of a great many services provided by ecosystems to human beings. Another is wild-caught fish, and Dalhousie University's Boris Worm, et al., find that human fishing is so diminishing populations and threatening species of ocean fish that we could see "the global collapse of all taxa currently fished by the mid-21st century." News reports have phrased this conclusion as "the end of commercial fishing," which stresses the economic impact on the lives of individual fishers, fish processing plants, and ports. There will also be a less economic impact on the well-being of peoples that depend on ocean fish for most of the protein in their diet.

When the Worm report appeared, critics were vocal. In response, Worm and one major critic, Ray Hilborn of the University of Washington, Seattle, WA, teamed up with other researchers and graduate students to study larger data sets in search of a better vision of the future. In "Rebuilding Global Fisheries," *Science* (July 31, 2009), Boris Worm, Ray Hilborn, et al., examine efforts to restore damaged marine ecosystems and fisheries and find that though there is some success, "63% of assessed fish stocks worldwide still require rebuilding, and even lower exploitation rates are needed to reverse the collapse of vulnerable species." Global fisheries remain in danger.

Key Concept: the value of biodiversity to human well-being

Human-dominated marine ecosystems are experiencing accelerating loss of populations and species, with largely unknown consequences. We analyzed local experiments, long-term regional time series, and global fisheries data to test how biodiversity loss affects marine ecosystem services across temporal and spatial scales. Overall, rates of resource collapse increased and recovery potential, stability, and water quality decreased exponentially with declining diversity. Restoration of biodiversity, in contrast, increased productivity fourfold and decreased variability by 21%, on average. We conclude that marine biodiversity loss is increasingly impairing the ocean's capacity to provide food, maintain water quality, and recover from perturbations. Yet available data suggest that at this point, these trends are still reversible.

What is the role of biodiversity in maintaining the ecosystem services on which a growing human population depends? Recent surveys of the terrestrial literature suggest that local species richness may enhance ecosystem productivity and stability. However, the importance of biodiversity changes at the landscape level is less clear, and the lessons from local experiments and theory do not seem to easily extend to long-term, large-scale management decisions. These issues are particularly enigmatic for the world's oceans, which are geographically large and taxonomically complex, making the scaling up from local to global scales potentially more difficult. Marine ecosystems provide a wide variety of goods and services, including vital food resources for millions of people. A large and increasing proportion of our population ives close to the coast; thus the loss of services such as flood control and waste detoxification can have disastrous consequences. Changes in marine biodiversity are directly caused by exploitation, pollution, and habitat destruction, or indirectly through climate change and related perturbations of ocean biogeochemistry. Although marine extinctions are only slowly uncovered at the global scale, regional ecosystems such as estuaries, coral reefs, and coastal and oceanic fish communities are rapidly losing populations, species, or entire functional groups. Although it is clear that particular species provide critical services to society, the role of biodiversity per se remains untested at the ecosystem level. We analyzed the effects of changes in marine biodiversity on fundamental ecosystem services by combining available data from sources ranging from small-scale experiments to global fisheries.

Experiments. We first used meta-analysis of published data to examine the effects of variation in marine diversity (genetic or species richness) on primary and secondary productivity, resource use, nutrient cycling, and ecosystem stability in 32 controlled experiments. Such effects have been contentiously debated, particularly in the marine realm, where high diversity and connectivity may blur any deterministic effect of local biodiversity on ecosystem functioning. Yet when the available experimental data are combined, they reveal a strikingly general picture. Increased diversity of both primary producers and consumers enhanced all examined ecosystem processes. Observed effect sizes corresponded to a 78 to 80% enhancement of primary and secondary production in diverse mixtures relative to monocultures and a 20 to 36% enhancement of resource use efficiency.

Experiments that manipulated species diversity or genetic diversity both found that diversity enhanced ecosystem stability, here defined as the ability to withstand recurrent perturbations. This effect was linked to either increased resistance to disturbance or enhanced recovery afterward. A number of experiments on diet mixing further demonstrated the importance of diverse food sources for secondary production and the channeling of that energy to higher levels in the food web. Different diet items were required to optimize different life-history processes (growth, survival, and fecundity), leading to maximum total production in the mixed diet. In summary, experimental results indicate robust positive linkages between biodiversity, productivity, and stability across trophic levels in marine ecosystems. Identified mechanisms from the original studies include complementary resource use, positive interactions, and increased selection of highly performing species at high diversity.

Coastal ecosystems. To test whether experimental results scale up in both space and time, we compiled long-term trends in regional biodiversity and services from a detailed database of 12 coastal and estuarine ecosystems and other sources. We examined trends in 30 to 80 (average, 48) economically and ecologically important species per ecosystem. Records over the past millennium revealed a rapid decline of native species diversity since the onset of industrialization. As predicted by experiments, systems with higher regional species richness appeared more stable, showing lower rates of collapse and extinction of commercially important fish and invertebrate taxa over time. Overall, historical trends led to the present depletion (here defined as > 50% decline over baseline abundance), collapse (>90% decline), or extinction (100% decline) of 91, 38, or 7% of species, on average. Only 14% recovered from collapse; these species were mostly protected birds and mammals.

These regional biodiversity losses impaired at least three critical ecosystem services: number of viable (non-collapsed) fisheries (−33%); provision of nursery habitats such as oyster reefs, seagrass beds, and wetlands (−69%); and filtering and detoxification services provided by suspension feeders, submerged vegetation, and wetlands (−63%). Loss of filtering services probably contributed to declining water quality and the increasing occurrence of harmful algal blooms, fish kills, shellfish and beach closures, and oxygen depletion. Increasing coastal flooding events are linked to sea level rise but were probably accelerated by historical losses of floodplains and erosion control provided by coastal wetlands, reefs, and submerged vegetation. An increased number of species invasions over time also coincided with the loss of native biodiversity; again, this is consistent with experimental results. Invasions did not compensate for the loss of native biodiversity and services, because they comprised other species groups, mostly microbial, plankton, and small invertebrate taxa. Although causal relationships are difficult to infer, these data suggest that substantial loss of biodiversity is closely associated with regional loss of ecosystem services and increasing risks for coastal inhabitants. Experimentally derived predictions that more species-rich systems should be more stable in delivering services are also supported at the regional scale.

Large marine ecosystems. At the largest scales, we analyzed relationships between biodiversity and ecosystem services using the global catch database from the United Nations Food and Agriculture Organization (FAO) and other sources. We extracted all data on fish and invertebrate catches from 1950 to 2003 within all 64 large marine ecosystems (LMEs) worldwide. LMEs are large (>150,000 km^2) ocean regions reaching from estuaries and coastal areas to the seaward boundaries of continental shelves and the outer margins of the major current systems. They are characterized by distinct bathymetry, hydrography, productivity, and food webs. Collectively, these areas produced 83% of global fisheries yields over the past 50 years. Fish diversity data for each LME were derived independently from a comprehensive fish taxonomic database.

Globally, the rate of fisheries collapses, defined here as catches dropping below 10% of the recorded maximum, has been accelerating over time, with 29% of currently fished species considered collapsed in 2003. This accelerating trend is best described by a power relation ($y = 0.0168x^{1.8992}$, $r = 0.96$ $P < 0.0001$), which predicts the percentage of currently collapsed taxa as a function of years elapsed since 1950. Cumulative collapses (including recovered species) amounted to 65% of recorded taxa. The data further revealed that despite large increases in global fishing effort, cumulative yields across all species and LMEs had declined by 13% (or 10.6 million metric tons) since passing a maximum in 1994.

Consistent with the results from estuaries and coastal seas, we observed that these collapses of LME fisheries occurred at a higher rate in species-poor ecosystems, as compared with species-rich ones. Fish diversity varied widely across LMEs, ranging from ~20 to 4000 species, and influenced fishery-related services in several ways. First, the proportion of collapsed fisheries decayed exponentially with increasing species richness. Furthermore, the average catches of non-collapsed fisheries were higher in species-rich systems. Diversity also seemed to increase robustness to overexploitation. Rates of recovery, here defined as any post-collapse increase above the 10% threshold, were positively correlated with fish diversity. This positive relationship between diversity and recovery became stronger with time after a collapse (5 years, $r = 0.10$; 10 years, $r = 0.39$; 15 years, $r = 0.48$). Higher taxonomic units (genus and family) produced very similar relationships as species richness; typically, relationships became stronger with increased taxonomic aggregation. This may suggest that taxonomically related species play complementary functional roles in supporting fisheries productivity and recovery.

A mechanism that may explain enhanced recovery at high diversity is that fishers can switch more readily among target species, potentially providing overfished taxa with a chance to recover. Indeed, the number of fished taxa was a log-linear function of species richness. Fished taxa richness was negatively related to the variation in catch from year to year and positively correlated with the total production of catch per year. This increased stability and productivity are likely due to the portfolio effect, whereby a more diverse array of species provides a larger number of ecological functions and economic opportunities, leading to a more stable trajectory and better performance over time. This portfolio effect has independently been confirmed by economic studies of multispecies harvesting relationships in marine ecosystems. Linear (or log-linear) relationships indicate steady increases in services up to the highest levels of biodiversity. This means that proportional species losses are predicted to have similar effects at low and high levels of native biodiversity.

Marine reserves and fishery closures. A pressing question for management is whether the loss of services can be reversed, once it has occurred. To address this question, we analyzed available data from 44 fully protected marine reserves and four large-scale fisheries closures. Reserves and closures have been used to reverse the decline of marine biodiversity on local and regional scales. As such, they can be viewed as replicated large-scale experiments. We used meta-analytic techniques to test for consistent trends in biodiversity and services across all studies.

We found that reserves and fisheries closures showed increased species diversity of target and nontarget species, averaging a 23% increase in species richness. These increases in biodiversity were associated with large increases in fisheries productivity, as seen in the fourfold average increase in catch per unit of effort in fished areas around the reserves. The difference in total catches was less pronounced, probably because of restrictions on fishing effort around many reserves. Resistance and recovery after natural disturbances from storms and thermal stress tended to increase in reserves, though not significantly in most cases. Community variability, as measured by the coefficient of variation in aggregate fish biomass, was reduced by 21% on average. Finally, tourism revenue measured as the relative increase in dive trips within 138 Caribbean protected areas strongly increased after they were established. For several variables, statistical significance depended on how studies were weighted. This is probably the result of large variation in sample sizes among studies. Despite the inherent variability, these results suggest that at this point it is still possible to recover lost biodiversity, at least on local to regional scales; and that such recovery is generally accompanied by increased productivity and decreased variability, which translates into extractive (fish catches around reserves) and nonextractive (tourism within reserves) revenue.

Conclusions. Positive relationships between diversity and ecosystem functions and services were found using experimental and correlative approaches along trajectories of diversity loss and recovery. Our data highlight the societal consequences of an ongoing erosion of diversity that appears to be accelerating on a global scale. This trend is of serious concern because it projects the global collapse of all taxa currently fished by the mid-21st century (based on the extrapolation of regression to 100% in the year 2048). Our findings further suggest that the elimination of locally adapted populations and species not only impairs the ability of marine ecosystems to feed a growing human population but also sabotages their stability and recovery potential in a rapidly changing marine environment.

We recognize limitations in each of our data sources, particularly the inherent problem of inferring causality from correlation in the larger-scale studies. The strength of these results rests on the consistent agreement of theory, experiments, and observations across widely different scales and ecosystems. Our analysis may provide a wider context for the interpretation of local biodiversity experiments that produced diverging and controversial outcomes. It suggests that very general patterns emerge on progressively larger scales. High-diversity systems consistently provided more services with less variability, which has economic and policy implications. First, there is no dichotomy between biodiversity conservation and long-term economic development; they must be viewed as interdependent societal goals. Second, there was no evidence for redundancy at high levels of diversity; the improvement of services was continuous on a log-linear scale. Third, the buffering impact of species diversity on the resistance and recovery of ecosystem services generates insurance value that must be incorporated into future economic valuations and management decisions. By restoring marine biodiversity through sustainable fisheries management, pollution control, maintenance of essential habitats, and the creation of marine reserves, we can invest in the productivity and reliability of the goods and services that the ocean provides to humanity. Our analyses suggest that business as usual would foreshadow serious threats to global food security, coastal water quality, and ecosystem stability, affecting current and future generations.

Critical Thinking

1. What can we do to prevent a catastrophic decline in ocean fisheries by the mid-21st century?
2. Why do collapsed fisheries recover better in high-diversity ecosystems?
3. Why are commercial fisheries in decline?

Rethinking Rain Forests: Biodiversity and Social Justice

John Vandermeer and Ivette Perfecto

There is no disagreement among ecologists about the importance of rain forests. These tropical ecosystems are home to more than three-quarters of all the species of plants and animals that inhabit the Earth. There is also broad agreement that human beings are rapidly destroying this vital natural resource and that there are many interacting causes for the destruction. There is much less agreement on what can be done to stop it.

John Vandermeer is a professor in the Department of Ecology and Evolutionary Biology at the University of Michigan. Ivette Perfecto is a professor of natural resources in the School of Natural Resources and Environment at the University of Michigan. They have both devoted many years to the study of the ecology of tropical rain forests as well as to the social, political, and economic factors that threaten their existence. They have written a book detailing their findings entitled *Breakfast of Biodiversity: The Truth About Rain Forest Destruction*, published by Food First Books. In the following selection from "Rethinking Rain Forests: Biodiversity and Social Justice," *Food First Backgrounder* (Summer 1995), Vandermeer and Perfecto seek to dispel the myths that attribute the assaults on the forests to any one of several contributing factors. Instead, they argue that there is a "web of causality" linking all these factors and that only by designing an action plan that takes into account all the components of this web can the forests be saved.

Key Concept: the interacting factors causing rain forest destruction

*T*he buzz is unmistakable. A huge chain saw cuts effortlessly through the wood of a beautiful rain forest tree, slicing up the trunk it has just felled into smaller bits to be taken away on giant lumber trucks. That image is fixed in our minds. It drives us to the same distraction it has driven so many before us. The rain forests are physically beautiful and contain the vast majority of our relatives on this planet. What sort of person would not be haunted by the sound of chain saws decimating them?

*Y*et another image is equally haunting. The bulldozed wooden shack, formerly the home of a poor family, constantly reminds us that lives as well as logs are being cut in most areas of tropical rain forests. Hungry children wander among the stumps of once majestic rain forest trees. Their mother cooks over an open fire, and their father fights the onslaught of weeds that continually threaten to choke out the crops the family needs for next

year's food. All live in fear that the bulldozers will come again to destroy their present home. What sort of person would not be haunted by the existence of such poverty in a world of plenty?

But for us the power of these two images lies in the way they are connected, a fact we are reminded of every morning we slice up bananas on our breakfast cereal. The banana cannot be grown in the United States, yet it is one of the most popular fruits here. As we all know, it is produced in the world's tropical regions, usually in the same areas where rain forests have flourished in the past. The link between the decimated forest and the hungry children is the banana. That is why it is so easy, as we slice up a banana in Michigan, for our thoughts to wander to the image of the chain saw slicing up the rain forest trees and the children who view the banana as a staple food rather than a luxury.

The majority of life on earth lives in the rain forest. Close to 80% of the terrestrial species of animals and plants are to be found there. And this cradle of life is disappearing at an enormous rate. This is what the popular press has labeled as the "biodiversity" crisis.

Some view the problem from only a utilitarian point of view. It is obvious that we depend on biodiversity for the most elementary aspects of existence—plants convert the sun's energy to a usable form, animals convert unusable plants to a product we can use, bacteria in our stomachs help digest our food. There are a host of other critical functions of life's diversity and furthermore, future utilitarian designs on biodiversity are most likely to follow the patterns of the past—medicines and genes for new crops being the obvious examples.

Yet even if these utilitarian concerns were absent, the spiritual concern that the world's biodiversity is being destroyed should be enough to drive us to action. Less than 50% of the original tropical rain forests of the world are left, and at the present rate of destruction almost all will be gone by 2025. Our families, our memories— indeed a piece of our humanness—will have been destroyed forever. For this reason many have sounded the alarm and called for action.

While we echo this same alarm, we are concerned that the calls for action may not be correctly placed. Indeed, many of these calls are based on one myth or another about what is causing rain forest destruction. We feel that these myths act to mask the true issue. [Here] we present arguments against the five main myths of rain forest destruction and argue that a more complex understanding is necessary to grasp what is causing the destruction of the world's rain forests. So we begin with an analysis of the five myths and conclude with a description of "the causal web," the *true* cause of rain forest destruction.

MYTH ONE: Loggers and logging companies are decimating the rain forest.

Certainly the most immediate and visually spectacular cause of tropical rain forest destruction is logging. Cutting trees is nothing new. The use of rain forest wood has been traditional for most human societies in contact with these ecosystems. But the European invasion of tropical lands accelerated wood cutting enormously, as tropical woods began contributing to the development of the modern industrial society.

The direct consequences of logging, apart from the obvious and dramatic visual effects, are largely unknown. Some facts are deducible from general ecological principles, and a handful of studies have actually measured a few of the consequences, but a detailed knowledge of the direct consequences of logging is lacking.

What can be deduced from ecological principles is not that tropical forests are irreparably damaged by logging, but quite the contrary: tropical forests are potentially quite resilient to disturbance. While this is a debatable deduction, most of the debate centers on how fast a forest will recover after a major disturbance, such as logging, not on whether it will. The process of ecological succession inevitably begins after logging, and the proper question to ask, then, is: how long will it take for the forest to recover?

In analyzing the effects of logging, we cannot assume a uniform process. There are a variety of logging techniques, some likely to lead to rapid forest recovery, others necessitating a longer period for recovery. For example, local residents frequently chop down trees for their own use as fence posts, charcoal, or dugout canoes. Forest recovery after such an intrusion can be thought of as virtually instantaneous, since the removal of a single tree is similar to a tree dying of natural causes, a perfectly natural process that happens regularly in all forests. At the other extreme is clear cutting, the extraction of all trees in an area. Though the physical nature of a clear cut forest is spectacularly different from the mature forest, from other perspectives the damage is not quite as dramatic as it appears. The process of secondary succession that begins immediately after such logging leads rapidly to the establishment of secondary forest. A great deal of biological diversity is contained in a secondary forest. Indeed, a late secondary forest is likely to appear indistinguishable from an old-growth forest to all but the most sophisticated observer, even though it may have been initiated from a clear cut. Large expanses of secondary forest may even contain more biological diversity than similar expanses of old-growth forest. No studies thus far have followed such an area to its return to a "mature" forest again, but a reasonable estimate is that it would take something on the order of 40 to 80 years before the area begins to regain the structure of an old growth forest.

Probably the most common type of commercial logging is not the clear cutting described above but, rather, selective logging. In an area of tropical forest that may contain 400 or more species of trees, only twenty or thirty will be of commercial importance. Thus, a logging company usually seeks out areas with particularly large concentrations of the valuable species and ignores the rest. Often the wood is so valuable that it makes economic sense to build a road to extract just a few trees. Yet these roads offer new access to the forest for hunters, miners, and peasant agriculturists. In most situations this aspect of selective logging contributes most egregiously to deforestation, but it is obviously an *indirect* consequence of the logging operation itself.

A selectively logged forest is damaged, but not destroyed. Even a single year after the selective logging the forest begins taking on the appearance of a "real" forest. If no further cutting occurs, the selectively logged forest may regain the structural features of old growth after ten or twenty years. Although the scars of selective logging will remain for decades to a trained eye, the general structure of the forest may rapidly return. But this is not to say selective logging is, in the end, benign. The roads and partial clearings are obvious entrance points for peasant agriculture, as described below.

MYTH TWO: Peasant farmers are increasing in numbers and cut down rain forests to make farms to feed their families.

This myth is especially popular among neo-Malthusians. The explosive growth in the population of poor people in most tropical countries of the world is seen as a consequence of the basic forces that cause populations to grow generally, and a simple extrapolation suggests that even if this is not the main problem now, it certainly will be if population growth is not somehow curtailed.

Debunking the neo-Malthusian myth is not our purpose here; that has been done well elsewhere. Rather, laying the blame for the destruction of the forest on the peasant farmer is really blaming the victim. Peasant farmers in most rain forest areas are forced to farm under circumstances that are unfavorable, to say the least, from both an ecological sociopolitical point of view.

At the outset, we must acknowledge the temptation to assume that, in rain forest areas, the potential for agriculture is great. Since there is neither winter nor lack of water, two of the main limiting factors for agriculture in other areas of the world, it is easy to conclude that production might very well be cornucopian. The tremendously lush vegetation of a tropical rain forest only heightens this impression, and indeed this perception may ultimately be valid. The ability to produce for twelve months of the year without worrying about irrigation is definitely a positive aspect to farming in such regions. But, so far at least, the woes are almost insurmountable, as most farmers forced to cultivate in rain forest areas can attest.

The first problem is the soils. Rain forest soils are usually acidic, made up of clay that cannot store nutrients well, and very low in organic matter. Even if nutrients are added to the soil they will be utilized relatively inefficiently because of the acidity, and then they will be washed out of the system because of its low storage capacity.

This problem is actually exacerbated by the forest itself. Because tropical rain forest plants have grown in these poor soil conditions for millions of years, they have evolved mechanisms for storing the system's nutrients in their vegetative matter (leaves, stems, roots, etc.) If they did not, much of the nutrient material would simply wash out of the system and no longer be available to them. This means that a vast majority of the nutrients in the ecosystem are stored in plant material rather than in the soil.

Consequently when a forest is cut down and burned, the nutrients in the vegetation are immediately made available to any crops that have been planted. The crops grow vigorously at first, but any nutrients unused during the first growing season will tend to leach out of the system. The "poverty" of the soil only becomes evident during the second growing season. This pattern is especially invidious when migrant farmers from areas with relatively stable soils arrive in a rain forest area. The first year they may produce a bumper crop, which creates a false sense of security. Then, if the second year is not a complete failure, almost certainly the third or fourth is, and the farmer is forced to move on to cut down another piece of forest.

A second problem is insects, diseases and weeds. The magnitude of the pest problem is often not fully anticipated by farmers or planners, and it is only after problems arise that the surprised agronomists become concerned. This is unfortunate, since one of the few things we can predict with confidence is that when rain forest is converted to agriculture, many pests arrive. The herbivores that used to eat the plants of the rain forest are not eliminated when the forest is cut. They are representatives of the massive biodiversity of tropical rain forests, and the potential number of them is enormous. Herbivores can devastate farmers' fields, and are able to destroy an entire crop in days.

A third problem is that because of the uniformly moist and warm environment, organisms that cause crop diseases find rain forest habitats quite hospitable. Consequently, the potential for losing crops to disease is far greater than in more temperate climates. Finally, just as the hot, wet environment is agreeable for crops, it is also agreeable for competitive plants. Since no two plants can occupy the same space, frequently the crop falls victim to the more aggressive vines and grasses that colonize open areas in tropical rain forest zones. Weeds are thus an especially difficult problem.

These, then, are some of the ecological problems faced by the peasant farmer seeking to establish a farm in a rain forest area. Sociopolitical forces, however, are far more devastating. And most of those sociopolitical forces are associated with a different form of agriculture—modern export agriculture.

When a modern export agricultural operation is set up, it tends to do two things regarding labor. First, it purchases, or sometimes steals, land from local peasant farmers, thus forcing them to move onto more marginal lands, with the kinds of problems we described above. Second, it frequently requires more labor than is locally available, thus acting as a magnet to attract unemployed people from other regions. Indeed, in most rain forest areas this magnet effect is a far more important factor leading to increased local populations than population growth.

But the modern agricultural operation, as detailed in the following section, is subject to dramatic fluctuations in production, since it is usually intimately connected with world agricultural commodity markets. Thus, there is a highly variable need for this labor, which means that today's workers always face the prospect of becoming tomorrow's peasant farmers.

In the contemporary world most peasant farmers find themselves in this precarious position. While it is true that many indigenous groups have lived and farmed in rain forest areas for hundreds of years and certainly deserve the world's attention and support in their attempts at preserving traditional ways of life, the vast majority of poor peasant farmers today are not indigenous.

Rather, they are people who have been marginalized by a politico-economic system that needs them to serve as laborers when times are good, and to take care of themselves when times are not. As long as times are good, the banana workers of Central America have jobs. But when economies sour, many of those banana workers suddenly become peasant farmers.

So in the end, the myth of the peasant farmers causing rain forest destruction is perhaps true in the narrow sense that a knitting needle causes yarn to form a sweater. But little understanding of what really drives the process is gained from the simple observation that a peasant's ax can chop a rain forest tree.

MYTH THREE: The transformation of rain forests into large-scale export agriculture is the main factor leading to deforestation.

Given the above description of how peasant agriculture is driven by industrialized agricultural activities, it is no wonder that many have concluded that the modern export agricultural system is the ultimate culprit. Furthermore, the images of large cattle ranchers purposefully burning Amazon rain forests to make cattle pastures fuels this interpretation. Again, there is some merit to this position. However, we feel that it, too, is an inappropriate window through which to view the problem of rain forest destruction.

The direct action of large modern agricultural enterprises is not really as involved in direct rain forest destruction as is popularly believed. Burning Amazon rain forests to replace them with cattle ranches is certainly an example of the direct destruction of rain forests by "big" agriculture. But the vast majority of modern agricultural transformations in tropical areas are confined to areas that had already been converted to agriculture. Developers of expanding banana plantations of Central America claim, for example, to be cutting no primary forest at all. While we doubt their full sincerity, it does seem that about 90% of the current expansion is into areas that had long ago been deforested. Attributing direct deforestation to them is, as they argue, probably quite unfair. On the other hand, their activities are not totally unrelated to the problem, as can be easily seen from a closer examination of their underlying structure.

The basic structure of modern agriculture is frequently misunderstood because of an overly romantic notion of agriculture—the small, independent, family farm, rich with tradition and a work ethic that even a Puritan could be impressed with. Such romanticism is fueled by a confusion between farming and agriculture.

Farming is a resource transformation process in which land, seed, and labor are converted into, for example, peanuts. It is Farmer Brown cultivating the land, sowing the seed, and harvesting the peanuts. Agriculture is the decision to invest money in this year's peanut production; the use of a tractor and cultivator to prepare the land; an automatic seeder for planting; application of herbicides, insecticides, fungicides, nematodes and bactericides to kill unwanted pieces of the ecology; automatic harvest of the commodity; sale of the commodity to a processing company where it is ground up and emulsifiers, taste enhancers, stabilizers and preservatives are added; packing in convenient "pleasing-to-the-consumer" jars; and, finally, marketing under a sexy brand name. In short, while farming is the production of peanuts from the land, agriculture is the production of peanut butter from petroleum. Over the last two hundred years, and especially in the last fifty, much farming has been transformed into agriculture.

The consequence of this evolution is that modern agriculture is remarkably intrusive on local ecologies. Take, for example, the establishment of a banana plantation. When the banana export business began, local peasant farmers grew most of the bananas and sold them to shipping companies. Gradually, the shipping companies turned into the banana producers, with huge areas devoted to the monocultural production of this single crop. To establish a modern banana plantation it is often necessary to construct a complex system of hydrological control wherein the soil is leveled and crisscrossed with drainage channels, significantly altering the physical nature of the soil. Contemporary banana production even includes burying plastic tubing in the ground to eliminate the natural variability in subsurface water depth. Metal monorails hang from braces placed into cement footings to haul the bunches of bananas. To avert fungal diseases, heavy use of fungicides is required, and because of the large scale of the operation chemical methods of pest control are the preferred option. The banana plants create an almost complete shade cover and thus replace all residual vegetation. Pesticide application is sometimes intense, other times almost absent, depending on conditions, but over the long run one can expect an enormous cumulative input of pesticides, the long-term consequences of which are unknown but likely to be unhealthy for both workers and the environment.

A major social transformation is also required. Banana production tends to promote a local "overpopulation crisis" by encouraging a great deal of migration into the area. As the international market for bananas ebbs and flows, workers are alternatively hired and fired. When fired, there is little alternative economic opportunity in banana zones, and displaced workers must either look for a piece of land to farm, or migrate to the cities to join the swelling ranks of shanty town dwellers.

Thus, the direct effect of most modern agricultural activities is not inexorably linked with the cutting and burning of rain forests, despite some obvious and spectacular examples of where it indeed is. More importantly, the overall operation of the modern agricultural system is integrated into a bigger picture. It is that bigger picture that we must examine to understand the causes of rain forest destruction.

MYTH FOUR: Local governments institute policies that cause rain forests to be destroyed.

Probably the most cited example of local government policy that promotes deforestation is that of the infamous transmigration programs of the Indonesian government, in which hundreds of thousands of Javanese farmers have been displaced to the exceedingly poor soils of Kalimantan. However, most local government programs in forestry and agriculture are frequently dictated by very specific economic and political forces that are effectively beyond the control of local governments. Once those forces are understood, it is difficult to lay the full blame on local governments. They may be corrupt, they may be inefficient, but in fact their hands are frequently tied by forces beyond their control.

Given today's global interconnectedness, in order to understand the Third World we must view it as embedded in the modern industrial system. In that system the people who provide the labor in the production processes are not the same people who provide the tools, machines and factories. The former are the workers in the factories, the latter are the owners of the factories. The owners of the machines and tools directly make the management decisions about all production processes. A good manager tries to minimize all production costs, including the cost of labor.

However, the owners of the factories face a complicated and contradictory task. While factory workers constitute a cost of production to be minimized, they also participate, along with the multitudes of other workers in society, in the consumption of products. In trying to maximize profits, factory owners are concerned that their factories' products sell for a high price. This can only happen if workers, in general, are making a lot of money. In contrast to what is desired at the level of the factory, the opposite goal is sought at the level of society. Factory owners must wear two hats, then: one as owners of the factories, and another as members of a social class. Owners wish the laborers to receive as little as possible, but members of the social class benefit if laborers in general receive as much as possible (to enable them to purchase the products produced in the factory). This has long been recognized as one of the classic contradictions of a modern economy.

The situation in much of the Third World appears superficially similar. For the most part we are dealing with agrarian economies in which there are two obvious social classes, those who produce crops for export like cotton, coffee, tea, rubber, bananas, chocolate, beef; and sugar; and those who produce food for their own consumption on their own small farms and, when necessary, provide the labor for export crop producers.

The typical arrangement in the Developed World is an articulated economy, while that in the Third World is disarticulated, in that the two main sectors of the economy are not articulated or connected with one another. The banana company does not really care whether its workers make enough money to buy bananas; that is not its market. The banana company cares that the workers of the Developed World have purchasing power, because those are the people expected to buy most of the bananas. This disarticulation, or dualism, helps to explain the differences between analogous classes in the First and Third Worlds. Flower producers in Colombia do not concern themselves much over the fact that their workers cannot buy their products. On the other hand, the factory owners in the U.S., whether they be private factories or government owned and/or subsidized industries, care quite a lot that the working class has purchasing power. General Motors "cares" that the general population in the U.S. can afford to buy cars. Naturally they aim to pay their own workers as little as possible, but that goal is balanced by their wish for the workers in general to be good consumers.

Seeing this structure at the national level in an underdeveloped country causes one to realize that one of the main, sometimes only, sources of capital to create a civil society is from agricultural exports. Because of the disarticulated nature of the economy the dream of development based on internally derived consumer demand is pie in the sky, and any realist must acknowledge that the only conceivable source of capital to invest in growth must come from exports. And most frequently agricultural exports are the only possibility.

Herein derives the need for Third World governments to continue expanding this export agriculture. This need is an inevitable consequence of the underlying structure of the general world system. Thus to blame local governments for initiating policies that are ultimately damaging to rain forests may be technically correct in that those policies frequently do just what the critics say they do—destroy rain forests. But taking a larger view we see that local governments are effectively constrained to do exactly that. Indeed, we predict that most of today's critics would wind up promoting the very same programs the local governments are currently promoting, if they were suddenly pushed into the same position the local governments currently find themselves.

MYTH FIVE: Decisions made by international agencies cause rain forest destruction.

As before, there is some truth to this position. As well documented, although not yet "retrospected" by Mr. McNamara, the World Bank has left a trail of rain forest destruction in the wake of its many socially and economically destructive programs in the Third World. From the point of view of the decision making agencies, they, along with other agencies involved in the overall problem, seem to be boxed in by circumstances.

The climate for investment is variable in the Developed World. There are times when it is difficult to find profitable investments at home. At such times it is useful to have alternatives to investment. The Third World is one source for those opportunities. The Developed World, because of its basic structure, tends to go through cycles of bust and boom, sometimes severe, other times merely annoying. During low economic times, where is an investor supposed to invest? The Third World provides a sink for investments during rough times in the First World. This is why the dualism of the Third World is a "functional dualism." It functions to provide an escape valve for investors from the Developed World. The West German entrepreneur who started an ornamental plant farm in Costa Rica, on which former peasant farmers work as night watchmen, invested his money in the Third World because opportunities in his native Germany were scarce at the time. What would he have done had there been no peasants willing to work for practically nothing, and no Costa Rica willing to accept his investments at very low taxation? Clearly Costa Rica is, for him, a place to make his capital work until the situation clears up in Germany. Union Carbide located its plant in Bhopal, India, and not in Grand Rapids, Michigan. U.S. pesticide companies export to Third World countries insecticides that have been banned at home. U.S. pharmaceutical companies pollute the ground water in Puerto Rico because they cannot do so (at least not so easily) in the United States. In all cases, Third World people are forced to accept such arrangements, largely because of their extremely underdeveloped economy.

With this analysis, the origin of the Third World as an outgrowth of European expansion (while a correct and useful historical point of view), can be seen as not the only factor to be considered. Even today, the maintenance of the Third World is a consequence of the way our world system operates. The Developed World remains successful at economic development for two reasons. First, because it has an articulated economy, and second, because it is able to weather the storm of economic crisis by seeking investment opportunities in the Third World. The Third World, in contrast, has been so unsuccessful because its economy is disarticulated, lacking the connections that would make it grow in the same way as the Developed World. Yet at a more macro scale, the dualism of the Third World is quite functional, in that it maintains the opportunity for investors from the Developed World to use the Third World as an escape valve in times of crisis. Indeed, it appears that the Developed World remains developed, at least in part, specifically because the Underdeveloped World is underdeveloped.

Given this structure, what really can be expected of international agencies? Their goal is usually stated in very humanitarian rhetoric. But their more basic goal has to be the preservation of the system that gives them their station in life. That is, above and beyond the stated goals of the World Bank or the IMF or the FAO, there must be a commitment to keep the world organized in its current state. Their activities can thus be viewed as trying to solve problems within the context of the current system. They thus become part of what preserves that system.

We should not expect large international agencies to be promoting such causes as land reform for peasant agriculture or labor standard regulations for export agriculture. Indeed, such proposals would be at odds with the manner in which the current world system is functioning, and would represent a legitimate challenge to the existence of the agency itself. Viewed from this perspective, the international agencies are just as boxed in as the local governments. The world system is functioning as well today as it was at the end of World War II—according to the standards adopted by those who benefit from its current structure.

The Web of Causality

In reading our demythologization of the above five myths, the reader has undoubtedly noted that there really is something valid about each of the myths. Loggers do cut trees down, peasant agriculturists do clear and burn forests, export agriculturists do cut down primary rain forests, local governments do encourage export agriculture, and international agencies do promote programs that destroy rain forests. But our attempt was not to disprove these myths per se, but rather to disprove the idea that any one of them could be the ultimate cause of rain forest destruction. Indeed, in each case we have emphasized not the direct consequences of the agency involved but, rather, the indirect connections that tie each of the agencies into a web of interaction.

We agree with Wallerstein's general assessment that the world is intricately connected, that it no longer makes sense to try understanding isolated pockets, such as nations, and, we add, that isolated thematic pockets are similarly incomprehensible unless embedded in this global framework. For this reason, attempts at understanding tropical rain forest destruction in isolation have largely failed. As should be clear by now, the fate of the rain forest is intimately tied to various agricultural activities, which are embedded in larger structures, some retaining a connection to agriculture, some not. Our position is that there is multiple causation of rain forest destruction, with logging, peasant agriculture, export agriculture, domestic sociopolitical forces, international socio-economic relations, and other factors intricately connected with one another in a "web of causality." This web is key to understanding why we face the problem of rain forest destruction in the first place. . . .

Furthermore, seeing the entire web of causality enables those engaged in highly focused political action to see their actions in relation to other actions, perhaps evoking an analysis of consequences that may be dramatic, even though quite indirect. For example, organizing consumer boycotts can be seen as clearly attacking the connection between consumers and modern export agriculture. But following through the logic of the web also suggests that a successful consumer boycott may likewise reduce the need of modern agriculture for workers, thus creating more peasant farmers, who will likely clear more forest. If a careful analysis of this situation reveals that the loss of jobs will be severe, the political action agenda might then be expanded to form alliances with a local farm worker union calling for job security or a political movement seeking secure land ownership for the increased number of peasant farmers that will surely be created if the boycott is a success.

The Political Action Plan

This analysis is meaningless without a program of political action. Political action must focus on the web of causality and eschew single issue foci. Calls for boycotts of tropical timbers or bananas need to be coupled with actions to change investment patterns and international banking pressures.

Above all, political action plans must be formulated so they do not make the situation worse—certainly a conceivable, perhaps even likely, consequence of any action, given the complex nature of the web of causality. It appears obvious that political action needs to be focused not only on rain forests and the subjects traditionally associated with them, but also on social justice. The same peasant farmers who formed the backbone of the Vietnamese liberation forces or the Salvadoran guerrillas are the ones who are forced into the marginal existence that compels them to continually move into the forests. So the same issues that compelled progressive organizers in the past to form solidarity committees and anti-war protests are the issues that must be addressed if the destruction of rain forests is to be stopped.

Just as the most effective political action in the past was organized in conjunction with and to some extent under the leadership of the people for whom social justice was being sought, so today political action should be coordinated with those same people. As that coordination proceeds, the alliances that grow will inevitably lead to the reformulation of goals, which the rain forest conservation activist must acknowledge and respect. Local people, quite obviously, must recognize something about the rain forest that is in their best interest to preserve, and it is the job of the progressive organizer to construct the political action so that such value is evident. In short, the alliance between the people who live in and around the rain forest and those from the outside who seek to stop the tide of rain forest destruction must be a two way alliance. If the people who live around the Lacandon forest in Mexico, for example, have as their major goal the reformulation of the Mexican political system, the rain forest conservationist must join the political movement to change that system—something that many would see as distant from the original goal of preserving rain forests. Political action to preserve rain forests, under the framework of the web of causality, will inevitably involve the serious preservationist in social justice issues, many of which initially may have seemed only marginally associated with the problem of rain forest destruction. Recalling the old slogan, "If you want peace, work for justice," we hope someday to hear, for example, "Save Mexican rain forests, support the Zapatistas," or "If you want to save Cuba's rain forests, break the illegal U.S. blockade."

Critical Thinking

1. Who is to blame for the ongoing destruction of tropical forests?
2. In what sense is it true that environmental problems such as the destruction of tropical forests have no single cause?

Pollution

Selection 19
JOHN EVELYN, from *Fumifugium: Or the Inconvenience of the Aer and Smoake of London Dissipated* (1661)

Selection 20
BEVERLY PAIGEN, from "Controversy at Love Canal," *Hastings Center Report* (June 1982)

Selection 21
MARGARET A. PALMER AND J. DAVID ALLEN, from "Restoring Rivers," *Issues in Science and Technology* (Winter 2006)

Selection 22
ORRIN H. PILKEY AND ROBERT S. YOUNG, from *The Rising Sea*

Learning Outcomes
After reading this unit, you should be able to:

1. Describe how air pollution affects the quality of urban life.
2. Describe at least one simple method of reducing urban air pollution.
3. Explain why hazardous wastes must be properly disposed of.
4. Explain why despite protective legislation rivers and streams are still being degraded.
5. Describe the threat posed by global warming to coastal settlements.
6. Describe necessary responses to the threat posed by rising sea level.

Fumifugium: Or the Inconvenience of the Aer and Smoake of London Dissipated

John Evelyn

Many environmental courses that include units on acid rain, photochemical smog, ozone depletion, and global warming may leave the impression that serious air pollution problems only became an issue in the second half of the twentieth century. This is certainly not the case. The earliest reports of noxious smoke in urban areas are associated with the introduction of high-sulfur coal from Newcastle as a fuel by cloth dyers and beer brewers in England's cities in the middle of the thirteenth century. By 1306 King Edward I had introduced restrictions on the burning of coal in London from nearby seashore outcroppings. During the fifteenth and sixteenth centuries increased populations in London and the shortage of firewood resulted in the rapid expansion in the use of coal as a domestic and industrial fuel.

John Evelyn (1620–1706) was a famous English writer whose historical diary is an important chronicle of life in seventeenth-century London. In 1661 Evelyn wrote the tract *Fumifugium: Or the Inconvenience of the Aer and Smoake of London Dissipated*, from which the following selection has been taken. This tract was addressed to King Charles II. In this work, which has become a much-cited document by historians of environmentalism, Evelyn describes the wretched effects of coal smoke on all aspects of life in the London of his day. His principal proposal for reducing the smoke pollution was to move those trades that burned large quantities of coal to a location five or six miles from London. Although King Charles ignored Evelyn's plea, subsequent government action did drastically reduce the burning of soft coal in London proper. However, by 1933 the industrial revolution had made smoke pollution such a widespread concern that the English National Smoke Abatement Society decided to reprint Evelyn's pamphlet as a means of publicizing their cause.

Key Concept: the long history of urban air pollution

TO THE KINGS MOST SACRED MAJESTY:

SIR,

It was one day, as I was Walking in Your MAJESTIES Palace at WHITEHALL, (where I have sometimes the honour to refresh myself with the Sight of Your Illustrious Presence, which is the Joy of Your Peoples hearts) that a presumptuous Smoake issuing from one or two Tunnels neer Northumberland-house, and not far from Scotland-yard, did so invade the Court; that all the Rooms, Galleries, and Places about it were filled and infested with it; and that to such a degree, as Men could hardly discern one another for the Clowd, and none could support, without manifest Inconveniency. It was not this which did first suggest to me what I had long since conceived against this pernicious Accident, upon frequent observation;

But it was this alone, and the trouble that it must needs procure to Your Sacred Majesty, as well as hazzard to Your Health, which kindled this Indignation of mine, against it, and was the occasion of what it has produced in these Papers.

Your Majesty, who is a Lover of noble Buildings, Gardens, Pictures, and all Royal Magnificences, must needs desire to be freed from this prodigious annoyance; and, which is so great an Enemy to their Lustre and Beauty, that where it once enters there can nothing remain long in its native Splendor and Perfection: Nor must I here forget that Illustrious and divine Princesse, Your Majesties only Sister, the now Dutchesse of Orleans, who at her Highnesse late being in this City, did in my hearing, complain of the Effects of this Smoake both in her Breast and Lungs, whilst She was in Your Majesties Palace. I cannot but greatly apprehend, that Your Majesty (who has been so long accustomed to the excellent Aer of other Countries) may be as much offended at it, in that regard also; especially since the Evil is so Epidemicall; indangering as well the Health of Your Subjects, as it sullies the Glory of this Your Imperial Seat.

Sir, I prepare in this short Discourse, an expedient how this pernicious Nuisance may be reformed. . . .

SIR,

Your Majesties ever Loyal, most obedient Subject and Servant, J. EVELYN. . . .

Part I

. . . I shall not here much descant upon the Nature of *Smoakes*, and other Exhalations from things burnt, which have obtained their several *Epithetes*, according to the quality of the Matter consumed, because they are generally accounted noxious and unwholsome; and I would not have it thought, that I doe here *Fumos vendere*, as the world is, or blot paper with insignificant remarks: It was yet haply no inept derivation of that *Critick*, who took our *English*, or rather, *Saxon* appellative, from the Greek word σμυχω *corrumpo* and *exuro*, as most agreeable to its destructive effects, especially of what we doe here so much declaim against, since this is certain, that of all the common and familiar materials which emit it, the immoderate use of, and indulgence to *Sea-coale* alone in the City of *London*, exposes it to one of the fowlest Inconveniences and reproaches, than possibly beffall so noble, and otherwise incomparable City: And that, not from the *Culinary* fires, which for being weak, and lesse often fed below, is with such ease dispelled and scattered above, as it is hardly at all discernible, but from some few particular Tunnells and Issues, belonging only to *Brewers*, *Diers*, *Lime-burners*, *Salt*, and *Sope-boylers*, and some other private Trades, *One* of whose *Spiracles* alone, does manifestly infect the *Aer*, more than all the Chimnies of *London* put together besides. And that this is not the least *Hyperbolic*, let the best of Judges decide it, which I take to be our senses: Whilst these are belching it forth their sooty jaws, the City of *London* resembles the face rather of Mount *Ætna*, the *Court of Vulcan, Stromboli*, or the Suburbs of *Hell*, than an Assembly of Rational Creatures, and the Imperial seat of our incomparable *Monarch*. For when in all other places the *Aer* is most Serene and Pure, it is here Ecclipsed with such a Cloud of Sulphure, as the Sun itself, which gives day to all the World besides, is hardly able to penetrate and impart it here; and the weary *Traveller*, at many Miles distance, sooner smells, than sees the City to which he repairs. This is that pernicious Smoake which sullyes all her Glory, superinducing a sooty Crust or Fur upon all that it lights, spoyling the moveables, tarnishing the Plate, Gildings and Furniture, and corroding the very Iron-bars and hardest Stones with these piercing and acrimonious Spirits which accompany its

Sulphure; and executing more in one year, than exposed to the pure *Aer* of the Country it could effect in some hundreds. . . .

It is this horrid Smoake which obscures our Churches, and makes our Palaces look old, which fouls our Clothes, and corrupts the Waters, so as the very Rain, and refreshing Dews which fall in the several Seasons, precipitate this impure vapour, which, with its black and tenacious quality, spots and contaminates whatever is exposed to it. . . .

The Consequences then of all this is, that (as was said) almost one half of them who perish in *London*, dye of *Phthisical* and *Pulmonic* distempers; That the *Inhabitants* are never free from *Coughs* and importunate *Rheumatisms* spitting of *Impostumated* and corrupt matter: for remedy whereof, there is none so infallible, as that, in time, the Patient change his *Aer*, and remove into the *Country*: Such as repair to *Paris* (where it is excellent) and other like Places, perfectly recovering of their health; which is a demonstration sufficient to confirm what we have asserted, concerning the perniciousnesse of that about this City, produc'd only, from this exital and intolerable Accident. . . .

Part II

We know (as the *Proverb* commonly speaks) that, as *there is no Smoake without Fire*; so neither is there hardly any *Fire* without *Smoake*, and the. . . materials which burn clear are very few, and but comparatively so tearmed: That to talk of serving this vast City (though *Paris* as great, be so supplied) with *Wood*, were madnesse; and yet doubtless it were possible, that much larger porportions of Wood might be brought to *London*, and sold at easier rates, if that were deligently observed, which both our *Laws* enjoyn, as faisible and practised in other places more remote, by Planting and preserving of *Woods* and *Copses*, and by what might by Sea, be brought out of the *Northern Countries*, where it so greatly abounds, and seems inexhaustible. But the *Remedy* which I would propose, has nothing in it of this difficulty, requiring only the Removal of such *Trades*, as are manifest *Nuisances* to the City, which, I would have placed at farther distances; especially, such as in their Works and Fournaces use great quantities of *Sea Coale*, the sole and only cause of those prodigious Clouds of *Smoake*, which so universally and so fatally infest the *Aer*, and would in no City of *Europe* be permitted, where Men had either respect to Health or Ornament. Such we named to be *Brewers, Diers, Sope* and *Salt-boylers, Lime-burners*, and the like: These I affirm, together with some few others of the same *Classe* removed at competent distance, would produce so considerable (though but partial) a Cure, as Men would even be found to breath a new life as it were, as well as *London* appear a new City, delivered from that, which alone renders it one of the most pernicious and insupportable abodes in the World, as subjecting her Inhabitants to so infamous an *Aer*, otherwise sweet and very healthful: For, (as we said) the *Culinary* fires (and which *charking* would greatly reform) contribute little, or nothing in comparison to these foul mouth'd Issues, and Curles of *Smoake*, which (as the Poet has it) do *Cælum* subtexere fumo, and draw a sable Curtain over Heaven. Let any man observe it, upon a *Sunday*, or such time as these Spiracles cease, that the Fires are generally extinguished, and he shall sensibly conclude, by the clearnesse of the Skie, and universal serenity of the *Aer* about it, that all the Chimnies in *London*, do not darken and poyson it so much, as *one* or two of those Tunnels of *Smoake*; and, that, because the most imperceptible transpirations, which *they* send forth, are ventilated, and dispersed with the least breath which is stirring: Whereas the *Columns* and Clowds of *Smoake*, which are belched forth from the sooty Throates of those Works, are so thick and plentiful, that rushing out with great impetuosity, they are capable even to resist the fiercest winds, and being

extremely surcharg'd with a fuliginous Body, fall down upon the City, before they can be dissipated, as the more thin and weak is; so as two or three of these *fumid vortices*, are able to whirle it about the whole City, rendring it in a few Moments like the Picture of *Troy* sacked by the *Greeks*, or the approaches of *Mount-Hecla*.

I propose therefore, that by an *Act* of this present *Parliament*, this infernal *Nuisance* be reformed; enjoyning, that all those *Works* be removed five or six miles distant from *London* below the River of *Thames*.

Critical Thinking

1. Industrial smoke can drastically impair the quality of urban life. How can such effects be prevented?
2. What makes some cities more vulnerable to air pollution than others?

Controversy at Love Canal

Beverly Paigen

The potentially disastrous consequences of improper hazardous waste disposal became a public issue in the late 1970s. The problem was dramatized by the evacuation from Niagara Falls, New York, of hundreds of residents, whose health was being threatened by chemicals leaking into their homes, backyards, and local school from the abandoned Love Canal, which had been used for many years as an industrial waste dump. The publicity given to this event led numerous other communities to discover and report local sites where toxic chemicals had been disposed of in open lagoons or were leaking from disintegrating steel drums. Such esoteric chemical names as dioxins and PCBs soon became part of the common lexicon, and numerous local citizens' groups were mobilized to demand the public's protection from exposure to hazardous wastes.

The Love Canal Homeowners Association was fortunate to have both an able leader, Lois Gibbs, who has since become a national organizer of the grassroots Citizens Committee on Hazardous Wastes (now The Center for Health, Environment and Justice), and a dedicated volunteer scientific adviser, Beverly Paigen. In 1978 Paigen, a research biologist now at The Jackson Laboratory in Bar Harbor, Maine, was studying differences in susceptibility to chemical toxins at the Roswell Park Memorial Institute of the New York State Department of Health, which is located in Buffalo, New York, a short distance from the Love Canal neighborhood. The developing controversy about the health effects of the chemicals to which the Love Canal residents had been exposed provided an opportunity for Paigen to make use of her professional expertise in a real world situation. She hoped to help answer the concerns of the homeowners while contributing to her own research interests. As she describes in the following selection from "Controversy at Love Canal," *Hastings Center Report* (June 1982), political and economic factors became serious obstacles in this endeavor and raised a series of important ethical questions about the behavior of various parties with conflicting interests in the resolution of a hazardous waste exposure issue.

Key Concept: the scientific and political dimensions of hazardous waste controversies

During the past four years, I have been involved in a controversy over the impact of hazardous waste buried at Love Canal on the health of the surrounding community. As a scientist employed by a research institute that is a division of the New York State Department of Health, I came to believe that additional studies should be done on the health of Love Canal residents. However, David Axelrod, who became Commissioner of Health in December 1978, maintained that the Health Department's studies were adequate and showed little or no health risk to the community. At the beginning, I thought our differences could be resolved in the traditional scientific manner by examining protocols, experimental design, and statistical analysis. But I was to learn that actual facts made little difference in resolving our disagreements—the Love Canal controversy was predominantly political in nature, and it raised a series of questions that had more to do with values than science.

The Environmental Protection Agency (EPA) estimates that 50,000 hazardous waste sites may exist in the United States, that 90 percent of these pose a potential health threat because they are improperly located or poorly managed, and that 2,000 are currently threatening the health of nearby communities. Thus the hazardous waste problem is large not only because of the number of sites but also because these sites are usually close to the places where people live and work. Because the issues faced at Love Canal will occur again and again, any lessons we can learn may be helpful in preventing or resolving controversy in similar situations and in protecting the public health.

Toxic Wastes at Love Canal

In 1942 Hooker Electrochemical Corporation (now Hooker Chemical and Plastics, a subsidiary of Occidental

Petroleum Corporation) began to fill an abandoned canal a half-mile long with toxic chemicals from the manufacture of chlorinated hydrocarbons and caustics. More than 21,000 tons of 200 or more chemicals had been deposited in the canal when the Niagara Falls Board of Education approached Hooker Chemical about purchasing the site for a school. Hooker claims that it warned the Board of Education that the site was not appropriate for a school. The company says it sold the property to the Board in 1953 for a token $1.00 only when the Board threatened to take the property by eminent domain. None of the people who were Board members at the time are living to confirm or deny that claim, and the minutes of the meetings do not bear out the claim. The deed transferring the property from Hooker to the Board of Education does contain a clause that, Hooker says, releases the company from liability. It states that the site was filled "with waste products resulting from the manufacturing of chemicals" and that "no claims, suit, action, or demand of any nature whatsoever" could be made against Hooker "for injury to a person or persons, including death resulting therefrom, . . . by reason of the presence of said industrial waste."

An elementary school was built in the center of the site and the north and south portions were sold to developers who built ninety-eight homes along the banks of the former canal. During the ensuing twenty-five years, chemicals from the dump site migrated as a thick black oily mixture through the topsoil into the surrounding community. As early as 1958 Hooker Chemical and city officials were informed that three children had suffered chemical burns from exposed wastes on the surface of the canal. The Niagara Falls Health Department and other local officials took no action on this and many other complaints.

Through the efforts of determined residents and newspaper reporter Michael Brown, the EPA and the New York State Department of Health eventually entered the picture. In 1978, these agencies identified many chemicals in the air of Love Canal homes, and the Department of Health documented an excess frequency of miscarriages in women living in homes immediately adjacent to Love Canal. On August 2, 1978, Robert Whalen, then the New York State Commissioner of Health, declared a health emergency. Shortly thereafter Hugh Carey, the governor, offered to purchase the 239 homes closest to the canal and to assist in relocating the families. A fence was placed around the homes and plans were made to construct trenches to intercept the flow of chemicals from the canal. The Health Department initiated health studies of 850 additional homes in the Love Canal neighborhood by distributing questionnaires and taking blood samples for analysis. In early fall of 1978 the department announced the preliminary results of these studies; officials assured the Love Canal residents that the neighborhood was a safe place to live and that the community beyond the homes that had already been evacuated was not at any increased health risk. This announcement was based on

data showing that the miscarriage rate in homes beyond the barrier was no higher than elsewhere.

The community was not reassured, citing visible seepage through basement walls, chemical odors in homes, and odors at storm sewer openings as evidence that chemicals had migrated beyond the fence. The residents also questioned why certain families living three to four blocks from the canal had multiple miscarriages and other illnesses.

History of the Controversy

At the time I was studying genetic variation in the metabolism of chemicals and trying to determine whether such differences might explain differing susceptibilities to environmental toxins. The exposure of 850 families to low levels of chemicals seemed to me an unusual opportunity to locate susceptible families with several sick members, match these with a healthy neighboring family, and compare the metabolism of chemicals in these two groups. I devised a health questionnaire, and a small group of residents systematically surveyed the neighborhood by telephone, collecting information about the number of people in the family, length of residence, and health.

I planned to plot the illnesses geographically with the following expectations: (1) if illnesses were clustered in families, that would indicate a possible genetic susceptibility to low-level chemical exposure and thus provide families for future study in my laboratory; or (2) if illnesses were geographically clustered, that would probably indicate migration of chemicals from the canal; or (3) if illnesses were randomly distributed, that would indicate no relationship to chemical exposure. Of course I realized that the sample size was small (1,140), so that any findings would only be a signal that further studies needed to be done. Plotting the results on a map revealed a strong geographical clustering of disease that appeared to be related to former stream beds and swales, which are low marshy areas that collect water but do not have a particular direction of flow. These streams or swales no longer exist because they were filled with building rubble as the neighborhood was developed. The predominant soil in the area is clay; thus a low-lying area filled with rubble could provide a permeable conduit for the flow of liquid chemicals from the canal.

The exact locations of these swales, as well as swamps and a pond, were determined by soil scientists under contract to the Health Department. Once these were defined, it was possible to divide the neighborhood into historically wet homes—those bordering streams and swales or built in swamps or the former pond— and historically dry homes. Comparing the incidence of disease in wet and dry homes provided an internal neighborhood control that randomized for reporting bias, occupational exposure and age.

We found a threefold increase in miscarriages in pregnant women who lived in wet homes compared to those who lived in dry homes. Particularly striking were the women in wet homes with three or more miscarriages. The frequency of such habitual aborters reported in the literature is between 0.4 to 0.7 percent. Of the sixty-four women living in wet homes at Love Canal who became pregnant, five (8 percent) had three or more miscarriages. The probability of this occurring by chance is less than 0.001.

Birth defects were also higher in wet homes. Of 120 children born in wet homes, 24 (20 percent) had birth defects compared to 6.8 percent of children born in dry homes. Some of these birth defects were minor, such as club feet, webbed toes, a missing ear, or an extra set of teeth; but many were serious, such as heart defects, missing or nonfunctional kidneys, deafness, absence of diaphragm, and mental retardation. Asthma was 3.5 times more frequent and urinary system disease was 2.8 times more frequent in wet homes than in dry. Various symptoms of central nervous system toxicity were reported by residents in wet homes: these included dizziness, fainting, seizures, blurred vision, depression, hyperactivity in children, suicides, suicide attempts, and nervous breakdowns. Since data included health effects only up to the declaration of the health emergency, the stress and anxiety that may have occurred after that date exerted no influence on the symptoms. Since central nervous system problems are very subjective, I evaluated only those that were severe and could be verified. I chose to group together admissions to a mental hospital and suicide attempts and found that these were five times more frequent in wet areas (p < .0005).

The survey, which was conducted without any funds, was adequate to locate sick and healthy families in order to study metabolism of chemicals, but it had never been intended as a full-scale epidemiological study. Once I had the data, however, I found myself in a very uncomfortable position. The data strongly indicated that chemicals might have migrated beyond the fence and that a health risk might be present in some parts of the Love Canal neighborhood. However, Commissioner Whalen was publicly stating that the chemicals had not migrated beyond the fence.

The department's claim arose from two factors. First, the department had analyzed miscarriage frequency by street (99th, 100th, 101st, etc.). Since the rate of miscarriage did not decrease as the streets grew more distant from the canal, the Health Department concluded that the miscarriages were not related to chemical exposure. However, if chemicals had migrated from the canal preferentially along swales, as I suspected, then miscarriage frequency should not necessarily decrease with perpendicular distance from the canal, but instead would be related to the location of the swales. Second, the Health Department's finding that miscarriage frequency in Love Canal women was no higher than elsewhere was based on the Warburton and Fraser study, which the department had selected as a literature control. However, that study

differed in several important ways from the Love Canal study conducted by the Health Department. The women in the Warburton and Fraser study had either previously had a child with a birth defect, or in a minority of cases, had given birth to twins. Thus they were not a representative population. In addition, unlike the Love Canal population, the women in the control study were poor and had little or no prenatal care. Finally Warburton and Fraser counted as valid any miscarriages reported by a woman, while the Health Department counted only those that could be independently confirmed by physicians' records. . . .

After reviewing the data analysis by the Health Department and the results of my own study, I felt I needed to convey my concerns to my superiors. On November 1, 1978, I suggested to the Health Department a hypothesis that needed to be tested: adverse pregnancy outcomes were more frequent in wet homes than in dry homes and adverse effects might be occurring in the urinary, respiratory, and central nervous system as well. The department then reevaluated their own data, and on February 8, 1979, the new health commissioner, David Axelrod, announced that women in wet homes were more likely to have miscarriages, babies of low birth weight, and children with congenital abnormalities. At that time families with children under two years of age or with women who could prove they were pregnant were evacuated. But evacuation was to continue only until the youngest child in the family reached the age of two. Women who wished to conceive were concerned because during most of the sensitive first trimester of pregnancy when the fetus is at highest risk they would be involved in proving they were pregnant and in processing their requests for evacuation. These women applied for evacuation but their requests were denied.

At the February 1979 meeting, the health commissioner publicly praised my contributions and promised the residents of Love Canal that studies would follow on the respiratory, urinary, and nervous systems. More than three years have elapsed, but these studies have not been done or even initiated. Still at the time I felt considerable relief. Since my superiors and I now agreed on the essential points, I thought the controversy was over. But I was wrong. The controversy continued for one-and-a-half years with increasing intensity, until the residents of Love Canal were given the opportunity to move out of the neighborhood.

That "scientific" controversy rolled on regardless of scientific facts was a hard lesson for me to learn. Shortly after the February 1979 meeting, I asked a New York state scientist in bewilderment exactly what we were disagreeing about since we both agreed that adverse pregnancy outcomes were elevated and the other diseases had not yet been examined by the state. He replied that the disagreement was now over the exact number of miscarriages and birth defects. Seeking to end that controversy, I said that I would accept their figures. I changed my slides and reports and even wrote to David Rall, chief of the National Institute of Environmental Health Science, that "the diagnosis of diseases in the State's data base is more reliable than those in my data base because the State has checked each disease with physician and hospital records while I have accepted the individual's self-report" (March 13, 1979). The state used this statement against me later, but the controversy I had hoped to end continued even though there was no longer any difference of opinion over facts.

These events caused me to think carefully about the nature of public health controversy, the factors that prevent resolution of controversy, and the ways in which controversy can be reduced or avoided. In discussing these ideas, I will illustrate with examples from Love Canal, but the events are characteristic of many such controversies.

The Elements of Controversy

The two opposing sides in the Love Canal controversy were the community and the New York State Department of Health. This was somewhat surprising, since the Health Department had declared the health emergency in the first place. However, when the community turned to the agency they regarded as their ally and protector, they felt that the response was inadequate.

Antagonism between a community and a health department is not unique to Love Canal. At a number of other hazardous waste sites, frustrated and angry communities have turned on the local health agency rather than on the industry that created the hazardous waste sites. Perhaps such a response is not so surprising after all. Hooker Chemical was a major industry in the region and employed many people in the Love Canal neighborhood. The community made some allowance for Hooker because the chemicals were buried many years before the chronic toxicity of chemicals was understood and before regulations concerning disposal of toxic waste existed. Hooker Chemical claimed that it had used state-of-the-art technology in burying the waste and that furthermore they had warned the Board of Education not to build a school on the site. The community also understood that the goal of industry is profits and that Hooker was acting in a manner consistent with its goals by using the cheapest method of disposal. So while many in the Love Canal community were inclined to make allowances for Hooker, they did not feel well disposed toward the Health Department, which was acting in the present with full knowledge of chemical toxicity. The stated goal of the department is to protect health, and the salaries of officials came from the community's tax money. So the community viewed the department as acting in a manner inconsistent with its goals and responsibilities when health effects were minimized or ignored.

Once the controversy was under way several factors impeded a resolution:

1. The failure to resolve any controversy may be advantageous to one side. In this case the state had much to gain from delay. Since over 600 other hazardous waste sites existed in New York, any action taken at Love Canal would set a precedent. Any state official who recommended positive action at Love Canal would have had to justify spending even more than the $42,000,000 the state had already allowed for construction to prevent further leakage and relocation of the families living closest to the canal. In contrast, an official who delayed was taking very little risk. If the decision to delay was later shown to be wrong, the community would suffer the risk of impaired health and some future official would probably have to worry about the results. . . .

This is not a scientific issue, nor can it be resolved by scientific methods. The issue is ethical, for it is a value judgment to decide whether to make errors on the side of protecting human health or on the side of conserving state resources. In science this problem is called avoiding type I and type II errors. In a type I error, a scientist accepts as true something that is actually false. The custom in science is to insist that we be 95 percent certain that something is true (for example, that the health of Love Canal residents has suffered) before we accept it as a fact. In contrast, a scientist who makes a type II error fails to recognize as true something that actually is true. . . .

Before Love Canal, I also needed to have 95 percent certainty before I was convinced of a result. But seeing this rigorously applied in a situation where the consequences of an error meant that pregnancies were resulting in miscarriages, stillbirths, and children with medical problems, I realized I was making a value judgment. In other issues of public health and safety—bomb threats; possible epidemics, etc.—we do not insist on 95 percent probability of harmful consequences before action is taken. Why is that the criterion in environmental health? This issue should be debated widely in the public health community so that choices can be made not out of habit but with full realization of the values involved.

2. Opponents may not agree on the question that needs to be answered. In this case, Commissioner Axelrod informed the residents that the epidemiologist would look for adverse health effects on the human fetus since that was the most vulnerable segment of the population. But the residents assumed that any increase in adverse pregnancy outcomes would indicate that toxic chemicals were present and that the entire population was at risk. They reacted angrily when the commissioner announced that the fetus was indeed at risk, but that the state would evacuate only pregnant women.

The design of a study also has a profound influence on the outcome. For example, in the spring of 1981, the Center for Disease Control (CDC) designed a study to answer the question of health impact at Love Canal. A great deal of money and effort went into answering the question of whether psychological harm occurred.

In contrast, few resources went into answering the question of whether there was chromosome damage. The CDC study was canceled due to lack of funds. However, had the design been carried out, after two years and several million dollars, the public would have been informed that Love Canal residents suffered psychological damage but not chromosomal damage—a conclusion dictated by the study design.

The most costly portion of the CDC study called for detailed health histories, physical examinations and laboratory tests for all Love Canal residents. However, these were not to be done on a comparable control population, nor were there plans for data compilation and analysis on the results of the Love Canal residents. . . .

3. In any controversy, since the type and quality of information gathered will influence the outcome, no one group should be in complete control of the information-gathering process. At Love Canal the state had the personnel and monies of a very good health department and an additional $1,500,000 specifically allocated by the New York legislature and Congress to gather information about health effects. In contrast, the residents had only their own energy and the help of a few scientists. Of the three small studies done outside the health department, two were done with no money (my survey and a study of nerve conduction by Dr. Stephen Barron) and one with $10,000 (a chromosome study by Dr. Dante Picciano under contract to the EPA). All had flaws in part attributable to the lack of funds. Since the data from these studies were widely available, it was easy for scientists to review and criticize them.

In contrast, the protocols and the data from the well-funded studies of the Health Department were—and still are—secret. According to a fact sheet released by the Department of Health on June 23, 1980, these studies cost $3,292,000 and 205,000 staff hours (122 staff years) and encompassed 4,386 blood samples, 11,138 field interviews, 5,924 soil samples, over 700 air, sump, and water samples, follow-up of 2,000 former residents, and 411 physical examinations of workers involved in remedial construction at Love Canal. With the exception of a provisional draft of the study on adverse pregnancy outcomes, which was released when a newspaper asked for it under the Freedom of Information Act, and the recently published Janerich study, the health studies are not available for scientific review or criticism even though conclusions are frequently quoted publicly. . . .

We must develop ways of providing communities with access to resources and expertise. This was done with remarkable success in one case at Love Canal. Shortly after the health emergency was declared, the New York State Department of Transportation planned some remedial construction work to prevent the further flow of chemicals from the canal into the community. A health and safety plan to protect the community was developed. During negotiations between the department and the community, the lawyer for the community arranged for a sum of money to be set aside for a toxicologist who would be on site during the construction to monitor the state's compliance with the safety plan and to report to the community. The cost was only $10,000 in a project that involved $3 to $4 million. . . .

Agencies that undertake health studies should certainly consider allocating money to the community so that residents can obtain their own expertise. But an even better solution would be to have information gathered and funded by neutral third parties. If a federally based and funded response team were established to deal with many hazardous waste sites, it might be in a position to evaluate the relative seriousness of each situation.

4. Beyond questions of money and expertise is the issue of full expression for dissident and minority opinions. Controversy is stifled, not resolved, by silencing the opposition. At Love Canal, scientists working for New York State who disagreed with the official stance were demoted, transferred, or harassed. For example, William Friedman, regional director of the Department of Environmental Conservation, was demoted to staff engineer in November 1978 (he subsequently left the agency) for "prodd[ing] the Albany hierarchy unrelentingly" (*Buffalo Evening News*, November 12, 1978). And Donald McKensie, a senior sanitary engineer in the regional office, Department of Environmental Conservation, wrote a letter to his superior raising questions about the manner in which the department was handling Love Canal and was promptly transferred from the Love Canal project to air quality projects (*Knickerbocker News*, May 12, 1980).

I too was harassed in a variety of ways. The most important involved my ability to raise money to support my research by grant applications. I first spoke publicly concerning the Love Canal problem in the summer of 1978. That September, the Department of Health administration withdrew one grant I had submitted without even informing me. That winter the administration refused to process papers so I could get funds from another grant that already had been awarded. This denial led the professional staff of Roswell Park Memorial Institute to charge the administration with scientific censorship. In June 1979, I was informed that, because of the "sensitive nature" of my work, all grants and research ideas had to go through a special review process. I was told to outline my research ideas and submit them for review "at the moment of conception before a single experiment was done." My professional mail arrived already opened and scotch-taped shut. My office was entered outside of working hours and my files were searched.

My income tax return was audited by New York State for the first time in over twenty years of filing returns and the auditor's file contained newspaper clippings, including such nonfactual material as "Letters to the Editor," about my role at Love Canal. Later James Tully, Jr., Commissioner of Taxation and Finance, apologized for several "errors in procedure," and wrote that the clippings resulted from a "a misunderstanding" of department policy by a local audit supervisor. Added to these serious problems were a variety of petty harassments, which continued until I announced in 1981 that I was leaving Roswell Park and accepting a position in another state.

In addressing the problem of whistleblowers, the Committee on Scientific Freedom and Responsibility of the American Association for the Advancement of Science (AAAS) stated in 1975 that scientists "must be assured of some form of due process in passing judgment on the issues that they raise. This would call for the presence of independent outside members on any board that passes judgment on the issues, and should also include some right of appeal." My experience showed that no such protection exists.

5. Scientists, who are no strangers to controversy, should follow the social controls on behavior that they have developed for the advancement of knowledge and the detection of error. These include openness of data, peer review and criticism, publication of data, and replication of experiments.

These norms of scientific behavior were violated at Love Canal. Secrecy of the state's data and protection from peer review and criticism prevailed. Of the repeated attempts to obtain state data, I will mention only a few. The Environmental Defense Fund requested the protocols of the Department of Health studies under the Freedom of Information Act, but this request was denied. On February 8, 1979, a news release from the Health Department made five claims regarding the results of the health studies (no liver disease, no benzene toxicity, no blood problems, no cancer, no epilepsy). My request for data supporting these claims was denied.

A short time later in a face-to-face meeting among the three of us, Governor Hugh Carey asked Commissioner Axelrod to give me the data. Though Dr. Axelrod agreed, he has never done so. Later toxicologist Steven Lester and I were promised by Dr. Axelrod access to the state's computerized data base. On an arranged date, Lester flew to Albany only to be denied access to any of the health data. On another occasion, Dr. Axelrod explained that he wanted to provide health data but could not do so ethically because confidentiality had been promised to the interviewed residents. He suggested that I get signed consent forms from each resident who provided health data to the state, releasing that data to me. The lawyer for the residents drew up the consent form, and the residents organized and obtained hundreds of signed consents. When I sent these to Dr. Axelrod with a cover letter, all I received back was a reprimand for writing my cover letter on Department of Health stationery. (See Archives Library of the State University of New York, Buffalo). The release of data is still "under review." Many more examples could be cited. . . .

6. In any attempt at controversy resolution, all parties to the conflict should agree on precisely what facts need resolving; all parties should agree on the composition of the body chosen to resolve the controversy; all parties should agree on the procedures by which that body will operate; and all parties must agree to abide by the decisions. Unfortunately, such a logical procedure rarely occurs.

There were at least two unsuccessful attempts to resolve controversy at Love Canal, and both failed because these requirements were not followed. The first occurred on February 22, 1979, when the local Congressman John LaFalce requested that representatives of the EPA and Department of Health, Education and Welfare meet with Dr. Axelrod and me "so that both sets of data can be reviewed by the federal government. I am hopeful," LaFalce wrote, "that such a meeting will resolve the differences that currently exist and that appropriate actions will then be taken to protect the health and welfare of my people." Dr. David Rall of the NIEHS was appointed chair, and he convened experts in biostatistics and epidemiology. Since Dr. Axelrod refused to have a face-to-face meeting with me, the federal committee met with me on April 12, 1979, and with State Health Department scientists on April 26. Their report agreed with my concerns and recommended that further studies be undertaken. In response to my suggestion that all women of childbearing age who wish more children should be evacuated, they recommended that exposure of Love Canal residents be "minimized to the extent feasible." Those familiar with Love Canal understood that the only way to minimize the exposure was to move the people out of the area. But, as the report pointed out, "The State appears to have a problem in terms of perceived conflict of interest." Thus the report recommended that "the involvement of outside scientists, both in the interpretation of data and the formulation of recommendations to the State policy makers, should be continued. In a manner such as this where there is much public concern it is wise that considerations of findings and of alternatives be conducted openly. To this end it might be wise to include representatives of the local population."

This attempt to resolve controversy failed because the Health Department chose to ignore the report. The department had nothing to gain by resolution of the controversy and had not agreed to abide by the decisions of the group.

The second attempt occurred about a year later. In May 1980, State Senator Thomas Bartosiewicz released a report on the hazardous waste problem in New York State. He then called for an investigation into the handling of Love Canal by the Department of Health. In a letter to Governor Carey and in a Senate resolution, he charged the state agencies (Department of Health and Department of Environmental Conservation) with unethical conduct. The detailed list of charges included:

- Appointment of a Blue Ribbon panel which had secret members and secret recommendations which were withheld from the public;

- Manipulation of health data. . . to minimize risks;
- Unexplained delays of up to eighteen months before the State was willing to admit a health problem existed;
- Demotion, transfers, and harassment of state employees sympathetic to Love Canal residents;
- An effort by the state to discourage and prevent independent professional health studies. . . .

The Next Decade

. . . When controversies arise in communities as they inevitably will, steps can be taken to ease the situation and protect the public health. First, scientists should scrupulously adhere to the norms of their profession such as openness of data, peer review and criticism, and publication of evidence. Second, community involvement should be sought and used at every level of the process. Third, funds should be provided so the community can hire its own experts.

It would also be useful to have a carefully reasoned process for conflict resolution, which could be published by scientific societies and incorporated into the ethical code of behavior for scientists. Such a process would help, even in cases like Love Canal, where the controversy was so political. The standards for conflict resolution should include guidelines for selecting an independent group of fact-finders, the rules of procedure for such a body, an agreement by both parties to abide by the decision, and adequate protection for whistle-blowers. Finally, since in conflicts between bureaucracy and community, the bureaucracy often does not gain by controversy resolution, third parties may have to try to apply pressure in order to get the bureaucracy to participate in such a procedure. . . .

Many of the controversies of Love Canal were stated as scientific issues, but they had their roots in ethical considerations. In such instances the controversy would have become easier to resolve if the ethical considerations and value judgments had been openly stated and understood.

Critical Thinking

1. Why might the people supposedly affected by an environmental problem (such as LoveCanal) and a government agency (such as the state Department of Health) disagree about what questions need to be answered about the problem?
2. What does "proper" hazardous waste disposal mean?
3. What political and economic factors can interfere with resolution of an environmental problem?

Restoring Rivers

Margaret A. Palmer and J. David Allen

Life on Earth is utterly dependent on water. Life on land requires clean, fresh water. So do ocean fish such as salmon, which breed in rivers and streams. Humans use huge amounts of fresh water for drinking, washing, flushing, food processing, industrial cooling, irrigation, recreation, and power generation. As a result, rivers have been fouled with chemicals and sewage, diverted to farms and cities to the point where some rivers no longer flow to the sea, and blocked with dams. Because a third of the world's people lack access to safe drinking water, they are subject to disease. Fights over water rights can take place in courtrooms and on battlefields.

What can be done? In the United States and Europe, water pollution is much less of a problem than it was half a century ago. Some dams have been removed to permit migratory fish to reach ancestral breeding grounds and restore populations. But many problems remain. Margaret A. Palmer, professor and director of the Chesapeake Biological Laboratory at the University of Maryland, and J. David Allan, professor at the School of Natural Resources and Environment at the University of Michigan, argue in this selection that rivers must be restored to "perform essential ecological and social functions such as mitigating floods, providing clean drinking water, removing excessive levels of nutrients and sediments. . . and supporting fisheries and wildlife." Present efforts are not sufficient.

Key Concept: human dependence on natural water systems

Between 1973 and 1998, U.S. fresh waters and rivers were getting cleaner. But that trend has reversed. If the reverse continues, U.S. rivers will be as dirty in 2016 as they were in the mid-1970s.

Water quality is not the only problem. In parts of the United States, the extraction of surface water and ground-water is so extreme that some major rivers no longer flow to the sea year round, and water shortages in local communities are a reality.

The damage and suffering wrought by Hurricane Katrina demonstrate that the restoration of waterways and wetlands is not a luxury. It is a national imperative. And the imperative does not apply only to hurricane-prone coastal waterways. More than one-third of rivers in the United States are impaired or polluted.

The flood-storage capacity of U.S. rivers is at an all-time low. Water shortages are increasingly common even in eastern states that historically have had plenty of it. Aquatic wildlife is going extinct at a rate much higher than organisms in either terrestrial or marine eco-systems. In June 2004, drinking water in Wisconsin was found to exceed levels of nitrate considered safe. In July 2005, scientists reported that high levels of nutrients and

sediments from river tributaries had created a dead zone that blanketed a third of the Chesapeake Bay. In October 2005, homes in Connecticut, New York, and New Jersey were flooded when rivers overflowed their banks after a week of rain. All three of these disasters are linked to the degradation of rivers and streams. And all three of them could have been prevented by ecological restoration.

River restoration means repairing waterways that can no longer perform essential ecological and social functions such as mitigating floods, providing clean drinking water, removing excessive levels of nutrients and sediments before they choke coastal zones, and supporting fisheries and wildlife. Healthy rivers and streams also enhance property values and are a hub for recreation. Clearly, degraded rivers and streams need to be repaired.

However, just as rivers are in need of restoration, so too are the art and practice of restoration itself. A recent study (in which we participated) published in *Science* documented a huge number of restoration projects being implemented in every region of the country at great cost and for a variety of reasons. The projects range from land acquisition (at a median cost of more than $800,000), to bank or channel reshaping to restore floodplains (median cost more than $200,000), to keeping livestock out of rivers and streams (median cost $15,000). However, distressingly few of them—just 10% of all restoration project records in the database put together by the National River Restoration Science Synthesis (NRRSS)—included any mention of assessment or evaluation. The study concluded that it is currently impossible to use existing databases to determine whether the desired environmental benefits of river restoration are being achieved. Even when monitoring was reported, it typically was an assessment of project implementation, not ecological outcomes.

The nation can do better. The United States needs regulatory and legislative federal policy reforms in order to improve the effectiveness of river restoration and thus the health of the nation's waterways.

How did the United States reach a point where the majority of our rivers are degraded and ecologically dysfunctional? People have always chosen to live and work near water. Cities and industrial facilities began to grow up along U.S. waterways centuries ago, and for most of U.S. history, dilution was the solution to pollution. U.S. streams and rivers were the dumping grounds for waste, and the hope was that the waste would be carried away.

Settlers also cut down riparian forests and filled in small tributaries and wetlands to make transportation and building easier. There was little understanding of the ecological roles that these forests and tributaries fill. In the first half of the 20th century, massive dams were erected with the goal of supplying power and minimizing floods. They did accomplish those objectives, but damming also led to the loss of water-starved native plants and animals downstream. They could not survive and reproduce without the seasonal changes in flow that the river had always brought them and that their life cycles depend on.

With increasing industrialization and population growth, cities and industries not only continued to dump raw sewage and other wastes into streams and rivers, but also "paved" many U.S. streams—lined them with concrete so they could convey water and pollutants more rapidly. Streams were viewed as pipelines, not the living entities we now know they are. Forgotten was their ability, when healthy, to cleanse water, store sediments, and provide materials essential to healthy coastal fisheries.

The crisis came in the 1960s, when it became known that two-thirds of U.S. waterways were polluted. In 1972, the Clean Water Act (CWA) was passed. Since then, U.S. rivers and streams have become healthier, largely because of controls over point-source pollution. Then in 2004, for the first time since the act was passed, the Environmental Protection Agency (EPA) reported that waterways were once again getting dirtier.

The primary reason why so many rivers and streams are still being degraded today is poor land stewardship. Human activities and alterations of the landscape have diverse and far-reaching effects. As land is cleared to build homes and shopping malls, entire watersheds are affected. Construction and the erosion of farmland introduce massive amounts of sediment into streams. Many streambeds are covered by heavy layers of silt. This silt suffocates fish eggs and invertebrates living on streambeds, destroys aquatic habitat, and even can interfere with the treatment of drinking water. Agriculture and urbanization move excessive amounts of nutrients and toxins from the land to rivers, streams, and coastal waters.

When land is cleared and replaced with hard surfaces such as parking lots and rooftops, stream flows are governed primarily by overland runoff or inputs from stormwater systems, not by the interaction of local climate, vegetation, and soil characteristics. Rainwater no longer soaks into the soil before it moves underground toward stream channels, the normal route for water flow in healthy temperate watersheds. Rapid runoff prevents replenishment of the groundwater table and results in "flashy" stream flows. Flashiness means more floods that can damage nearby property as well as flows so low in the summer that entire channels dry up. Stream and river banks begin to scour and slump. Channels widen and deepen. If the riparian vegetation has been removed, which is true for a huge number of streams and most rivers in the United States, erosion worsens. Moreover, the large amount of pavement in urban watersheds retains a lot of heat in the summer, so the flashy flows that follow heavy rains inundate the streams with pulses of warm water, destroying fish and bottom-dwelling organisms.

How has the United States tried to solve these problems? Are the solutions working? It is not as if nothing has been done. The CWA went a long way toward minimizing point-source inputs of pollutants to rivers. Unfortunately, rapid changes in land use and the many effects that urbanization and agriculture have had on rivers and streams are not as easy to remedy as point-source discharges.

Attempts have been made to minimize those effects. For example, the Conservation Reserve Enhancement Program of the Department of Agriculture's Farm Service Agency paid farmers to participate in long-term conservation projects such as planting riparian buffers on their property or keeping land out of agriculture. In recognition that more diffuse sources of pollution of waterways have become increasingly common, the EPA recently adopted stricter standards for control of stormwater runoff. This is Phase II of the National Pollutant Discharge Elimination System, which extends permitting regulations to smaller population centers. These regulations require communities and public entities to develop, implement, and enforce a stormwater program designed to reduce the discharge of pollutants. In addition, many cities and towns enacted their own regulations to slow the clearing of land, to require that only a minimal number of trees be removed during construction, or to mitigate damage to forests and wetlands.

Despite these and many other efforts to minimize the environmental impact of developing the land or extracting natural resources (such as mining), streams and rivers have continued to degrade. The controls have simply not been able to keep up with the rate of development and associated watershed damage. Moreover, many rivers and streams were suffering years before conservation programs were enacted.

River and stream restoration thus grew out of the recognition that active interventions and aggressive programs were needed to improve the health of U.S. waterways. There are many ways to restore rivers, and they vary depending on the underlying problem. The most common goals of river and stream restoration are to improve water quality, manage or replant riparian vegetation, enhance in-stream habitat, provide for fish passage, and stabilize banks. Practices for accomplishing these goals are diverse and overlapping.

For example, bank stabilization can be achieved in a number of ways, including riparian plantings, wire baskets filled with stones, large slabs of concrete, and rope netting. Improving water quality may involve enhancing upstream stormwater treatment and planting vegetation along stream banks. Enhancing habitat and improving fisheries may require adding logs or boulders to streams, constructing ladders for migrating fish that cannot pass dams, or reconnecting a river floodplain to its channel to provide spawning habitat or nursery grounds for young fish.

Restoration activities such as these are now common in the United States. . . . [But] the fact that the health of coastal areas such as the Chesapeake Bay continues to decline despite thousands of stream and river restoration projects demonstrate that something is wrong with U.S. restoration policies. The health of U.S. waterways is not improving fast enough despite the fact that the number of projects in the United States is increasing rapidly. The country is now spending well over $1 billion per year on river and stream restoration, and it is not getting its money's worth.

The problem is that there are no policies to support restoration standards, to promote the use of proven methods, or to provide basic data needed for planning and implementing restoration. Although much is known about effective restoration, this information has not been used in most projects or policies. For this reason, many restoration efforts fail: Stream banks collapse, pollutants from upstream reaches that were never considered for restoration overwhelm newly restored sites downstream, channel reconfiguration projects that were over engineered are buried in sediment, and flood waters flow over river banks.

What to do? First, government officials must deal with the fact that most restoration projects are being done piecemeal, with little or no assessment of ecological effectiveness. They do not even know which of the various restoration approaches are most effective. In addition, there is little coordination among restoration plans and projects in most watersheds.

Second, because there are no national standards for measuring success in restoration, there is no system for evaluating how effective projects are. Watershed managers and restoration practitioners have little guidance for choosing among various restoration methods to ensure ecological improvements.

Third, the United States has no national tracking system that gathers basic information on what is done where and when. Thus, there is no way to prioritize projects based on what is being done elsewhere or what is known to work.

The solution to pollution is to reform federal, state, and local policies. Here we address the federal level because of the critical role that federal policies play in funding and permitting restoration projects. Different regulations and laws are needed in four areas.

Federal agencies must be directed to adopt and abide by standards for successful river and stream restoration. Progress in the science and practice of river restoration has been hampered by the lack of agreed-on criteria for judging ecological success. The restoration community—which includes legislators, agencies, practitioners, and citizen groups—should adopt common criteria for defining and assessing ecological success in restoration. Success can also be achieved through the involvement of stakeholders and learning from experience, but it is ecological success that will improve the health of U.S. waterways.

Five basic standards have been recommended by an eminent team of U.S. scientists and engineers and endorsed by an international group of river scientists as well as by restoration practitioners. The standards are:

- The design of a river restoration project should be based on a specific guiding image of a more dynamic, healthy river.
- The river's ecological condition must show measurable improvement.
- The river system must be more self-sustaining and resilient to external perturbations, so that only minimal follow up maintenance is needed.
- During the construction phase, no lasting harm should be inflicted on the ecosystem.
- Both pre- and post-assessments must be completed and data made publicly available.

Simple metrics are already available that can be applied to each standard, so implementing the standards would not be difficult. If federal agencies involved in funding river restoration adopted these standards, it would go a long way toward ensuring that projects meet their stated ecological goals. . . .

A coordinated tracking system for restoration projects must be implemented. If the restoration of aquatic ecosystems is to be effective, restorers must be able to learn from past efforts. At present there is no coordinated tracking of river restoration projects. Existing federal databases are highly fragmented, often relying on ad hoc or volunteer data entry. They are inadequate for evaluating even the most basic trends in river restoration. Bernhardt et al. found that these national databases cataloged less than 8% of the projects in the NRRSS database. They could not be used to evaluate regional differences in restoration goals, expenditures, and assessment.

Agencies at all levels of government that share jurisdiction over rivers and streams are engaging in restoration projects of various kinds. Legislators who authorize projects and appropriate funds are also taking greater notice of the opportunities presented by river restoration projects that can drive funding toward their districts and constituents. Thus, there is an urgent need for a centralized tracking system that catalogs every stream and river restoration project implemented in the United States. This should include at least the following information for every project: Geographic Information Systems coordinates; spatial extent, intent, and goals; a catalog of project actions; implementation year; contact information; cost; and monitoring results.

Restoration monitoring presents its own suite of challenges. Implementation monitoring—determining whether a project was built as designed, is in compliance with permit requirements, and was implemented using practices that had been vetted by effectiveness-monitoring research programs—is not carried out routinely. This monitoring should be required of all projects and included in the centralized database. This ensures compliance with permits and with stated intents and requires that monitoring be part of project planning and designed in light of project goals.

Effectiveness monitoring involves an in-depth research evaluation of ecological and physical performance to determine whether a particular type of restoration or method of implementation provides the desired environmental benefits. Because effectiveness monitoring is time-consuming and expensive, it is unrealistic to make this a routine expectation. However, it is necessary when comparing the effectiveness of different restoration approaches, when evaluating unproven restoration practices, or when the ecological risks of a project are considered high. . . .

Undertake a national study to evaluate the effectiveness of restoration projects. Because restoration effectiveness has not received adequate attention, it is not always clear which restoration methods are most appropriate or most likely to lead to ecological improvements. Agency practitioners often rely on best professional judgment that their projects are meeting intended ecological goals, rather than undertaking scientific measurement and

evaluation. Only a small fraction of projects are currently being monitored to determine their relative success, so little is known about the environmental benefits. Something must be done to ensure that projects are doing what they set out to do and that money is well spent.

Although monitoring of project effectiveness in meeting ecological goals is always desirable, not every project requires sophisticated and costly effectiveness monitoring. In fact, many people worry that such an expectation would diminish the number of restoration projects on the ground by siphoning off available resources.

One way to balance the need for evaluation and accountability with limited resources is to conduct detailed monitoring of a sample of projects. The information gained would provide an efficient means of understanding project effectiveness and help restorers learn from the experience of others. Such a program could involve detailed monitoring of a sample of all projects within each of the major categories of river and stream restoration, perhaps beginning with the most interventionist restoration practices (such as channel reconfiguration) or the most costly forms of restoration (such as floodplain reconnection). . . .

Use existing funding for river restoration more efficiently and supplement funding. Although there are many areas in which Congress and federal agencies could make improvements in the policy and practice of river restoration, it is first necessary to ensure that existing funding is wisely allocated so that projects are successful. That means developing a mechanism to authorize and fund restoration projects in a much more coordinated fashion than the balkanized system that supports them today. There are more than 40 federal programs that fund stream and river restoration projects. Although large-scale high-profile projects such as those in the Everglades receive a great deal of attention, most projects in the United States are small in spatial extent. The cumulative costs and benefits of the many small restoration projects can be very high, which argues for better coordination.

We suggest that a Water Resources Restoration Act (WRRA) could serve as a mechanism to authorize and fund river restoration projects. Like the Water Resources Development Act (WRDA); WRRA would support projects of various shapes and sizes, all for the purpose of making federal investments in natural capital and infrastructure. Money would still flow through individual agencies, but prioritization and coordination would be achieved through an administrative body with representation from all agencies that fund river restoration activities. That body would ensure that restoration funds are spent efficiently as well as address unmet needs. These projects would yield enormous benefits in the form of ecosystem services, including flood control, protection of infrastructure, and maintenance of water quality. They also would have benefits similar to those of more traditional infrastructure projects. They create jobs in member districts, but if they are carefully chosen and designed, they can also save taxpayer money. By including interagency tracking mechanisms and building in compliance monitoring requirements to each project, Congress can ensure the necessary feedback and accountability to make these projects wise investments. Instead of funds being allocated independently by 40-plus federal programs, WRRA would ensure that project prioritization occurs on watershed scales and is based on criteria that are consistent across the nation.

In addition to the need for better coordination and thus more efficient use of existing funding, current funding falls short of what is needed. The magnitude of the problems and the demands that citizens are making for healthier waters require additional funds for cleaning U.S. rivers and streams. Aging sewer and stormwater infrastructure combined with increased development of the land make it imperative that a combined approach involving better coordination and an increase in funding be a priority. Additional funding will not only make possible more recovery of damaged river ecosystems, but will enable inter- or intraagency mechanisms for tracking projects and allow more pre- and post-project monitoring of their effectiveness. New funding will not be easy to come by in the current budget climate and with increased competition for investments in water quality. Many federal programs that involve river restoration are being cut, not increased. The growing need for upgrading stormwater and sewer infrastructure goes hand in hand with river restoration; one cannot replace the other. Only together will they accomplish the goals of improved water quality, more productive fisheries, and the restoration of other services that rivers provide.

River restoration is a necessity, not a luxury. U.S. citizens depend on the services that healthy streams and rivers provide. People from all walks of life are demanding cleaner, restored waterways. Replacing the services that healthy streams provide with human-made alternatives is extremely expensive, so river restoration is akin to investments such as highways, municipal works, or electric transmission. Congress already commits billions of taxpayer dollars in public infrastructure through the transportation bill or WRDA. It should make similar investments in natural capital.

Much can be accomplished by allocating scarce resources and prioritizing efforts based on sound policies that ensure that the most effective methods are applied and that agreed-on standards are adhered to. Changes in agency policies and practices require overcoming bureaucratic inertia and confronting competing constituencies. Instituting tracking systems and comprehensive studies of project effectiveness requires cooperation among multiple agencies, scientists, environmental groups, and affected industries. But with congressional oversight and wise appropriation of scarce dollars, U.S. rivers and streams can once again flow clear and clean.

Critical Thinking

1. We passed water-protecting legislation decades ago. Why are rivers and streams still being degraded?
2. Why should rivers and streams be restored?

The Rising Sea

Orrin H. Pilkey and Robert S. Young

Orrin H. Pilkey (b. 1934) is the James B. Duke Professor Emeritus of Geology at Duke University. A winner of the Society for Sedimentary Geology's Francis Shepard Medal for excellence in marine geology (1987), the American Geological Institute's award for outstanding contributions to public understanding of geology (1993), and the Priestley Medal, the Dickinson College Award in Memory of Joseph Priestley, for distinguished research in coastal geology and public service in policy formulation and education about America's coastal resources (2003), Pilkey has earned a reputation as one of the most outspoken critics of efforts to manage erosion along America's coastline. Robert S. Young is an associate professor of geology in the Department of Geosciences and Natural Resources Management of Western Carolina University.

In their recent book, *The Rising Sea*, excerpted here, Pilkey and Young argue that sea level rise (largely due to global warming) poses such a threat to coastal lands that coastal settlements and island nations will vanish beneath the waves. Governments should no longer encourage coastal development or finance rebuilding after storm damage. "Relocation of buildings and infrastructure [and even populations] should be a guiding philosophy." As we saw after Hurricane Katrina severely damaged New Orleans, such recommendations are not welcome to those who have invested money and sentiment in coastal zones. But to ignore them is to waste money and effort and even to increase the damage to coastal ecosystems.

Key Concept: efforts to protect coastal beaches and infrastructure in the face of rising seas due to global warming are economically and ecologically counterproductive

Chapter 1: Living on the Edge

A rising sea is not something that may happen in the future. It is already upon us. Planners turned down construction of a large residential development on the Yorke Peninsula, South Australia, because it would be flooded by rising seas. In England, regulators declared that six small villages on the Norfolk Broads northeast of London will need to be abandoned as sea level rises. To avoid the rising sea, the 580 Inupiat Eskimo inhabitants of Shishmaref, Alaska, will likely be moved to the mainland at a cost of several hundred thousand dollars per resident. On barrier islands along the Pacific coast of Colombia where the sea level is rising with particular rapidity (because the land is also sinking), moving buildings and entire villages to higher ground is already a routine matter. In rural Cape Town, South Africa, a "blue line" may be established seaward of which nothing can be built because it would lie within an expected flood zone from sea level rise. Plans to abandon many Pacific atolls are now on the drawing board because they will

soon be flooded by the expanding oceans. And in South Carolina, retreat from the shoreline in response to the sea level rise is now official state policy.

Still, despite strong evidence of global warming and attendant sea level rise, many communities, governments, and developers continue to ignore the inevitability of a continued rise in sea level and the corresponding increase in shoreline erosion. Singapore continues to fill in its bays to create more low-elevation land for development. In a stunning act of developmental hubris, the government of Dubai has constructed spectacular, palm-shaped artificial islands along the Persian Gulf providing space for hundreds of homes, all at low elevation and immediately susceptible to even modest sea level rise. In the United States, the State of Florida seems content to spend billions of dollars in a losing battle to hold the shoreline in place with artificial beaches, breakwaters, and seawalls while high-rise beachfront construction continues apace. And state and federal officials continue to insist that most Mississippi Delta communities can be maintained in their present location despite recent rapid sea level rise augmented by subsidence (sinking) of the land.

Global warming is changing many things: the extent of ice on the surface of the Arctic Ocean, the extent of mountain glaciers, patterns of rainfall and drought around the world, and routes of ocean currents. As the oceans warm, wide swaths of coral reefs, responsible for much of the diversity of marine life, may be degraded as human activities prevent their natural expansion to the north or to the south away from the equator. Shoreline-hugging and biologically important salt marshes and their warm-water equivalent, mangroves, already seriously reduced by the activities of humans, will further degrade as sea level rises. The distribution and the migration pathways of land mammals, birds, and insects will change, and some species will disappear entirely. Mosquitoes will appear in the high Arctic. . . .

Substantial sea level change will play a critical role in humankind's future just as it sometimes has in the past. . . .

It has been established that during a period forty thousand to sixty thousand years ago in which sea level was considerably lower than it is today, Aborigines walked across an exposed land bridge between New Guinea and Australia, and some of these early Australians walked on from Australia to Tasmania. The first Americans may also have taken advantage of a more recent comparatively low sea level, perhaps eleven thousand years ago, to cross the Bering Land Bridge from Asia to North America. Stone Age people at the same time must have crossed back and forth between the British Isles and Europe. Eventually, as we know, the sea rose and covered these ancient access routes, changing the face of the earth and the lives of its people.

Many societies are candidates to be the first in our time to suffer catastrophic impacts from impending sea level changes. Eventually, every nation with a coast will feel the effects of sea level rise. . . . It's a cruel irony that many of these societies, for the most part non-industrial, have played almost no role in the global warming that lies behind so much of current sea level rise; in that sense, they are truly innocent victims of the industrialized world. . . .

The Rising Sea Revisited

Here we give a summary of our most salient conclusions and recommendations and our views on decisions we believe need to be made to begin to address the challenge of sea level rise.

The causes of sea level rise are changing. The relative importance of the various drivers of sea volume change (eustatic sea level change) will likely be different in this century than they were in the twentieth century. According to the IPCC, in the twentieth century, the major contributors were, in descending order of importance,

- Thermal expansion of the oceans
- Mountain glacier melting
- Melting of the Greenland ice sheet

Recent data from a variety of sources suggest that in the twenty-first century, the West Antarctic ice sheet will likely become the major source of meltwater, possibly replacing thermal expansion as the most important cause of eustatic rise, making the main drivers these:

- Melting of the West Antarctic ice sheet
- Melting of the Greenland ice sheet
- Thermal expansion of the oceans
- Mountain glacier melting

In the short term, reduction of carbon dioxide emissions will not halt sea level rise. Many researchers have pointed out that global warming and sea level rise have a certain momentum. Even if we slow the input of CO_2 into the atmosphere, there will likely not be a direct reduction in the rate of sea level rise. The potential collapse of additional ice shelves in Antarctica, for example, is an unpredictable wild card that could raise sea level even as CO_2 falls. We must expect that rapidly rising sea level will be with us for a long time. By no means does this imply that reducing anthropogenic CO_2 should not be a critical long-term goal for our society.

It's not just sea level rise. Storm surge, storm waves, shoreline erosion, groundwater salinization, and infrastructure destruction will force a retreat from the shoreline long before actual inundation occurs. Simple maps showing the areas that will be slowly inundated by a given amount of sea level rise should be viewed with great skepticism. They do not truly consider the inland reach of the changes to be wrought by rising sea level.

Assume a minimal sea level rise of 7 feet (2 m) by 2100 for planning purposes. This is not a prediction; it is a scenario, a recommendation. But a rise of this magnitude (7 ft; 2 m) is a real possibility. Seven feet is a catastrophic sea level rise. At a bare minimum, we recommend using a minimal 3 feet (0.9 m) for a 50- to 100-year planning horizon in communities where the politics would not permit the consideration of more forward-looking coastal management.

Three feet of sea level rise will doom much, if not most, barrier island development. Maintaining the static shoreline required to keep sandy barrier islands in place is economically, environmentally, and oceanographically impossible with a 3-foot (0.9 m) rise in sea level.

Immediately prohibit the construction of high-rise buildings in areas vulnerable to future sea level rise. Decisions concerning community planning and development should be based on minimizing or avoiding altogether the damage from the expanding ocean. This means, first and foremost, no more high-rises near the beachfront. Buildings placed in future hazardous zones should be small and movable or disposable.

Relocation of buildings and infrastructure should be a guiding philosophy. Instead of making major repairs on infrastructure such as bridges, water supply, and sewer and drainage systems, when major maintenance is needed, go the extra mile and place them out of reach of the sea. In our view, no new sewer and water lines should be introduced to zones that will be adversely affected by sea level rise in the next fifty years. Relocation of some beach buildings could be implemented after severe storms or with financial incentives.

Stop government assistance for oceanfront rebuilding. The guarantee of recovery is perhaps the biggest obstacle to a sensible response to sea level rise. The goal in the past has always been to restore conditions to what they were before a storm or flood. In the United States, hurricanes have become urban renewal programs. The replacement houses become larger and larger and even more costly to replace again in the future. The problem is compounded when even people whose rental investment houses were destroyed are considered victims; maybe people who insist on building adjacent to eroding shorelines facing the open ocean should be considered fools rather than victims?

Underlying this problem in the United States is the Stafford Act. Passed by the U.S. Congress in 1988, the act allows the expenditure of money to restore community infrastructure once the president issues a disaster declaration. It is an unquestioning, automatic response to a disaster. No environmental impact statement is required, and money may even be spent on coastal areas where other expenditures of federal money are prohibited. Those who invest in vulnerable coastal areas need to assume responsibility for that decision. If you stay, you pay.

Stop asking only coastal engineers for a solution to coastal erosion. If a coastal community asks a coastal engineer for a solution to a coastal erosion problem, that community gets a coastal engineering response (e.g., build a seawall, build a groin, renourish the beach). Coastal engineers are selling a product. They are not likely to suggest that the community relocate property. This would put them out of business. Beach nourishment, the currently preferred method of fighting coastal erosion, is becoming increasingly expensive. In the future, beaches will need more sand, more frequently. In most cases, the sand resources are not available to fight this battle into the last half of the twenty-first century. In light of this, relocation may soon begin to seem like a more reasonable option, we hope. Coastal communities need to include a broader circle of experts in their quest to seek solutions to coastal erosion and global sea level rise.

Get the Corps off the shore. The U.S. Army Corps of Engineers, more or less by default, is the government agency in charge of much of the planning and the funding for the nation's response to sea level rise. It is an agency ill-suited for the job. It has too long a history of checkered competence, high-cost construction, and inefficiency due in significant part to its close dependence on Congress for pork barrel funding, as many critics have pointed out. Part of the problem is that the engineers' "we can fix it" mentality is the wrong mentality for a sensible approach to changing sea level (the agency's motto is "Essayons," a French term meaning "Let us try"). A fundamental reorganization of the Corps is needed.

Take the reins from local government. The problems created by sea level rise are international and national, not local, in scope. Local governments of towns (understandably) follow the self-interests of coastal property owners and developers, so preservation of buildings is inevitably a very high priority. In the debate over which is more important, buildings or beaches, buildings always win. Many tourist beach communities are controlled by a small number of year-round resident voters who make a living from the coastal environment that is utilized by many thousands of nonvoting people during the summer. In Spain, many of the coastal communities are inhabited primarily by nonvoting foreigners. Sea level rise is a national crisis, an issue of national and international interest that must not be solved simply for the benefit of those unwise enough to build at low elevation near an eroding ocean shoreline. In addition, the resources needed to respond to the sea level rise are far beyond those of the local communities.

Sea level rise will threaten coastal ecosystems. Direct destruction of coastal marshes, mangroves, and coral reefs by human activities is currently having a greater impact on these ecosystems than is sea level rise. If we do not immediately act to better protect these critical ecosystems, the combination of rising sea level and human development will destroy them.

Sea level rise provides an opportunity and a challenge for all. Sea level rise does not have to be a natural catastrophe. It could be seen as an opportunity for society to redesign with nature, to anticipate the changes that will occur in the future and to respond in such a fashion as to maintain a coast that future generations will find both useful and enjoyable. It provides a challenge to scientists, planners, environmentalists, politicians, and other citizens alike to stretch the limits of their imagination to respond with flexibility and with careful foresight to development challenges that our society has not faced before. Opportunities will abound for entrepreneurs with fresh ideas on how to live with a rising sea.

The science tells us that the world's shorelines will look different a hundred years from now. These changes need not end the coastal economy as we know it. But preserving our coastal resources and the businesses that depend on them will require insightful and long-term planning. Beginning an honest assessment of how we may deal with inevitable future sea level rise can help ensure that our coastal communities remain the vibrant places that they are today.

Critical Thinking

1. Which level of government—local, state, or federal—is best placed to govern how local communities prepare for and respond to sea level rise?
2. In the face of rising sea level, should damaged coastal infrastructure such as roads, bridges, and sewers be repaired or moved inland?
3. Just how great is the threat posed by global warming to coastal settlements?

Global Warming and Ozone Depletion

Selection 23
THE INTERGOVERNMENTAL PANEL ON CLIMATE CHANGE, from *Summary for policymakers: Climate Change 2007: The Physical Science Basis: IPCC Forth Assessment Report* (February 2007)

Selection 24
BILL MCKIBBEN, from "The Most Important Number on Earth," *Mother Jones* (November/December 2008)

Learning Outcomes
After reading this unit, you should be able to:

1. Describe how air pollution affects the quality of urban life.
2. Describe at least one simple method of reducing urban air pollution.
3. Explain why hazardous wastes must be properly disposed of.
4. Explain why despite protective legislation rivers and streams are still being degraded.
5. Describe the threat posed by global warming to coastal settlements.
6. Describe necessary responses to the threat posed by rising sea level.

Summary for Policymakers: Climate Change 2007: The Physical Science Basis

The Intergovernmental Panel on Climate Change (IPCC)

The idea that the heat-trapping ability of infrared-absorbing gases in the atmosphere is similar to that of the glass panes in a greenhouse (hence the "greenhouse effect") was first proposed by the French mathematical physicist Jean-Baptiste-Joseph Fourier in 1827. In 1896, the Swedish chemist Svante Arrhenius, who later won the 1903 Nobel prize in chemistry, predicted that if atmospheric carbon dioxide (CO_2) levels doubled due to the burning of fossil fuels, the resulting increase in the average temperature at the Earth's surface would amount to four to six degrees Celsius (seven to ten degrees Fahrenheit).

The Arrhenius prediction about global warming was all but forgotten for more than half a century until direct observations and historical data demonstrated that by 1960, atmospheric CO_2 levels had risen to 315 ppm from the preindustrial level of 280 ppm. Careful measurements since then have shown that the present CO_2 level is approaching 380 ppm, and rising. The Arrhenius prediction that the average temperature on Earth will rise four to six degrees Celsius is likely to come true before the end of the twenty-first century if present fossil fuel use and forest destruction trends continue. Most atmospheric scientists agree that such a warming will be accompanied by changes in the world's weather patterns and a significant increase in sea levels. The data on which these conclusions are based, as well as the conclusions themselves, have been vigorously debated for years.

In 1988, due to concern about the potentially serious disruptive effects that would result from significant, short-term changes in world climate, the United Nations Environment Programme joined with the World Meteorological Organization to establish the Intergovernmental Panel on Climate Change (IPCC) to assess the available scientific, technical, and socioeconomic information regarding greenhouse gas-induced climate change. Thousands of meteorologists and other atmospheric and climate scientists have participated in periodic reviews of the data. The Fourth Assessment Report of the IPCC appeared in 2007. The following selection is taken from the first section of that report, "Summary for Policymakers: Climate Change 2007: The Physical Science Basis." It clearly states that "Warming of the climate system is unequivocal, as is now evident from observations of increases in global average air and ocean temperatures, widespread melting of snow and ice, and rising global average sea level." Later sections of the report (see http://www.ipcc.ch/) make it clear that the impacts on ecosystems and human well-being (especially in developing nations) will be serious and outline the steps that must be taken to prevent, ease, or cope with these impacts. Other reports (see Nicholas Stern, *Stern Review: The Economics of Climate Change*, Executive Summary, October 30, 2006 [http://www.hm-treasury.gov.uk/independent_reviews/stern_review_ economics_climate_change/sternreview_index.cfm]) make it clear that although taking steps now to limit future impacts of global warming would be very expensive, "the benefits of strong, early action considerably outweigh the costs . . . Ignoring climate change will eventually damage economic growth. . . . Tackling climate change is the pro-growth strategy for the longer term, and it can be done in a way that does not cap the aspirations for growth of rich or poor countries. The earlier effective action is taken, the less costly it will be."

Almost all the few remaining critics of the reality of global warming are either employed by or funded by industries and nations that have a financial stake in resisting proposals for significant reductions in the release of greenhouse gases.

Key Concept: the causes and effects of greenhouse gas-induced climate change

Human and Natural Drivers of Climate Change

Global atmospheric concentrations of carbon dioxide, methane and nitrous oxide have increased markedly as a result of human activities since 1750 and now far exceed pre-industrial values determined from ice cores spanning many thousands of years (see Figure 1). The global increases in carbon dioxide concentration are due primarily to fossil fuel use and land-use change, while those of methane and nitrous oxide are primarily due to agriculture.

- Carbon dioxide is the most important anthropogenic greenhouse gas. The global atmospheric concentration of carbon dioxide has increased from a pre-industrial value of about 280 ppm to 379 ppm in 2005. The atmospheric concentration of carbon dioxide in 2005 exceeds by far the natural range over the last 650,000 years (180 to 300 ppm) as determined from ice cores. The annual carbon dioxide concentration growth-rate was larger during the last 10 years (1995–2005 average: 1.9 ppm per year), than it has been since the beginning of continuous direct atmospheric measurements (1960–2005 average: 1.4 ppm per year) although there is year-to-year variability in growth rates.

- The primary source of the increased atmospheric concentration of carbon dioxide since the pre-industrial period results from fossil fuel use, with land use change providing another significant but smaller contribution. Annual fossil carbon dioxide emissions increased from an average of 6.4 [6.0 to 6.8] GtC [giga Ton of Carbon; equivalent to 3.67 $GtCO_2$] (23.5 [22.0 to 25.0] $GtCO_2$) per year in the 1990s, to 7.2 [6.9 to 7.5] GtC (26.4 [25.3 to 27.5] $GtCO_2$) per year in 2000–2005 (2004 and 2005 data are interim estimates). Carbon dioxide emissions associated with land-use change are estimated to be 1.6 [0.5 to 2.7] GtC (5.9 [1.8 to 9.9] $GtCO_2$) per year over the 1990s, although these estimates have a large uncertainty.

- The global atmospheric concentration of methane has increased from a pre-industrial value of about 715 ppb to 1732 ppb in the early 1990s, and is 1774 ppb in 2005. The atmospheric concentration of methane in 2005 exceeds by far the natural range of the last 650,000 years (320 to 790 ppb) as determined from ice cores. Growth rates have declined since the early 1990s, consistent with total emissions (sum of anthropogenic and natural sources) being nearly constant during this period. It is *very likely* that the observed increase in methane concentration is due to anthropogenic activities, predominantly agriculture and fossil fuel use, but relative contributions from different source types are not well determined.

- The global atmospheric nitrous oxide concentration increased from a pre-industrial value of about

CHANGES IN GREENHOUSE GASES FROM ICE-CORE AND MODERN DATA

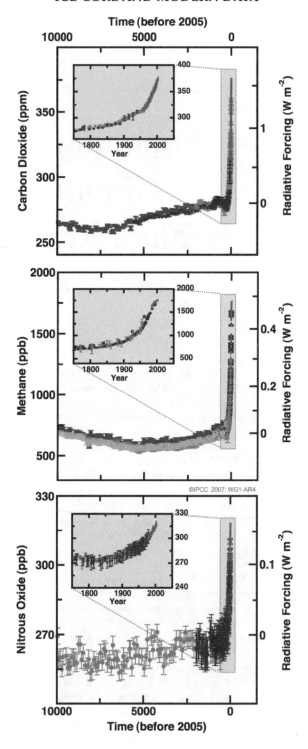

Figure 1 Atmospheric concentrations of carbon dioxide, methane and nitrous oxide over the last 10,000 years (large panels) and since 1750 (inset panels). Measurements are shown from ice cores (symbols with different colours for different studies) and atmospheric samples (red lines). The corresponding radiative forcings are shown on the right hand axes of the large panels.

270 ppb to 319 ppb in 2005. The growth rate has been approximately constant since 1980. More than a third of all nitrous oxide emissions are anthropogenic and are primarily due to agriculture.

The understanding of anthropogenic warming and cooling influences on climate has improved since the Third Assessment Report (TAR), leading to *very high confidence* that the globally averaged net effect of human activities since 1750 has been one of warming, with a radiative forcing of +1.6 [+0.6 to +2.4] W m⁻².

- The combined radiative forcing due to increases in carbon dioxide, methane, and nitrous oxide is +2.30 [+2.07 to +2.53] W m⁻², and its rate of increase during the industrial era is *very likely* to have been unprecedented in more than 10,000 years. The carbon dioxide radiative forcing increased by 20% from 1995 to 2005, the largest change for any decade in at least the last 200 years.
- Anthropogenic contributions to aerosols (primarily sulphate, organic carbon, black carbon, nitrate and dust) together produce a cooling effect, with a total direct radiative forcing of −0.5 [−0.9 to −0.1] W m⁻² and an indirect cloud albedo forcing of −0.7 [−1.8 to −0.3] W m⁻². These forcings are now better understood than at the time of the TAR due to improved *in situ*, satellite and ground-based measurements and more comprehensive modelling, but remain the dominant uncertainty in radiative forcing. Aerosols also influence cloud lifetime and precipitation.
- Significant anthropogenic contributions to radiative forcing come from several other sources. Tropospheric ozone changes due to emissions of ozone-forming chemicals (nitrogen oxides, carbon monoxide, and hydrocarbons) contribute +0.35 [+0.25 to +0.65] W m⁻². The direct radiative forcing due to changes in halocarbons is +0.34 [+0.31 to +0.37] W m⁻². Changes in surface albedo, due to land-cover changes and deposition of black carbon aerosols on snow, exert respective forcings of −0.2 [−0.4 to 0.0] and +0.1 [0.0 to +0.2] W m⁻². Additional terms smaller than ±0.1 W m⁻² are shown in Figure 2.
- Changes in solar irradiance since 1750 are estimated to cause a radiative forcing of +0.12 [+0.06 to +0.30] W m⁻², is less than half the estimate given in the TAR.

Direct Observations of Recent Climate Change

Warming of the climate system is unequivocal, as is now evident from observations of increases in global average air and ocean temperatures, widespread melting of snow and ice, and rising global average sea level (see Figure 2).

- Eleven of the last twelve years (1995–2006) rank among the 12 warmest years in the instrumental

CHANGES IN TEMPERATURE, SEA LEVEL AND NORTHERN HEMISPHERE SNOW COVER

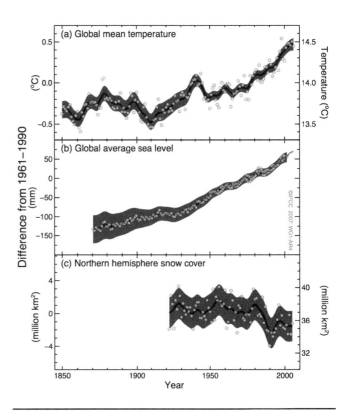

Figure 2 Observed changes in (a) global average surface temperature; (b) global average sea level rise from tide gauge (blue) and satellite (red) data and (c) Northern Hemisphere snow cover for March–April. All changes are relative to corresponding averages for the period 1961–1990. Smoothed curves represent decadal averaged values while circles show yearly values. The shaded areas are the uncertainty intervals estimated from a comprehensive analysis of known uncertainties (a and b) and from the time series (c).

record of global surface temperature (since 1850). The updated 100-year linear trend (1906–2005) of 0.74 [0.56 to 0.92]°C is therefore larger than the corresponding trend for 1901–2000 given in the TAR of 0.6 [0.4 to 0.8]°C. The linear warming trend over the last 50 years (0.13 [0.10 to 0.16]°C per decade) is nearly twice that for the last 100 years. The total temperature increase from 1850–1899 to 2001–2005 is 0.76 [0.57 to 0.95]°C. Urban heat island effects are real but local, and have a negligible influence (less than 0.006°C per decade over land and zero over the oceans) on these values.

- New analyses of balloon-borne and satellite measurements of lower- and mid-tropospheric temperature show warming rates that are similar to those of the surface temperature record and are consistent within their respective uncertainties, largely reconciling a discrepancy noted in the TAR.

- The average atmospheric water vapour content has increased since at least the 1980s over land and ocean as well as in the upper troposphere. The increase is broadly consistent with the extra water vapour that warmer air can hold.

- Observations since 1961 show that the average temperature of the global ocean has increased to depths of at least 3000 m and that the ocean has been absorbing more than 80% of the heat added to the climate system. Such warming causes seawater to expand, contributing to sea level rise (Table 1).

- Mountain glaciers and snow cover have declined on average in both hemispheres. Widespread decreases in glaciers and ice caps have contributed to sea level rise (ice caps do not include contributions from the Greenland and Antarctic ice sheets). (see Table 1.)

- New data since the TAR now show that losses from the ice sheets of Greenland and Antarctica have *very likely* contributed to sea level rise over 1993 to 2003 (Table 1). Flow speed has increased for some Greenland and Antarctic outlet glaciers, which drain ice from the interior of the ice sheets. The corresponding increased ice sheet mass loss has often followed thinning, reduction or loss of ice shelves or loss of floating glacier tongues. Such dynamical ice loss is sufficient to explain most of the Antarctic net mass loss and approximately half of the Greenland net mass loss. The remainder of the ice loss from Greenland has occurred because losses due to melting have exceeded accumulation due to snowfall.

Table 1 Observed rate of sea level rise and estimated contributions from different sources

Source of sea level rise	Rate of sea level rise (mm per year)	
	1961–2003	1993–2003
Thermal expansion	0.42 ± 0.12	1.6 ± 0.5
Glaciers and ice caps	0.50 ± 0.18	0.77 ± 0.22
Greenland ice sheet	0.05 ± 0.12	0.21 ± 0.07
Antarctic ice sheet	0.14 ± 0.41	0.21 ± 0.35
Sum of individual climate contributions to sea level rise	1.1 ± 0.5	2.8 ± 0.7
Observed total sea level rise	1.8 ± 0.5[a]	3.1 ± 0.7[a]
Difference (Observed minus sum of estimated climate contributions)	0.7 ± 0.7	0.3 ± 1.0

[a]Data prior to 1993 are from tide gauges and after 1993 are from satellite altimetry.

- Global average sea level rose at an average rate of 1.8 [1.3 to 2.3] mm per year over 1961 to 2003. The rate was faster over 1993 to 2003, about 3.1 [2.4 to 3.8] mm per year. Whether the faster rate for 1993 to 2003 reflects decadal variability or an increase in the longer-term trend is unclear. There is *high confidence* that the rate of observed sea level rise increased from the 19th to the 20th century. The total 20th century rise is estimated to be 0.17 [0.12 to 0.22] m.

- For 1993–2003, the sum of the climate contributions is consistent within uncertainties with the total sea level rise that is directly observed (see Table 1). These estimates are based on improved satellite and *in-situ* data now available. For the period of 1961 to 2003, the sum of climate contributions is estimated to be smaller than the observed sea level rise. The TAR reported a similar discrepancy for 1910 to 1990.

At continental, regional, and ocean basin scales, numerous long-term changes in climate have been observed. These include changes in Arctic temperatures and ice, widespread changes in precipitation amounts, ocean salinity, wind patterns and aspects of extreme weather including droughts, heavy precipitation, heat waves and the intensity of tropical cyclones.

- Average Arctic temperatures increased at almost twice the global average rate in the past 100 years. Arctic temperatures have high decadal variability, and a warm period was also observed from 1925 to 1945.

- Satellite data since 1978 show that annual average Arctic sea ice extent has shrunk by 2.7 [2.1 to 3.3]% per decade, with larger decreases in summer of 7.4 [5.0 to 9.8]% per decade. These values are consistent with those reported in the TAR.

- Temperatures at the top of the permafrost layer have generally increased since the 1980s in the Arctic (by up to 3°C). The maximum area covered by seasonally frozen ground has decreased by about 7% in the Northern Hemisphere since 1900, with a decrease in spring of up to 15%.

- Long-term trends from 1900 to 2005 have been observed in precipitation amount over many large regions. Significantly increased precipitation has been observed in eastern parts of North and South America, northern Europe and northern and central Asia. Drying has been observed in the Sahel, the Mediterranean, southern Africa and parts of southern Asia. Precipitation is highly variable spatially and temporally, and data are limited in some regions. Long-term trends have not been observed for the other large regions assessed.

- Changes in precipitation and evaporation over the oceans are suggested by freshening of mid and high latitude waters together with increased salinity in low latitude waters.

- Mid-latitude westerly winds have strengthened in both hemispheres since the 1960s.
- More intense and longer droughts have been observed over wider areas since the 1970s, particularly in the tropics and subtropics. Increased drying linked with higher temperatures and decreased precipitation have contributed to changes in drought. Changes in sea surface temperatures (SST), wind patterns, and decreased snowpack and snow cover have also been linked to droughts.
- The frequency of heavy precipitation events has increased over most land areas, consistent with warming and observed increases of atmospheric water vapour.
- Widespread changes in extreme temperatures have been observed over the last 50 years. Cold days, cold nights and frost have become less frequent, while hot days, hot nights, and heat waves have become more frequent (see Table 2).
- There is observational evidence for an increase of intense tropical cyclone activity in the North Atlantic since about 1970, correlated with increases of tropical sea surface temperatures. There are also suggestions of increased intense tropical cyclone activity in some other regions where concerns over data quality are greater. Multi-decadal variability and the quality of the tropical cyclone records prior to routine satellite observations in about 1970 complicate the detection of long-term trends in tropical cyclone activity. There is no clear trend in the annual numbers of tropical cyclones. . . .

Understanding and Attributing Climate Change

Most of the observed increase in globally averaged temperatures since the mid-20th century is *very likely* due to the observed increase in anthropogenic greenhouse gas concentrations. This is an advance since the TAR's conclusion that "most of the observed warming over the last 50 years is *likely* to have been due to the increase in greenhouse gas concentrations". Discernible human influences now extend to other aspects of climate, including ocean warming, continental-average temperatures, temperature extremes and wind patterns.

- It is *likely* that increases in greenhouse gas concentrations alone would have caused more warming than observed because volcanic and anthropogenic aerosols have offset some warming that would otherwise have taken place.
- The observed widespread warming of the atmosphere and ocean, together with ice mass loss, support the conclusion that it is *extremely unlikely* that global climate change of the past fifty years can be explained without external forcing, and *very likely* that it is not due to known natural causes alone.
- Warming of the climate system has been detected in changes of surface and atmospheric temperatures, temperatures in the upper several hundred metres of the ocean and in contributions to sea level rise. Attribution studies have established anthropogenic contributions to all of these changes. The observed pattern of tropospheric warming and stratospheric cooling is *very likely* due to the combined influences of greenhouse gas increases and stratospheric ozone depletion.
- It is *likely* that there has been significant anthropogenic warming over the past 50 years averaged over each continent except Antarctica. The observed patterns of warming, including greater warming over land than over the ocean, and their changes over time, are only simulated by models that include anthropogenic

Table 2   Projected globally averaged surface warming and sea level rise at the end of the 21st century

	Temperature Change (°C at 2090–2099 relative to 1980–1999)[a]		Sea Level Rise (m at 2090–2099 relative to 1980–1999)
Case	Best estimate	*Likely* range	Model-based range excluding future rapid dynamical changes in ice flow
Constant Year 2000 concentrations[b]	0.6	0.3–0.9	NA
B1 scenario	1.8	1.1–2.9	0.18–0.38
A1T scenario	2.4	1.4–3.8	0.20–0.45
B2 scenario	2.4	1.4–3.8	0.20–0.43
A1B scenario	2.8	1.7–4.4	0.21–0.48
A2 scenario	3.4	2.0–5.4	0.23–0.51
A1F1 scenario	4.0	2.4–6.4	0.26–0.59

[a]These estimates are assessed from a hierarchy of models that encompass a simple climate model, several Earth Models of Intermediate Complexity(EMICs), and a large number of Atmosphere-Ocean Global Circulation Models (AOGCMs).
[b]Year 2000 constant composition is derived from AOGCMs only.

forcing. The ability of coupled climate models to simulate the observed temperature evolution on each of six continents provides stronger evidence of human influence on climate than was available in the TAR.

- Difficulties remain in reliably simulating and attributing observed temperature changes at smaller scales. On these scales, natural climate variability is relatively larger making it harder to distinguish changes expected due to external forcings. Uncertainties in local forcings and feedbacks also make it difficult to estimate the contribution of greenhouse gas increases to observed small-scale temperature changes.
- Anthropogenic forcing is *likely* to have contributed to changes in wind patterns, affecting extra-tropical storm tracks and temperature patterns in both hemispheres. However, the observed changes in the Northern Hemisphere circulation are larger than simulated in response to 20th century forcing change.
- Temperatures of the most extreme hot nights, cold nights and cold days are *likely* to have increased due to anthropogenic forcing. It is *more likely than not* that anthropogenic forcing has increased the risk of heat waves.

Analysis of climate models together with constraints from observations enables an assessed *likely* range to be given for climate sensitivity for the first time and provides increased confidence in the understanding of the climate system response to radiative forcing.

- The equilibrium climate sensitivity is a measure of the climate system response to sustained radiative forcing. It is not a projection but is defined as the global average surface warming following a doubling of carbon dioxide concentrations. It is *likely* to be in the range 2 to 4.5°C with a best estimate of about 3°C, and is *very unlikely* to be less than 1.5°C. Values substantially higher than 4.5°C cannot be excluded, but agreement of models with observations is not as good for those values. Water vapour changes represent the largest feedback affecting climate sensitivity and are now better understood than in the TAR. Cloud feedbacks remain the largest source of uncertainty.
- It is *very unlikely* that climate changes of at least the seven centuries prior to 1950 were due to variability generated within the climate system alone. A significant fraction of the reconstructed Northern Hemisphere interdecadal temperature variability over those centuries is *very likely* attributable to volcanic eruptions and changes in solar irradiance, and it is *likely* that anthropogenic forcing contributed to the early 20th century warming evident in these records.

Projections of Future Changes in Climate

For the next two decades a warming of about 0.2°C per decade is projected for a range of SRES emission scenarios. Even if the concentrations of all greenhouse gases and aerosols had been kept constant at year 2000 levels, a further warming of about 0.1°C per decade would be expected.

- Since IPCC's first report in 1990, assessed projections have suggested global averaged temperature increases between about 0.15 and 0.3°C per decade for 1990 to 2005. This can now be compared with observed values of about 0.2°C per decade, strengthening confidence in near-term projections.
- Model experiments show that even if all radiative forcing agents are held constant at year 2000 levels, a further warming trend would occur in the next two decades at a rate of about 0.1°C per decade, due mainly to the slow response of the oceans. About twice as much warming (0.2°C per decade) would be expected if emissions are within the range of the SRES scenarios. Best-estimate projections from models indicate that decadal-average warming over each inhabited continent by 2030 is insensitive to the choice among SRES scenarios and is *very likely* to be at least twice as large as the corresponding model-estimated natural variability during the 20th century.

Continued greenhouse gas emissions at or above current rates would cause further warming and induce many changes in the global climate system during the 21st century that would *very likely* be larger than those observed during the 20th century.

- Advances in climate change modelling now enable best estimates and *likely* assessed uncertainty ranges to be given for projected warming for different emission scenarios. Results for different emission scenarios are provided explicitly in this report to avoid loss of this policy-relevant information. Projected globally-averaged surface warmings for the end of the 21st century (2090–2099) relative to 1980–1999 are shown in Table 2. These illustrate the differences between lower to higher SRES emission scenarios and the projected warming uncertainty associated with these scenarios.
- Best estimates and *likely* ranges for globally average surface air warming for six SRES emissions marker scenarios are given in this assessment and are shown in Table 2. For example, the best estimate for the low scenario (B1) is 1.8°C (*likely* range is 1.1°C to 2.9°C), and the best estimate for the high scenario (A1FI) is 4.0°C (*likely* range is 2.4°C to 6.4°C). Although these projections are broadly consistent with the span quoted in the TAR (1.4 to 5.8°C), they are not directly comparable (See Figure 3). The AR4 is more advanced as it provides best estimates and an assessed likelihood range for each of the marker scenarios. The new assessment of the *likely* ranges now relies on a larger number of climate models of increasing complexity and realism, as well as new information regarding the

MULTI-MODEL AVERAGES AND ASSESSED RANGES FOR SURFACE WARMING

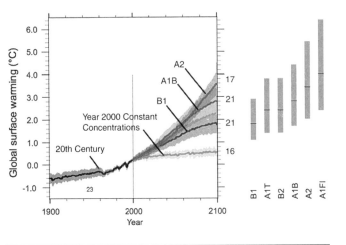

Figure 3 Solid lines are multi-model global averages of surface warming (relative to 1980–99) for the scenarios A2, A1B and B1, shown as continuations of the 20th century simulations. Shading denotes the plus/minus one standard deviation range of individual model annual averages. The orange line is for the experiment where concentrations were held constant at year 2000 values. The gray bars at right indicate the best estimate (solid line within each bar) and the *likely* range assessed for the six SRES marker scenarios. The assessment of the best estimate and *likely* ranges in the gray bars includes the AOGCMs in the left part of the figure, as well as results from a hierarchy of independent models and observational constraints.

nature of feedbacks from the carbon cycle and constraints on climate response from observations.

- Warming tends to reduce land and ocean uptake of atmospheric carbon dioxide, increasing the fraction of anthropogenic emissions that remains in the atmosphere. For the A2 scenario, for example, the climate-carbon cycle feedback increases the corresponding global average warming at 2100 by more than 1°C. Assessed upper ranges for temperature projections are larger than in the TAR (see Table 2) mainly because the broader range of models now available suggests stronger climate-carbon cycle feedbacks.
- Model-based projections of global average sea level rise at the end of the 21st century (2090–2099) are shown in Table 2. For each scenario, the midpoint of the range in Table 2 is within 10% of the TAR model average for 2090–2099. The ranges are narrower than in the TAR mainly because of improved information about some uncertainties in the projected contributions.
- Models used to date do not include uncertainties in climate-carbon cycle feedback nor do they include the full effects of changes in ice sheet flow,

because a basis in published literature is lacking. The projections include a contribution due to increased ice flow from Greenland and Antarctica at the rates observed for 1993–2003, but these flow rates could increase or decrease in the future. For example, if this contribution were to grow linearly with global average temperature change, the upper ranges of sea level rise for SRES scenarios shown in Table 2 would increase by 0.1 m to 0.2 m. Larger values cannot be excluded, but understanding of these effects is too limited to assess their likelihood or provide a best estimate or an upper bound for sea level rise.

- Increasing atmospheric carbon dioxide concentrations lead to increasing acidification of the ocean. Projections based on SRES scenarios give reductions in average global surface ocean pH of between 0.14 and 0.35 units over the 21st century, adding to the present decrease of 0.1 units since pre-industrial times.

There is now higher confidence in projected patterns of warming and other regional-scale features, including changes in wind patterns, precipitation, and some aspects of extremes and of ice.

- Projected warming in the 21st century shows scenario-independent geographical patterns similar to those observed over the past several decades. Warming is expected to be greatest over land and at most high northern latitudes, and least over the Southern Ocean and parts of the North Atlantic ocean.
- Snow cover is projected to contract. Widespread increases in thaw depth are projected over most permafrost regions.
- Sea ice is projected to shrink in both the Arctic and Antarctic under all SRES scenarios. In some projections, Arctic late-summer sea ice disappears almost entirely by the latter part of the 21st century.
- It is *very likely* that hot extremes, heat waves, and heavy precipitation events will continue to become more frequent.
- Based on a range of models, it is *likely* that future tropical cyclones (typhoons and hurricanes) will become more intense, with larger peak wind speeds and more heavy precipitation associated with ongoing increases of tropical SSTs. There is less confidence in projections of a global decrease in numbers of tropical cyclones. The apparent increase in the proportion of very intense storms since 1970 in some regions is much larger than simulated by current models for that period.
- Extra-tropical storm tracks are projected to move poleward, with consequent changes in wind, precipitation, and temperature patterns, continuing the broad pattern of observed trends over the last half-century.
- Since the TAR there is an improving understanding of projected patterns of precipitation. Increases in the amount of precipitation are *very likely* in

---

THE EMISSION SCENARIOS OF THE IPCC SPECIAL REPORT
ON EMISSION SCENARIOS (SRES)

A1. The A1 storyline and scenario family describes a future world of very rapid economic growth, global population that peaks in mid-century and declines thereafter, and the rapid introduction of new and more efficient technologies. Major underlying themes are convergence among regions, capacity building and increased cultural and social interactions, with a substantial reduction in regional differences in per capita income. The A1 scenario family develops into three groups that describe alternative directions of technological change in the energy system. The three A1 groups are distinguished by their technological emphasis: fossil intensive (A1FI), non-fossil energy sources (A1T), or a balance across all sources (A1B) (where balanced is defined as not relying too heavily on one particular energy source, on the assumption that similar improvement rates apply to all energy supply and end use technologies).

A2. The A2 storyline and scenario family describes a very heterogeneous world. The underlying theme is self reliance and preservation of local identities. Fertility patterns across regions converge very slowly, which results in continuously increasing population. Economic development is primarily regionally oriented and per capita economic growth and technological change more fragmented and slower than other storylines.

B1. The B1 storyline and scenario family describes a convergent world with the same global population, that peaks in mid-century and declines thereafter, as in the A1 storyline, but with rapid change in economic structures toward a service and information economy, with reductions in material intensity and the introduction of clean and resource efficient technologies. The emphasis is on global solutions to economic, social and environmental sustainability, including improved equity, but without additional climate initiatives.

B2. The B2 storyline and scenario family describes a world in which the emphasis is on local solutions to economic, social and environmental sustainability. It is a world with continuously increasing global population, at a rate lower than A2, intermediate levels of economic development, and less rapid and more diverse technological change than in the B1 and A1 storylines. While the scenario is also oriented towards environmental protection and social equity, it focuses on local and regional levels.

An illustrative scenario was chosen for each of the six scenario groups A1B, A1FI, A1T, A2, B1 and B2. All should be considered equally sound.

The SRES scenarios do not include additional climate initiatives, which means that no scenarios are included that explicitly assume implementation of the United Nations Framework Convention on Climate Change or the emissions targets of the Kyoto Protocol.

---

high-latitudes, while decreases are *likely* in most subtropical land regions (by as much as about 20%), continuing observed patterns in recent trends.

- Based on current model simulations, it is *very likely* that the meridional overturning circulation (MOC) of the Atlantic Ocean will slow down during the 21st century. Temperatures in the Atlantic region are projected to increase despite such changes due to the much larger warming associated with projected increases of greenhouse gases. It is *very unlikely* that the MOC will undergo a large abrupt transition during the 21st century. Longer-term changes in the MOC cannot be assessed with confidence.

Anthropogenic warming and sea level rise would continue for centuries due to the timescales associated with climate processes and feedbacks, even if greenhouse gas concentrations were to be stabilized.

- Climate carbon cycle coupling is expected to add carbon dioxide to the atmosphere as the climate system warms, but the magnitude of this feedback is uncertain. This increases the uncertainty in the trajectory of carbon dioxide emissions required to achieve a particular stabilisation level of atmospheric carbon dioxide concentration. Based on

current understanding of climate carbon cycle feedback, model studies suggest that to stabilise at 450 ppm carbon dioxide, could require that cumulative emissions over the 21st century be reduced from an average of approximately 670 [630 to 710] GtC (2460 [2310 to 2600] $GtCO_2$) to approximately 490 [375 to 600] GtC (1800 [1370 to 2200] $GtCO_2$). Similarly, to stabilise at 1000 ppm this feedback could require that cumulative emissions be reduced from a model average of approximately 1415 [1340 to 1490] GtC (5190 [4910 to 5460] $GtCO_2$) to approximately 1100 [980 to 1250] GtC (4030 [3590 to 4580] $GtCO_2$).

- If radiative forcing were to be stabilized in 2100 at B1 or A1B levels a further increase in global average temperature of about 0.5°C would still be expected, mostly by 2200.

- If radiative forcing were to be stabilized in 2100 at A1B levels, thermal expansion alone would lead to 0.3 to 0.8 m of sea level rise by 2300 (relative to 1980–1999). Thermal expansion would continue for many centuries, due to the time required to transport heat into the deep ocean.

- Contraction of the Greenland ice sheet is projected to continue to contribute to sea level rise after 2100. Current models suggest ice mass losses increase with temperature more rapidly than gains due to precipitation and that the surface mass balance becomes negative at a global average warming (relative to pre-industrial values) in excess of 1.9 to 4.6°C. If a negative surface mass balance were sustained for millennia, that would lead to virtually complete elimination of the Greenland ice sheet and a resulting contribution to sea level rise of about 7 m. The corresponding future temperatures in Greenland are comparable to those inferred for the last interglacial period 125,000 years ago, when paleoclimatic information suggests reductions of polar land ice extent and 4 to 6 m of sea level rise.

- Dynamical processes related to ice flow not included in current models but suggested by recent observations could increase the vulnerability of the ice sheets to warming, increasing future sea level rise. Understanding of these processes is limited and there is no consensus on their magnitude.

- Current global model studies project that the Antarctic ice sheet will remain too cold for widespread surface melting and is expected to gain in mass due to increased snowfall. However, net loss of ice mass could occur if dynamical ice discharge dominates the ice sheet mass balance.

- Both past and future anthropogenic carbon dioxide emissions will continue to contribute to warming and sea level rise for more than a millennium, due to the timescales required for removal of this gas from the atmosphere.

Critical Thinking

1. Given that projections of global climate change are not certain, should we act now? If not, how long should we wait?
2. What consequences can we expect from global warming?
3. Are the steps we are urged to take to deal with global warming likely to be beneficial in other ways? In what ways?

The Most Important Number on Earth
We're way past the carbon tipping point. It's time to freak out—and get to work.

Bill McKibben

It is clear that the more carbon dioxide and other greenhouse gases are in the atmosphere, the greater will be the warming effect on climate and hence the greater will be sea level rise and other effects. The big question is how high a level of carbon dioxide can we—human society as well as the Earth's ecosystems—tolerate before the effects become catastrophic. Before the industrial revolution, the atmosphere held about 280 ppm of carbon dioxide. Today the level is about 380 ppm, and it is rising. Some people think we have already passed the "tipping point" and catastrophic effects have become inevitable.

Is it too late to prevent such effects? Can we stop burning fossil fuels quickly enough to bring the carbon dioxide back to a safe level? If we can, what is that safe level? Bill McKibben is an environmentalist and writer whose first book, *The End of Nature* (Random House, 1989), is regarded as the first book for a general audience about climate change. In 2008, he took the conclusion by climate scientist James Hansen and his colleagues that the maximum safe level of carbon dioxide is 350 ppm and founded 350.org, an international campaign to increase awareness and stimulate action to achieve that level.

Key Concept: the need to reduce atmospheric carbon dioxide levels

Sooner or later, you have to draw a line. We've spent the last 20 years in the opening scenes of what historians will one day call the Global Warming Era—the preamble to the biggest drama that humans have ever staged, the overture that hints at the themes that will follow for centuries to come. But none of the notes have resolved, none of the story lines yet come into clear view. And that's largely because until recently we didn't know quite where we were. From the moment in 1988 when a Nasa scientist named James Hansen told Congress that burning coal and gas and oil was warming the earth, we've struggled to absorb this one truth: The central fact of our economic lives (the ubiquitous fossil fuel that developed the developed world) is wrecking the central fact of our physical lives (the stable climate and sea level on which civilization rests). For a while, and much longer in the US than elsewhere, we battled over whether this was true. But warm year succeeded warm year and that fight began to subside. Instead, the real question became, is this a future peril, the kind of thing you take out a reasonably priced insurance policy to guard against?

Or is it the oh-my-lord crisis you drop everything else to deal with? Will Hitler be happy with the Sudetenland, or is the world going to spend every cent it has, not to mention tens of millions of lives, fighting him off? Trouble, or TROUBLE? These last 12 months, we've found out.

It was September 2007 that the tide began to turn. Every summer Arctic sea ice melts, and every fall it refreezes. The amount of open water has been steadily increasing for three decades, a percent or two every year—it's been going at about the pace that the hairline recedes on a middle-aged man. It was worrisome, and scientists said all the summer ice could be gone by 2070 or so, which is an eyeblink in geologic time but an eternity in politician time. In late summer of last year, though, the melt turned into a rout—it was like those stories of people whose hair turns gray overnight. An area the size of Colorado was disappearing every week; the Northwest Passage was staying wide open all September, for the first time in history. Before long the Arctic night mercifully descended and the ice began to refreeze, but scientists were using words like "astounding." They were recalculating—by one Nasa scientist's estimate the summer Arctic might now be free of ice by 2012. Which in politician years is "beginning of my second term."

The key phrase, really, was "tipping point." As in "I'd say we are reaching a tipping point or are past it for the ice. This is a strong indication that there is an amplifying mechanism here." That's Pål Prestrud of the Center for International Climate and Environmental Research-Oslo. Or this, from Mark Serreze, of the National Snow and Ice Data Center at the University of Colorado: "When the ice thins to a vulnerable state, the bottom will drop out. . . I think there is some evidence that we may have reached that tipping point, and the impacts will not be confined to the Arctic region."

"Tipping point" is not, in this context, an idle buzzword. It means that the physical world is taking over the process that humans began. We poured carbon into the atmosphere, trapping excess heat; that excess heat began to melt ice. When that ice was melted, there was less white up north to reflect the sun's rays back out to space, and more blue ocean to absorb them. Events began to feed upon themselves. And in the course of the last year, we've seen the same thing happening in other systems. In April, the National Oceanic and Atmospheric Administration released a report showing that 2007 had seen a sudden and dramatic surge in the amount of methane, another heat-trapping gas, in the atmosphere. Apparently, one reason is that when we burned all that fossil fuel and began raising the temperature, we also started melting the permafrost—melting eight times more of it in some places over two decades than had thawed for the previous 1,000 years. And as that frozen soil thaws, it releases methane; enough of it now bubbles out to make "hot spots" in lakes and ponds that don't freeze during the deepest part of the Siberian winter. The more methane, the more heat, the more methane. Wash, rinse, repeat.

The final piece of the puzzle came early this year, and again from James Hansen. Twenty years after his crucial testimony, he published a paper with several coauthors called "Target Atmospheric CO_2" (*.pdf*). It put, finally, a number on the table—indeed it did so in the boldest of terms. "If humanity wishes to preserve a planet similar to that on which civilization developed and to which life on Earth is adapted," it said, "paleoclimate evidence and ongoing climate change suggest that CO_2 will need to be reduced from its current 385 ppm to at most 350 ppm."

Get that? Let me break it down for you. For most of the period we call human civilization, the amount of carbon dioxide in the atmosphere hovered at about 275 parts per million. Let's call that the Genesis number, or depending on your icons, the Buddha number, the Confucius number, the Shakespeare number. Then, in the late 18th century, we started burning fossil fuel in appreciable quantities, and that number started to rise. The first time we actually measured it, in the late 1950s, it was already about 315. Now it's at 385, and growing by more than 2 parts per million annually.

And it turns out that that's too high. We never had a number before, so we never knew whether we'd crossed a red line. We half guessed and half hoped that the danger zone might be 450 or 550 parts per million—those were still a little ways in the distance. Therefore we could get away with thinking like the young Augustine: "Lord, make me chaste, but not yet." Not anymore. We have been told by science that we're already over the line.

And so we're now in the land of tipping points. We know that we've passed some of them—Arctic sea ice is melting, and so is the permafrost that guards those carbon stores. But the logic of Hansen's paper was clear. Above 350, we are at constant risk of crossing other, even worse, thresholds, the ones that govern the reliability of monsoons, the availability of water from alpine glaciers, the acidification of the ocean, and, perhaps most spectacularly, the very level of the seas. It is at least conceivable that instead of a slow, steady rise in the height of the oceans, we could see rapid melt in Greenland and the West Antarctic, where much of the world's frozen water resides. We can't rule out, warns Hansen, a sea level rise of up to 20 feet this century. Plug that into Google Earth and watch waterfront developments turn into high-priced reefs. We can't rule out, in other words, the collapse of human society as we've known it. "If humanity wishes to preserve a planet similar to that on which civilization developed and to which life on Earth is adapted. . ." We should add the phrase to the oath of office for every politico on the third planet.

So what does this mean? If you took 350 to be the most important number on the planet, what would it imply?

In essence, it means that we've got to transform the world's economy far more quickly than we'd hoped. Almost everyone knows that this transformation is coming—that by century's end we won't be relying on fossil fuel, both because the oil will have run out and because the environmental damage will be intense. But the question is how quickly. The kind of change envisioned before last year was still a little leisurely—maybe the developed world cutting its carbon emissions 15 or 20 percent by 2020. That's far more than the Bush administration or its energy-industry cronies would go for, of course—at ExxonMobil's annual meeting last spring, ceo Rex Tillerson said he envisioned a world that still used fossil fuel for two-thirds of its power in 2030. A world where change came slowly enough that everyone could make every last penny off their sunk investments in coal mines and oil platforms. And a world where politicians didn't need to raise the price of carbon steeply, and hence didn't need to arouse voters.

But the 350 world looks different. We're not worried we might have a weight problem. We've been to the doctor and the doctor has said, "Your cholesterol is too high. Scaring me. You're in the danger zone. You need to change your diet and then you need to pray that you get back down where you're supposed to be before the stroke that's coming at you." When that happens, you clean the cheese out of the refrigerator and go cold turkey.

In energy terms, that would look like this:

1 No more new coal plants, because although the world still has immense amounts of coal, it's immensely dirty. And the people who tell you about clean coal are blowing smoke—literally.

2 A cap on the amount of carbon the country can produce—which, in essence, is a tax. America would say, just as it does now with sulfur from coal plants, "We're only going to release so much carbon every year." CO_2 would stop being free; in fact, it would become expensive. In order to simplify the process, the upstream producer who mines, imports, or sells the fossil fuel would get the tab. ExxonMobil would have to pay dearly for a permit to release x amount of carbon, a cost it would pass on to consumers. Then those consumers would use less, and markets would go to work figuring out all the possible ways to cut demand and boost renewables.

3 An international agreement, including China and India, to do the same thing around the world.

Now, these are three of the hardest tasks we've even thought about since we took on Hitler. They go to the very heart of the way our economy operates: We get most of our electricity from fossil fuels, any increase in the price of energy affects every single part of the economy, and China and India are pulling people out of poverty largely by burning cheap coal. If you're a person who uses a lot of fossil fuel, i.e. an American, then they're unappealing. If you're a person who would like to use even a little energy, i.e. almost anyone in the developing world, then they're maddening. And yet they are what the physics and chemistry of the situation dictate. So the question becomes, how to make them happen?

The logic imposed by 350 is fairly straightforward. In order to keep Americans from rebelling, we need to take the money we're charging ExxonMobil for those pollution permits and return it to the taxpayers—everyone needs to get a check every month to, in essence, buy us all off. To help make us whole for the price rises that will inevitably come, the price rises that will do the work of wringing fossil fuel out of the economy. ExxonMobil would pay, then we'd pay—but we'd get some of the money back in the mail. We've got to make the switch so fast that it's going to be brutally expensive—think $10 gas—and our democracy will never support it for long without that monthly check.

But we can't give ourselves back all the money. Because some of it is needed to make the rest of the world whole—to build windmills for the Indians so they won't use the same cheap coal that we used for 200 years in order to get rich. That is, we're going to need a Marshall Plan for carbon—with the same mix of idealism and self-interest that motivated the Marshall Plan in Hitler's wake.

We also need serious investment in infrastructure, both technological and human. For instance, concepts like concentrated solar power—those big mirror arrays in the desert—have gained real momentum in the last 18 months. Former Clinton administration energy analyst Joseph Romm recently calculated that such arrays could provide America with all of its electricity from a 92-square-mile grid in the Southwest desert—but only if promoted via loan guarantees for the entrepreneurs who build them and a new generation of transcontinental transmission lines. Meanwhile, demand is skyrocketing for small rooftop solar panels, but increasingly there's a shortage of trained installers, which means our community colleges need money to start training them. No matter what the price of energy, homes aren't going to insulate themselves—this is the great opening for a green-jobs revolution. (See "The Truth About Green Jobs.")

You'll note here I'm talking more about what we should do in the US House (and Senate) in the next year or two than which bulbs you should be changing in *your* house. diy conservation makes great practical sense, but we won't save the planet that way. One by one, trying to do the right thing, we add up to . . . not nearly enough. You cannot make the math work that way—there are too many sockets and too many tailpipes and most of all too much inertia for voluntary action to do the trick. It didn't work when President Bush made voluntary reduction by corporations his global warming "policy," and it won't work fast enough with individuals either.

Which is not to say that life at home doesn't need to change. It does—and it will, once we've taken the political step of making the price of carbon reflect the damage it does to the environment. Look at what happened this past year when the price of gas finally rose far enough to get our attention. We began riding trains and buses in record numbers. Total miles driven fell, sharply, for the first time since we started keeping records in 1942. We groused and moaned and we started to change. General Motors decided to sell its Hummer factory.

If we get that check every month to cover some of the damage, it will help attenuate the very real heat-or-eat dilemma that will grip many people this coming winter, but the incentive to change will still be there. Buses and bikes. Smaller homes that are easier to heat. Solar panels, bought on the installment plan with loans paid off from the power generated on your roof. Local food (and lots more local farmers). Vacations in the neighborhood—no more jetting off for the weekend.

You can see every one of these trends in embryo already, driven by the run-up in energy prices that we've seen so far. The quick contraction of the airline industry. The collapse in home values in the distant suburbs, while homes along the commuter rail lines fare better. Again the question is all about pace—what will make them happen fast enough, across a wide enough swath of the planet. Al Gore set the example with his call for a 10-year conversion to noncarbon electricity. It's at the outer edge of doable, and the outer edge is where we need to be. We'll have plug-in hybrid electric vehicles on sale by 2010. The question is, can we have nothing else on sale by 2020? We built more than half of the interstate highway system in a decade. Would rebuilding our rail networks to a European standard be all that much harder? Can we get the price of energy up quickly enough to get markets on the task of finding a low-carbon way of life that works? And by works, I mean reverses the flow of carbon into the atmosphere. Because physics and chemistry won't reward good intentions. Methane is seriously uninterested in compromise. Permafrost, notoriously, refuses to bargain. Even the absolute political power represented by King Canute couldn't hold back the rising seas. Those forces will only pay attention if we can scramble back below 350.

Forcing that pace requires a new kind of politics. It requires forging a consensus that this toughest of all changes must happen. The consensus must be broad, it must come quickly, and it must encompass the whole earth—they don't call it global warming for nothing.

The list of things on which we've achieved a broad and deep global consensus is pretty much limited to . . . Coke Is It. And that took billions of dollars and several decades, and it involved inducing people to drink sugar water. The odds against a strong global movement about anything tougher than that are low, with language barriers, religious barriers, cultural barriers. And we start from such incredibly different places—Americans use 12 times the energy of sub-Saharan Africans.

And yet we do have this one tool that at least offers the possibility, a tool that wasn't fully there even a few years ago. The Internet—and its attendant technologies, like cell phones and texting—does link up most of the known world at this point. You can get pretty far back of beyond in most of the world, and someone in that village has a mobile.

And we have a number—350. The most important number on earth. If the Internet has a cosmic purpose, this could be it—to take that number and spread it everywhere on the planet, so that everyone, even if they knew little else about climate change, understood that it represented a kind of safety, a bulwark against the monsoon turning erratic, the sea rising over their fields, the mosquito spreading up their mountain.

I'm part of a group of people calling ourselves 350.org. Our goal is simple—to try to get people everywhere to spread that number. We've started finding musicians and artists, athletes and video makers, and most of all activists, the kinds of people who are working to save watersheds or babies, or to educate girls or to block dams, or any of the other thousand lovely things that won't happen if we allow the basic physical stability of the planet to come unglued. We need a lot of noise, and we need it fast, in the scant months—14 now—before the world meets in Copenhagen next December to draw up a new climate treaty. Because one clear implication of 350 is that that treaty is our last real chance to get it right. If we don't, then all we'll be dealing with is the consequences. Once the ocean really starts to rise, dike building is pretty much the only project.

It's not clear if a vocal world citizenry will be enough to beat inertia and vested interest. If 350 emerges as the clear bar for success or failure, then the odds of the international community taking effective action increase, though the odds are still long. Still, these are the lines it is our turn to speak. To be human in 2008 is to rise in defense of the planet we have known and the civilization it has spawned.

Critical Thinking

1. In what sense is fossil fuel dependence the central fact of our economic lives?
2. In what ways does fossil fuel dependence threaten our physical lives?
3. Can global warming be properly addressed without getting ordinary people involved?

Part 4: Human Health and the Environment

Food

Selection 25
LESTER BROWN, from "Food Scarcity: An Environmental Wakeup Call," *The Futurist* (January/February 1998)

Selection 26
N. V. FEDOROFF, ET AL., from "Radically Rethinking Agriculture for the 21st Century," *Science* (February 12, 2010)

Learning Outcomes
After reading this unit, you should be able to:

1. Explain how food shortages may threaten global political stability.
2. Describe the major threats to the global food supply.
3. Describe measures that could improve global food security.
4. Describe the advantages of organic farming.

Could Food Shortages Bring Down Civilization?[1]

Lester R. Brown

Lester Brown started his career as a tomato farmer in southern New Jersey. In 1974, he founded the World Watch Institute in Washington, D.C. In 2001, he founded the Earth Policy Institute, which he currently serves as president and senior researcher. The *Washington Post* has called Brown "one of the world's most influential thinkers." He has authored or coauthored many highly influential books, among which are *Outgrowing the Earth: The Food Security Challenge in an Age of Falling Water Tables and Rising Temperatures* (Earthscan, 2005), *Plan B 4.0: Mobilizing to Save Civilization* (W. W. Norton, 2009), and *World on the Edge: How to Prevent Environmental and Economic Collapse* (W. W. Norton, 2011).

In this selection, Brown warns that demand for food is growing faster than the supply, and the ability to maintain supply at current levels, much less increase it to meet demand, is threatened by shortages of fresh water, loss of topsoil, and global warming. The likely results include malnourished populations, rising prices that limit access to food by the poor, and political conflict (even including widespread war). However, the worst results can be prevented by cutting carbon emissions, stabilizing population, eradicating poverty, and restoring forests, soils, and aquifers (a program Brown calls "Plan B").

Key Concept: the effect of environmental degradation and population growth on food supply, and hence on political stability

One of the toughest things for people to do is to anticipate sudden change. Typically we project the future by extrapolating from trends in the past. Much of the time this approach works well. But sometimes it fails spectacularly, and people are simply blindsided by events such as today's economic crisis.

For most of us, the idea that civilization itself could disintegrate probably seems preposterous. Who would not find it hard to think seriously about such a complete departure from what we expect of ordinary life? What evidence could make us heed a warning so dire— and how would we go about responding to it? We are so inured to a long list of highly unlikely catastrophes that we are virtually programmed to dismiss them all with a wave of the hand: Sure, our civilization might devolve into chaos—and Earth might collide with an asteroid, too!

For many years I have studied global agricultural, population, environmental, and economic trends and their interactions. The combined effects of those trends and the political tensions they generate point to the breakdown of governments and societies. Yet I, too, have resisted the idea that food shortages could bring down not only individual governments but also our global civilization.

I can no longer ignore that risk. Our continuing failure to deal with the environmental declines that are undermining the world food economy—most important, falling water tables, eroding soils, and rising temperatures— forces me to conclude that such a collapse is possible.

The Problem of Failed States

Even a cursory look at the vital signs of our current world order lends unwelcome support to my conclusion. And those of us in the environmental field are well into our third decade of charting trends of environmental decline without seeing any significant effort to reverse a single one.

In six of the past nine years world grain production has fallen short of consumption, forcing a steady drawdown in stocks. When the 2008 harvest began, world carryover stocks of grain (the amount in the bin when the new harvest begins) were at 62 days of consumption, a near record low. In response, world grain prices in the spring and summer of last year climbed to the highest level ever.

As demand for food rises faster than supplies are growing, the resulting food-price inflation puts severe stress on the governments of countries already teetering on the edge of chaos. Unable to buy grain or grow their own, hungry people take to the streets. Indeed, even before the steep climb in grain prices in 2008, the number of failing states was expanding. Many of their problems stem from a failure to slow the growth of their populations. But if the food situation continues to deteriorate, entire nations will break down at an ever increasing rate. We have entered a new era in geopolitics. In the 20th century the main threat to international security was superpower conflict; today it is failing states. It is not the concentration of power but its absence that puts us at risk.

States fail when national governments can no longer provide personal security, food security, and basic social services such as education and health care. They often lose control of part or all of their territory. When governments lose their monopoly on power, law and order begin to disintegrate. After a point, countries can become so dangerous that food relief workers are no longer safe and their programs are halted; in Somalia and Afghanistan, deteriorating conditions have already put such programs in jeopardy.

Failing states are of international concern because they are a source of terrorists, drugs, weapons, and refugees, threatening political stability everywhere. Somalia, number one on the 2008 list of failing states, has become a base for piracy. Iraq, number five, is a hotbed for terrorist training. Afghanistan, number seven, is the world's leading supplier of heroin. Following the massive genocide of 1994 in Rwanda, refugees from that troubled state, thousands of armed soldiers among them, helped to destabilize neighboring Democratic Republic of the Congo (number six).

Our global civilization depends on a functioning network of politically healthy nation-states to control the spread of infectious disease, to manage the international monetary system, to control international terrorism and to reach scores of other common goals. If the system for controlling infectious diseases—such as polio, SARS, or avian flu—breaks down, humanity will be in trouble. Once states fail, no one assumes responsibility for their debt to outside lenders. If enough states disintegrate, their fall will threaten the stability of global civilization itself.

A New Kind of Food Shortage

The surge in world grain prices in 2007 and 2008—and the threat they pose to food security—has a different, more troubling quality than the increases of the past.

During the second half of the 20th century, grain prices rose dramatically several times. In 1972, for instance, the Soviets, recognizing their poor harvest early, quietly cornered the world wheat market. As a result, wheat prices elsewhere more than doubled, pulling rice and corn prices up with them. But this and other price shocks were event-driven—drought in the Soviet Union, a monsoon failure in India, crop-shrinking heat in the U.S. Corn Belt. And the rises were short-lived: Prices typically returned to normal with the next harvest.

In contrast, the recent surge in world grain prices is trend-driven, making it unlikely to reverse without a reversal in the trends themselves. On the demand side, those trends include the ongoing addition of more than 70 million people a year; a growing number of people wanting to move up the food chain to consume highly grain-intensive livestock products [see "The Greenhouse Hamburger," by Nathan Fiala; *Scientific American,* February 2009]; and the massive diversion of U.S. grain to ethanol-fuel distilleries.

The extra demand for grain associated with rising affluence varies widely among countries. People in low-income countries where grain supplies 60 percent of calories, such as India, directly consume a bit more than a pound of grain a day. In affluent countries such as the U.S. and Canada, grain consumption per person is nearly four times that much, though perhaps 90 percent of it is consumed indirectly as meat, milk, and eggs from grain-fed animals.

The potential for further grain consumption as incomes rise among low-income consumers is huge. But that potential pales beside the insatiable demand for crop-based automotive fuels. A fourth of this year's U.S. grain harvest—enough to feed 125 million Americans or half a billion Indians at current consumption levels—will go to fuel cars. Yet even if the entire U.S. grain harvest were diverted into making ethanol, it would meet at most 18 percent of U.S. automotive fuel needs. The grain required to fill a 25-gallon SUV tank with ethanol could feed one person for a year.

The recent merging of the food and energy economies implies that if the food value of grain is less than its fuel value, the market will move the grain into the energy economy. That double demand is leading to an epic competition between cars and people for the grain supply and to a political and moral issue of unprecedented dimensions. The U.S., in a misguided effort to reduce its dependence on foreign oil by substituting grain-based fuels, is generating global food insecurity on a scale not seen before.

Water Shortages Mean Food Shortages

What about supply? The three environmental trends I mentioned earlier—the shortage of freshwater, the loss of topsoil, and the rising temperatures (and other effects) of global warming—are making it increasingly hard to expand the world's grain supply fast enough to keep up with demand. Of all those trends, however, the spread of water shortages poses the most immediate threat. The biggest challenge here is irrigation, which consumes 70 percent of the world's freshwater. Millions of irrigation wells in many countries are now pumping water out of underground sources faster than rainfall can recharge them. The result is falling water tables in countries populated by half the world's people, including the three big grain producers—China, India, and the U.S.

Usually aquifers are replenishable, but some of the most important ones are not: The "fossil" aquifers, so called because they store ancient water and are not recharged by precipitation. For these—including the vast Ogallala Aquifer that underlies the U.S. Great Plains, the Saudi aquifer and the deep aquifer under the North China Plain—depletion would spell the end of pumping. In arid regions such a loss could also bring an end to agriculture altogether.

In China the water table under the North China Plain, an area that produces more than half of the country's wheat and a third of its corn, is falling fast. Overpumping has used up most of the water in a shallow aquifer there, forcing well drillers to turn to the region's deep aquifer, which is not replenishable. A report by the World Bank foresees "catastrophic consequences for future generations" unless water use and supply can quickly be brought back into balance.

As water tables have fallen and irrigation wells have gone dry, China's wheat crop, the world's largest, has declined by 8 percent since it peaked at 123 million tons in 1997. In that same period China's rice production dropped 4 percent. The world's most populous nation may soon be importing massive quantities of grain.

But water shortages are even more worrying in India. There the margin between food consumption and survival is more precarious. Millions of irrigation wells have dropped water tables in almost every state. As Fred Pearce reported in New Scientist:

> Half of India's traditional hand-dug wells and millions of shallower tube wells have already dried up, bringing a spate of suicides among those who rely on them. Electricity blackouts are reaching epidemic proportions in states where half of the electricity is used to pump water from depths of up to a kilometer [3,300 feet].

A World Bank study reports that 15 percent of India's food supply is produced by mining groundwater. Stated otherwise, 175 million Indians consume grain produced with water from irrigation wells that will soon be exhausted. The continued shrinking of water supplies could lead to unmanageable food shortages and social conflict.

Less Soil, More Hunger

The scope of the second worrisome trend—the loss of topsoil—is also startling. Topsoil is eroding faster than new soil forms on perhaps a third of the world's cropland. This thin layer of essential plant nutrients, the very foundation of civilization, took long stretches of geologic time to build up, yet it is typically only about six inches deep. Its loss from wind and water erosion doomed earlier civilizations.

In 2002 a U.N. team assessed the food situation in Lesotho, the small, landlocked home of two million people embedded within South Africa. The team's finding was straightforward: "Agriculture in Lesotho faces a catastrophic future; crop production is declining and could cease altogether over large tracts of the country if steps are not taken to reverse soil erosion, degradation and the decline in soil fertility."

In the Western Hemisphere, Haiti—one of the first states to be recognized as failing—was largely self-sufficient in grain 40 years ago. In the years since, though, it has lost nearly all its forests and much of its topsoil, forcing the country to import more than half of its grain.

The third and perhaps most pervasive environmental threat to food security—rising surface temperature—can affect crop yields everywhere. In many countries crops are grown at or near their thermal optimum, so even a minor temperature rise during the growing season can shrink the harvest. A study published by the U.S. National Academy of Sciences has confirmed a rule of thumb among crop ecologists: For every rise of one degree Celsius (1.8 degrees Fahrenheit) above the norm, wheat, rice, and corn yields fall by 10 percent.

In the past, most famously when the innovations in the use of fertilizer, irrigation, and high-yield varieties of wheat and rice created the "green revolution" of the 1960s and 1970s, the response to the growing demand for food was the successful application of scientific agriculture: the technological fix. This time, regrettably, many of the most productive advances in agricultural technology have already been put into practice, and so the long-term rise in land productivity is slowing down. Between 1950 and 1990 the world's farmers increased the grain yield per acre by more than 2 percent a year, exceeding the growth of population. But since then, the annual growth in yield has slowed to slightly more than 1 percent. In some countries the yields appear to be near their practical limits, including rice yields in Japan and China.

Some commentators point to genetically modified crop strains as a way out of our predicament. Unfortunately, however, no genetically modified crops have led to dramatically higher yields, comparable to the doubling or tripling of wheat and rice yields that took place during the green revolution. Nor do they seem likely to do so, simply because conventional plant-breeding techniques have already tapped most of the potential for raising crop yields.

Jockeying for Food

As the world's food security unravels, a dangerous politics of food scarcity is coming into play: Individual countries acting in their narrowly defined self-interest are actually worsening the plight of the many. The trend began in 2007, when leading wheat-exporting countries such as Russia and Argentina limited or banned their exports, in hopes of increasing locally available food supplies and thereby bringing down food prices domestically. Vietnam, the world's second-biggest rice exporter after Thailand, banned its exports for several months for the same reason. Such moves may reassure those living in the exporting countries, but they are creating panic in importing countries that must rely on what is then left of the world's exportable grain.

In response to those restrictions, grain importers are trying to nail down long-term bilateral trade agreements that would lock up future grain supplies. The Phillippines, no longer able to count on getting rice from the world market, recently negotiated a three-year deal with Vietnam for a guaranteed 1.5 million tons of rice each year. Food-import anxiety is even spawning entirely new efforts by food-importing countries to buy or lease farmland in other countries.

In spite of such stopgap measures, soaring food prices and spreading hunger in many other countries are beginning to break down the social order. In several provinces of Thailand the predations of "rice rustlers" have forced villagers to guard their rice fields at night with loaded shotguns. In Pakistan an armed soldier escorts each grain truck. During the first half of 2008, 83 trucks carrying grain in Sudan were hijacked before reaching the Darfur relief camps.

No country is immune to the effects of tightening food supplies, not even the U.S., the world's breadbasket. If China turns to the world market for massive quantities of grain, as it has recently done for soybeans, it will have to buy from the U.S. For U.S. consumers, that would mean competing for the U.S. grain harvest with 1.3 billion Chinese consumers with fast-rising incomes—a nightmare scenario. In such circumstances, it would be tempting for the U.S. to restrict exports, as it did, for instance, with grain and soybeans in the 1970s when domestic prices soared. But that is not an option with China. Chinese investors now hold well over a trillion U.S. dollars, and they have often been the leading international buyers of U.S. Treasury securities issued to finance the fiscal deficit. Like it or not, U.S. consumers will share their grain with Chinese consumers, no matter how high food prices rise.

Plan B: Our Only Option

Since the current world food shortage is trend-driven, the environmental trends that cause it must be reversed. To do so requires extraordinarily demanding measures, a monumental shift away from business as usual—what we at the Earth Policy Institute call Plan A—to a civilization-saving Plan B. [see "Plan B 3.0: Mobilizing to Save Civilization," at www.earthpolicy.org/Books/PB3/]

Similar in scale and urgency to the U.S. mobilization for World War II, Plan B has four components: a massive effort to cut carbon emissions by 80 percent from their 2006 levels by 2020; the stabilization of the world's population at eight billion by 2040; the eradication of poverty; and the restoration of forests, soils, and aquifers.

Net carbon dioxide emissions can be cut by systematically raising energy efficiency and investing massively in the development of renewable sources of energy. We must also ban deforestation worldwide, as several countries already have done, and plant billions of trees to sequester carbon. The transition from fossil fuels to renewable forms of energy can be driven by imposing a tax on carbon, while offsetting it with a reduction in income taxes.

Stabilizing population and eradicating poverty go hand in hand. In fact, the key to accelerating the shift to smaller families is eradicating poverty—and vice versa. One way is to ensure at least a primary school education for all children, girls as well as boys. Another is to provide rudimentary, village-level health care, so that people can be confident that their children will survive to adulthood. Women everywhere need access to reproductive health care and family-planning services.

The fourth component, restoring the earth's natural systems and resources, incorporates a worldwide initiative to arrest the fall in water tables by raising water productivity: the useful activity that can be wrung from each drop. That implies shifting to more efficient irrigation systems and to more water-efficient crops. In some countries, it implies growing (and eating) more wheat and less rice, a water-intensive crop. And for industries and cities, it implies doing what some are doing already, namely, continuously recycling water.

At the same time, we must launch a worldwide effort to conserve soil, similar to the U.S. response to the Dust Bowl of the 1930s. Terracing the ground, planting trees as shelterbelts against windblown soil erosion, and practicing minimum tillage—in which the soil is not plowed and crop residues are left on the field—are among the most important soil-conservation measures.

There is nothing new about our four interrelated objectives. They have been discussed individually for years. Indeed, we have created entire institutions intended to tackle some of them, such as the World Bank to alleviate poverty. And we have made substantial progress in some parts of the world on at least one of them—the distribution of family-planning services and the associated shift to smaller families that brings population stability.

For many in the development community, the four objectives of Plan B were seen as positive, promoting development as long as they did not cost too much. Others saw them as humanitarian goals—politically correct and morally appropriate. Now a third and far more momentous rationale presents itself: Meeting these goals may be necessary to prevent the collapse of our civilization. Yet the cost we project for saving civilization would amount to less than $200 billion a year, a sixth of current global military spending. In effect, Plan B is the new security budget.

Time: Our Scarcest Resource

Our challenge is not only to implement Plan B but also to do it quickly. The world is in a race between political tipping points and natural ones. Can we close coal-fired power plants fast enough to prevent the Greenland ice sheet from slipping into the sea and inundating our coastlines? Can we cut carbon emissions fast enough to save the mountain glaciers of Asia? During the dry season their meltwaters sustain the major rivers of India and China—and by extension, hundreds of millions of people. Can we stabilize population before countries such as India, Pakistan, and Yemen are overwhelmed by shortages of the water they need to irrigate their crops?

It is hard to overstate the urgency of our predicament. [For the most thorough and authoritative scientific assessment of global climate change, see "Climate Change 2007. Fourth Assessment Report of the Intergovernmental Panel on Climate Change," available at www.ipcc.ch]

Every day counts. Unfortunately, we do not know how long we can light our cities with coal, for instance, before Greenland's ice sheet can no longer be saved. Nature sets the deadlines; nature is the timekeeper. But we human beings cannot see the clock.

We desperately need a new way of thinking, a new mind-set. The thinking that got us into this bind will not get us out. When Elizabeth Kolbert, a writer for the *New Yorker*, asked energy guru Amory Lovins about thinking outside the box, Lovins responded: "There is no box."

There is no box. That is the mind-set we need if civilization is to survive.

Critical Thinking

1. According to Lester Brown, what is the greatest threat to global political stability?
2. According to Lester Brown, demand for food is growing faster than the supply. What are the effects of this trend likely to be? How can we prevent the worst effects?

Radically Rethinking Agriculture for the 21st Century

N. V. Fedoroff, et al.

Population growth, arable land and fresh water limits, and climate change have profound implications for the ability of agriculture to meet this century's demands for food, feed, fiber, and fuel while reducing the environmental impact of their production. Success depends on the acceptance and use of contemporary molecular techniques, as well as the increasing development of farming systems that use saline water and integrate nutrient flows.

Many people expect that feeding the world's billions will become difficult in the twenty-first century. Yet many people also raise objections to innovative technologies that may be able to raise food production significantly. These technologies are largely grouped under the heading of genetic engineering or genetic modification. Already in use are modifications to crop plants that give them resistance to pests and to herbicides that control weeds. Future modifications may improve the ability of crops to tolerate drought, flooding, and saltwater, give crops the ability to turn atmospheric nitrogen into usable nitrogen compounds, and improve the efficiency of nutrient utilization. Other technologies center on integrating the production of multiple plant crops and animals in order to enhance nutrient cycling and minimize waste and pollution.

In September 2009, an international group of researchers held a workshop for the U.S. Department of State to discuss these issues. Their conclusion was that it is time to reevaluate existing regulatory frameworks based on data rather than fear. It is also time to scale up and build on existing innovations, and to do so immediately. The lead author of the resulting report, reprinted here, is Nina V. Federoff, a professor of life sciences at Pennsylvania State University and Science and Technology Adviser to the Secretary of State and to the Administrator of USAID, U.S. Department of State.

Key Concept: the need for new approaches to agriculture in order to meet growing demand for food

Population experts anticipate the addition of another roughly 3 billion people to the planet's population by the mid-21st century. However, the amount of arable land has not changed appreciably in more than half a century. It is unlikely to increase much in the future because we are losing it to urbanization, salinization, and desertification as fast as or faster than we are adding it[1]. Water scarcity is already a critical concern in parts of the world[2].

Climate change also has important implications for agriculture. The European heat wave of 2003 killed some 30,000 to 50,000 people[3]. The average temperature that summer was only about 3.5°C above the average for the last century. The 20 to 36% decrease in the yields of grains and fruits that summer drew little attention. But if the climate scientists are right, summers will be that hot on average by midcentury, and by 2090 much of the world will be experiencing summers hotter than the hottest summer now on record.

The yields of our most important food, feed, and fiber crops decline precipitously at temperatures much above 30°C[4]. Among other reasons, this is because photosynthesis has a temperature optimum in the range of 20° to 25°C for our major temperate crops, and plants develop faster as temperature increases, leaving less time to accumulate the carbohydrates, fats, and proteins that constitute the bulk of fruits and grains[5]. Widespread adoption of more effective and sustainable agronomic practices can help buffer crops against warmer and drier environments[6], but it will be increasingly difficult to maintain, much less increase, yields of our current major crops as temperatures rise and drylands expand[7].

Climate change will further affect agriculture as the sea level rises, submerging low-lying cropland, and as glaciers melt, causing river systems to experience shorter and more intense seasonal flows, as well as more flooding[7].

Recent reports on food security emphasize the gains that can be made by bringing existing agronomic and food science technology and know-how to people who do not yet have it[8, 9], as well as by exploring the genetic variability in our existing food crops and developing more ecologically sound farming practices[10]. This requires building local educational, technical, and research capacity, food processing capability, storage capacity, and other aspects of agribusiness, as well as rural transportation and water and communications infrastructure. It also necessitates addressing the many trade, subsidy, intellectual property, and regulatory issues that interfere with trade and inhibit the use of technology.

What people are talking about today, both in the private and public research sectors, is the use and improvement of conventional and molecular breeding, as well as molecular genetic modification (GM), to adapt our existing food crops to increasing temperatures, decreased water availability in some places and flooding in others, rising salinity[8, 9], and changing pathogen and insect threats[11]. Another important goal of such research is increasing crops' nitrogen uptake and use efficiency, because nitrogenous compounds in fertilizers are major contributors to waterway eutrophication and greenhouse gas emissions.

There is a critical need to get beyond popular biases against the use of agricultural biotechnology and develop forward-looking regulatory frameworks based on scientific evidence. In 2008, the most recent year for which statistics are available, GM crops were grown on almost 300 million acres in 25 countries, of which 15 were developing countries[12]. The world has consumed GM crops for 13 years without incident. The first few GM crops that have been grown very widely, including insect-resistant and herbicide-tolerant corn, cotton, canola, and soybeans, have increased agricultural productivity and farmers' incomes. They have also had environmental and health benefits, such as decreased use of pesticides and herbicides and increased use of no-till farming[13].

Despite the excellent safety and efficacy record of GM crops, regulatory policies remain almost as restrictive as they were when GM crops were first introduced. In the United States, case-by-case review by at least two and sometimes three regulatory agencies (USDA, EPA, and FDA) is still commonly the rule rather than the exception. Perhaps the most detrimental effect of this complex, costly, and time-intensive regulatory apparatus is the virtual exclusion of public-sector researchers from the use of molecular methods to improve crops for farmers. As a result, there are still only a few GM crops, primarily those for which there is a large seed market[12], and the benefits of biotechnology have not been realized for the vast majority of food crops.

What is needed is a serious reevaluation of the existing regulatory framework in the light of accumulated evidence and experience. An authoritative assessment of existing data on GM crop safety is timely and should encompass protein safety, gene stability, acute toxicity, composition, nutritional value, allergenicity, gene flow, and effects on nontarget organisms. This would establish a foundation for reducing the complexity of the regulatory process without affecting the integrity of the safety assessment. Such an evolution of the regulatory process in the United States would be a welcome precedent globally.

It is also critically important to develop a public facility within the USDA with the mission of conducting the requisite safety testing of GM crops developed in the public sector. This would make it possible for university and other public-sector researchers to use contemporary molecular knowledge and techniques to improve local crops for farmers.

However, it is not at all a foregone conclusion that our current crops can be pushed to perform as well as they do now at much higher temperatures and with much less water and other agricultural inputs. It will take new approaches, new methods, new technology—indeed, perhaps even new crops and new agricultural systems.

Aquaculture is part of the answer. A kilogram of fish can be produced in as little as 50 liters of water[14], although the total water requirements depend on the feed source. Feed is now commonly derived from wild-caught fish, increasing pressure on marine fisheries. As well, much of the growing aquaculture industry is a source of nutrient pollution of coastal waters, but self-contained and isolated systems are increasingly used to buffer aquaculture from pathogens and minimize its impact on the environment[15].

Another part of the answer is in the scale-up of dryland and saline agriculture[16]. Among the research leaders are several centers of the Consultative Group on International Agricultural Research, the International Center for Biosaline Agriculture, and the Jacob Blaustein Institutes for Desert Research of the Ben-Gurion University of the Negev.

Systems that integrate agriculture and aquaculture are rapidly developing in scope and sophistication. A 2001 United Nations Food and Agriculture Organization report[17] describes the development of such systems in many Asian countries. Today, such systems increasingly integrate organisms from multiple trophic levels[18]. An approach particularly well suited for coastal deserts includes inland seawater ponds that support aquaculture, the nutrient efflux from which fertilizes the growth of halophytes, seaweed, salt-tolerant grasses, and mangroves useful for animal feed, human food, and biofuels, and as carbon sinks[19]. Such integrated systems can eliminate today's flow of agricultural nutrients from land to sea. If done on a sufficient scale, inland seawater systems could also compensate for rising sea levels.

The heart of new agricultural paradigms for a hotter and more populous world must be systems that close the loop of nutrient flows from microorganisms and plants to animals and back, powered and irrigated as much as possible by sunlight and seawater. This has the potential to decrease the land, energy, and freshwater demands of agriculture, while at the same time ameliorating the pollution currently associated with agricultural chemicals and animal waste. The design and large-scale implementation of farms based on nontraditional species in arid places will undoubtedly pose new research, engineering, monitoring, and regulatory challenges, with respect to food safety and ecological impacts as well as control of pests and pathogens. But if we are to resume progress toward eliminating hunger, we must scale up and further build on the innovative approaches already under development, and we must do so immediately.

References and Notes

1. *The Land Commodities Global Agriculture & Farmland Investment Report 2009* (Land Commodities Asset Management AG, Baar, Switzerland, 2009; www.landcommodities.com).
2. *Water for Food, Water for Life: A Comprehensive Assessment of Water Management* (International Water Management Institute, Colombo, Sri Lanka, 2007).
3. D. S. Battisti, R. L. Naylor, Historical warnings of future food insecurity with unprecedented seasonal heat. *Science* **323**, 240 (2009). [Abstract/Free Full Text]
4. W. Schlenker, M. J. Roberts, Nonlinear temperature effects indicate severe damages to U.S. crop yields under climate change. *Proc. Natl. Acad. Sci. U.S.A.* **106**, 15594 (2009). [Abstract/Free Full Text]
5. M. M. Qaderi, D. M. Reid, in *Climate Change and Crops*, S. N. Singh, Ed. (Springer-Verlag, Berlin, 2009), pp. 1–9.
6. J. I. L. Morison, N. R. Baker, P. M. Mullineaux, W. J. Davies, Improving water use in crop production. *Philos. Trans. R. Soc. London Ser. B* **363**, 639 (2008). [Abstract/Free Full Text]
7. Intergovernmental Panel on Climate Change, *Climate Change 2007: Impacts, Adaptation and Vulnerability* (Cambridge Univ. Press, Cambridge, 2007; www.ipcc.ch/publications_and_data/publications_ipcc_fourth_assessment_report_wg2_report_impacts_adaptation_and_vulnerability.htm).
8. *Agriculture for Development* (World Bank, Washington, DC, 2008; http://siteresources.worldbank.org/INTWDR2008/Resources/WDR_00_book.pdf).
9. *Reaping the Benefits: Science and the Sustainable Intensification of Global Agriculture* (Royal Society, London, 2009; http://royalsociety.org/Reapingthebenefits).
10. *The Conservation of Global Crop Genetic Resources in the Face of Climate Change* (Summary Statement from a Bellagio Meeting, 2007; http://iis-db.stanford.edu/pubs/22065/Bellagio_final1.pdf).
11. P. J. Gregory, S. N. Johnson, A. C. Newton, J. S. I. Ingram, Integrating pests and pathogens into the climate change/food security debate. *J. Exp. Bot.* **60**, 2827 (2009). [Abstract/Free Full Text]
12. C. James, *Global Status of Commercialized Biotech/GM Crops: 2008* (International Service for the Acquisition of Agri-biotech Applications, Ithaca, NY, 2008).
13. G. Brookes, P. Barfoot, *AgBioForum* **11**, 21 (2008).
14. S. Rothbard, Y. Peretz, in *Tilapia Farming in the 21st Century*, R. D. Guerrero III, R. Guerrero-del Castillo, Eds. (Philippines Fisheries Associations, Los Baños, Philippines, 2002), pp. 60–65.
15. *The State of World Fisheries and Aquaculture 2008* (United Nations Food and Agriculture Organization, Rome, 2009; www.fao.org/docrep/011/i0250e/i0250e00.HTM).
16. M. A. Lantican, P. L. Pingali, S. Rajaram, Is research on marginal lands catching up? The case of unfavourable wheat growing environments. *Agric. Econ.* **29**, 353 (2003). [CrossRef]
17. *Integrated Agriculture-Aquaculture* (United Nations Food and Agriculture Organization, Rome, 2001; www.fao.org/DOCREP/005/Y1187E/y1187e00.htm).
18. T. Chopin et al., in *Encyclopedia of Ecology*, S. E. Jorgensen, B. Fath, Eds. (Elsevier, Amsterdam, 2008), pp. 2463–2475.
19. The Seawater Foundation, www.seawaterfoundation.org.
20. The authors were speakers in a workshop titled "Adapting Agriculture to Climate Change: What Will It Take?" held 14 September 2009 under the auspices of the Office of the Science and Technology Adviser to the Secretary of State. The views expressed here should not be construed as representing those of the U.S. government. N.V.F. is on leave from Pennsylvania State University. C.N.H. is co-chair of Global Seawater, which promotes creation of Integrated Seawater Farms.

Critical Thinking

1. What are the major current threats to global food security?
2. How can the agricultural system be changed to improve global food security?

Chemicals

Selection 27
ROBERT VAN DEN BOSCH, from *The Pesticide Conspiracy* (Doubleday, 1978)

Selection 28
SANDRA STEINGRABER, from *Living Downstream: An Ecologist Looks at Cancer and the Environment* (Addison-Wesley, 1997)

Selection 29
THEO COLBORN, DIANNE DUMANOSKI, AND JOHN PETERSON MYERS, from *Our Stolen Future* (Dutton, 1996)

Learning Outcomes
After reading this unit, you should be able to:

1. Explain how natural selection ensures the eventual failure of chemical pest control methods.
2. Explain how natural enemies of insects can be used to control their numbers.
3. Describe the difficulty of evaluating the carcinogenicity in humans of chemicals.
4. Explain why more research into the environmental causes of cancer is needed.
5. Describe the hazards posed by environmental hormone mimics.

The Pesticide Conspiracy

Robert Van Den Bosch

The attempts by the supporters of uncontrolled pesticide use to discredit ecologist Rachel Carson, author of *Silent Spring* (Houghton Mifflin, 1962), in which she warns of the dangers of indiscriminate pesticide use, emphasize the fact that although she was a scientist, her training and expertise were not specifically in the field of pest management. No such claim could be made about Robert van den Bosch (1922–1978), a highly accomplished scientist whose research career was devoted to the problem of controlling pest populations. Although van den Bosch's prose lacks the grace and literary merit that helped to make *Silent Spring* a best-seller, his well-documented book *The Pesticide Conspiracy* (Doubleday, 1978) is an even more potent indictment of the pesticide industry. In particular, he exposes the industry's influence in turning the U.S. Department of Agriculture into a willing accomplice in promoting the irresponsible, ecologically disastrous, and ineffective overuse of chemical poisons.

In the following excerpt from *The Pesticide Conspiracy,* van den Bosch not only details the reasons why chemical poisons alone have not, and cannot, control agricultural pests, but he emphasizes that a much more effective alternative exists. This alternative is integrated pest management (IPM), which relies on a wide variety of technologies (including judicious use of appropriate chemicals), based on detailed knowledge of the pest in question. Van den Bosch was one of the developers and early advocates of IPM. The fact that it has taken more than three decades since *Silent Spring* for government agricultural policymakers to embrace IPM can in some part be attributed to the untimely death of van den Bosch shortly after the publication of *The Pesticide Conspiracy.*

Key Concept: integrated pest management

Introduction

A Can of Worms

In the early summer of 1976, a popular California radio station broadcast to growers an insecticide advertisement prepared for a major chemical company by a New York ad agency. The broadcast warned the growers of the imminent appearance of a "menacing" pest in one of their major crops and advised that as soon as the bugs "first appear" in the fields the growers should start a regular spray program, using, of course, the advertised insecticide. The broadcast also claimed that the material was *the one* insecticide the growers in the area could depend on for effective and economical control of the threatening pest, and further told the growers that through its use they would get a cleaner crop and more profit at harvest time.

The advertisement epitomizes what is wrong with the American way of killing bugs, a practice more often concerned with merchandising gimmickry than it is with applied science. In connection with this gimmickry, much of modern chemical pest control is dishonest, irresponsible, and dangerous. This was true of the radio advertisement just described. It was *dishonest* in its claim that the touted insecticide was *the one* material that growers could depend upon, for in actuality there are several equally effective insecticides and none will assure a cleaner crop and more profit. The advertisement was *irresponsible* in advising growers to initiate a regular spray program upon "first appearance" of the pest. Intensive research has shown that spraying of the crop should be undertaken only when the pest population reaches and maintains a prescribed level during the budding season and that sprays should never be applied on a regular schedule. Finally, the advertisement was *dangerous*, because the advised spraying, if widely adopted by the growers, would have resulted in the senseless dumping of huge amounts of a highly hazardous poison into the environment.

As a veteran researcher in insect control, I have long been disturbed by the dishonest, irresponsible, and dangerous nature of our prevailing chemical control strategy, but I am even more distressed by the knowledge that this simplistic strategy cannot possibly contain the versatile, prolific, and adaptable insects. For a third of a century following the emergence of DDT, we have been locked onto this costly and hazardous insect control strategy, which for biological and ecological reasons, never had a chance to succeed.

What is most disturbing of all is our inability to clean up the mess by shifting to the workable, ecologically based, alternative strategy that modern pest-control specialists term *integrated control* (also termed integrated pest management). Integrated control, as the name implies, is a holistic strategy that utilizes technical information, continuous pest-population monitoring, resource (crop) assessment, control-action criteria, materials, and methods, in concert with natural mortality factors, to manage pest populations in a safe, economical, and effective way. Integrated control is the only strategy that will work effectively against the insects, because it systematically utilizes all possible tactics in such a way that they attain full individual impact, function collectively for maximum mutual effect, and cause minimum detriment to the surrounding environment. In other words, unlike the prevailing chemical control strategy, with its emphasis on product merchandising, integrated control is a technology. It is scientific pest control and, as such, the only way we can hope to gain the upper hand in our battle with the insects. In every respect, integrated control makes sense, and it works. Despite this, our swing to this better pest-management strategy has been painfully slow, and for a clear reason. The impediment has been a powerful coalition of individuals, corporations, and agencies that profit from the prevailing chemical control strategy and brook no interference with the status quo. This power consortium has been unrelenting in its efforts to keep things as they are and as so frequently happens in our society, the games it plays to maintain the status quo are often corruptive, coercive, and sinister.

This book, then, is a tale of a contemporary technology gone sour under the pressures generated by a powerful vested interest. Bugs provide the theme, but politics, deceit, corruption, and treachery are its substance. I feel that the story is a most timely one, for it describes an ecological rip-off and how this atrocity is being perpetuated by tacticians of pure Watergate stripe. The book is largely based on personal experiences and insights gained from more than a quarter century of battling the bugs and their human allies who devised and maintain the inadequate chemical control strategy. It is a tale of personal outrage that I hope proves highly infectious. . . .

Integrated Control— A Better Way to Battle the Bugs

The 1975 meeting of the Entomological Society of America was the scene of an interesting comparison between the contrasting insect-control strategies of two of the world's great nations, the People's Republic of China and the United States of America. And from what transpired, it appears as though the Chinese pest-control system has more going for it than does ours. I would like to dwell on this matter a bit, for not only does it cast light on the right and wrong ways to combat insects but also because, if we are willing to read the signals honestly, it gives us considerable insight into what is going wrong with the American way of doing things. There may be something of value in such an exercise.

Insect control in China was described, to an audience of two thousand attending the opening plenary session of the Entomological Society, by a panel of America's leading entomologists who earlier in the year had visited China under the China-U.S. cultural exchange. I know most of the panelists, some intimately, and would characterize them largely as politically moderate Middle Americans. In other words, they had no ax to grind on behalf of China and its Marxist political ideology but reported things as they witnessed and recorded them. From what they had to say, it seems that China's entomologists constantly sift the world's literature and other information sources for relevant techniques, methods, and materials, and integrate them along with their own technical developments into a highly effective national integrated pest-management system. Under this system there is continuous monitoring of pest populations, use of action-precipitating pest-population thresholds (economic thresholds), and the implementation of a variety of tactics, including chemical, cultural, and biological controls, as circumstances dictate.

This program is serving China well. For example: using this pest-control system, China grows 39 percent of the world's rice, which not only feeds her 900 million citizens but enables her to be a major rice exporter. China also utilizes her pest-management system against disease-transmitting and nuisance insects such as mosquitoes and flies. It is interesting that in mosquito control she employs virtually no DDT, apparently relying instead on reduction of mosquito breeding sources, mosquito exclusion tactics, natural controls, and the judicious use of "safe" insecticides. In this latter connection it is especially noteworthy that China, though producing about one hundred insecticides, relies heavily on seven organo-phosphates because of their limited hazard to warm-blooded animals. And under her insect-control system, she uses these materials judiciously.

Now let's see how we do things in the U.S.A. Two days after the China report, the Entomological Society heard Assistant Agriculture Secretary Robert Long tell us all about it. On this occasion we were a captive audience, since the convention registration fee included the price of a ticket to the Society's annual awards luncheon, before which industry's spokesman Long performed as "distinguished" guest speaker. In reading the fine print of the meeting program I had earlier discovered that Long's visit to New Orleans was arranged at the behest of the agri-chemical industry. And it didn't take long for him to burst into his expected song as he unleashed a vicious attack on industry's great tormenter, the Environmental Protection Agency [EPA]. In his speech, Long first chortled over the recently enacted, politically inspired amendments to the Federal Insecticide, Fungicide, and Rodenticide Act (FIFRA), which give USDA [the U.S. Department of Agriculture] considerable veto power over EPA pesticide decisions. But then he made it abundantly clear that this was not enough. Despite the FIFRA amendments, Long left little doubt that in his mind EPA still had too much control over the registration and regulation of pesticides, particularly as regards EPA's intentions to seek re-registration of America's fourteen hundred pesticide species and their thirty thousand formulations. Here he ran up the alarm pennant by maintaining that EPA's protocols were so deeply mired in bureaucratic stickum that the agri-chemical industry simply would not make the effort to re-register their materials. In other words, he flatly told us that we were about to lose our thirty thousand pesticides, and he painted a terrifying picture of impending starvation, pestilence, and disease in the wake of this loss.

This rhetoric, as it was intended to do, quite probably frightened the naïve in the crowd while bringing joy to the hearts of Long's chemical-company sponsors. Robert Long, a glib spellbinder, well knew that his prediction of an imminent pesticide wipe-out was complete nonsense. Legal road-blocks and political gamesmanship make this a virtual impossibility. What Long was actually telling us was that the U.S. Department of Agriculture, with powerful political backing, intended to hound EPA into loosening its control over pesticide registration and regulation, to the point where the agri-chemical industry would have things just about as they were in the days before passage of the National Environmental Policy Act. The speech was simply a trial run, with Long using the entomologists to perfect the pitch with which he and other USDA brass planned to bushwhack EPA in forthcoming political jousting.

What he and his sponsors hoped to accomplish, then, was an easing of the way for the American agri-chemical industry to unload its fourteen hundred pesticides in their thirty thousand varieties onto the environment, with USDA bulldozing the path. Fortunately, the 1976 presidential election aborted this plan, which, if it had unfolded, would have permitted the interests of the American chemical industry to transcend environmental quality, public health, and the economic well-being of the farmer and consumer. Madison Avenue would have predominated, while scientific pest control would have remained a fuzzy dream in the minds of a few radical researchers.

But let's return to China. How can she feed, and protect from pestilence, 900 million people, with just a handful of insecticides, while we are led to believe that we must have thousands of poisons or otherwise be overwhelmed by an insect avalanche? Is it that we have a vastly more severe pest problem? I hardly think so. Malaria is nowhere endemic in the United States, but it is in China, as are other horrible, insect-borne diseases. Nor do we have 900 million mouths to feed. What, indeed, has happened is that China has used her intelligence to invoke a national *integrated pest-management strategy*, while our strategy is chemical control dominated by the marketing thrust of the agri-chemical industry. Result: pest-control chaos, and if we care to look about us, we will find that similar chaos characterizes many of the other things that we do.

But it isn't too late to change our ways in pest control or, for that matter, in other aspects of applied technology. As I have mentioned several times, it was a mistake to challenge the insects head on with crude chemical weapons. The bugs are too diverse, adaptable, and prolific to be beaten by such a simple strategy. But we were so dazzled by DDT's great killing efficiency and, perhaps, our cleverness in concocting the stuff, that we ignored the possibility of a bug backlash and plunged full blast into the chemical "extermination" campaign. And once we had made our move, we were hooked onto an insecticide treadmill just like an addict on junk.

Now, suddenly, in the midst of the nightmare, when our addiction demands heavier doses and more frequent fixes, the chemicals are hard to get and very expensive. Alarmingly, with famine an increasing global concern, many of the chemical eggs in our bug-control basket are no longer effective. The insects, our great rivals for the earthly bounty, are gearing up to march through our gardens, groves, forests, and fields largely immune to our chemical weapons and freed from natural controls. And in the disease area, too, the breakdown is having a disturbing effect, as malaria makes its dreadful resurgence largely because of mosquito resistance to DDT and other insecticides.

The situation would be much more frightening but for a handful of pest-control radicals who never tumbled to the chemical strategy. These are the renegades who quietly worked away on integrated control programs while most in the pest-control arena were on the chemical kick. Though integrated control is still limited in scope, there are enough programs in operation or under development to offer encouragement that there is indeed a better way to battle the bugs.

What is Integrated Control?

Integrated control is simply rational pest control: the fitting together of information, decision-making criteria, methods, and materials with naturally occurring pest mortality into effective and redeeming pest-management systems.

Under integrated control, natural enemies, cultural practices, resistant crop and livestock varieties, microbial agents, genetic manipulation, messenger chemicals, and yes, even pesticides become mutually augmentative instead of individually operative or even antagonistic, as is often the case under prevailing practice (e.g., insecticides versus natural enemies). An integrated control program entails six basic elements: (1) man, (2) knowledge/information, (3) monitoring, (4) the setting of action levels, (5) methods, and (6) materials.

Man conceives the program and makes it work. *Knowledge* and *information* are used to develop a system and are vital in its day-to-day operation. *Monitoring* is the continuous assessment of the pest-resource system. *Action levels* are the pest densities at which control methods are invoked. *Methods* are the pathways of action taken to manipulate pest populations. *Materials* are the tools of manipulation.

Sounds like what's going on in China, doesn't it!

Integrated control systems are dynamic, involving continuous information gathering and evaluation, which in turn permit flexibility in decision-making, alteration of the pathways of action, and variation in the agents used. It is the pest-control adviser who gives integrated control its dynamism. By constantly "reading" the situation and invoking tactics and materials as conditions dictate, he acts as a surrogate insecticide, "killing" insects with knowledge and information as well as pesticides, pathogens, parasites, and predators. Integrated control's dynamism is a major factor that sets it off from conventional pest control. Thus, though the latter involves some of the same elements, it lacks dynamism in that it is essentially preprogrammed to the prophylactic or therapeutic use of pesticides. In other words, pesticides dominate the system and constitute its rigid backbone. Where a crop is involved, there is little or no on-going assessment of the crop ecosystem and the dynamic interplay of plant, pests, climate, and natural enemies. This pest-control pattern prevails even in California, our most advanced agro-technology, where over one hundred research entomologists busily at work killing bugs for more than a quarter century have developed fewer than half a dozen valid economic thresholds for the hundreds of pest species. A perusal of the stack of official University of California pest-control recommendations reveals the following kinds of pest-control action criteria:

- when damaging plants
- when present
- when damage occurs
- when they first appear
- when colonies easily found
- when abundant
- when needed
- early season
- when present in large numbers before damage occurs
- anytime when present
- early, mid, and late season
- on small plants as needed
- when present and injuring the plants
- when feeding on the pods
- throughout the season
- when infestation spotty
- when plants are three feet tall.

What this long menu of senseless gobbledygook implies is that in California the insecticide folks have a wide-open field in which to hustle their chemicals, and this they do with greater success than anywhere else in the world.

Under the prevailing chemical control strategy, there is virtually no flexibility in decision-making, particularly as regards alternative pathways of action. The game plan is set at the start and it is stubbornly followed. Result, the familiar case of the fruit grower who year after year automatically sprays his orchard a dozen times or more with the calendar as his main decision-making guide. Or the cotton grower who typically sprays when a chemical-company fieldman drops around and tells him that a few stinkbugs, bollworms, or army worms are showing up in the south forty.

In conventional pest control, one turns on the chemical switch, sits back, and lets the insecticides do the job. It is the lazy man's approach, which characterizes so many aspects of modern life and for which society and the environment pay dearly. A measure of this cost can be gained from a brief analysis of pest control in California.

California's pest control is locked to chemical pesticides. The state is the country's greatest user of these materials, and as stated earlier, receives about 5 percent of the world's pesticide load. It appears that along with its primacy in smog and earthquakes, California has another distinction: leadership in pesticide pollution. Little wonder! More than fourteen hundred chemical company fieldmen (salesmen) prowl the state, servicing the prevailing pest-control system. They assure a sustained chemical blizzard as well as a fat market for the agri-chemical industry. And at what a cost! Not only does this horde of hustling polluters dump hundreds of tons of unneeded pesticides into the environment, but in the bargain they annually cost California's economy about $50 million to support their huckstering. The chemical companies and many of the major pesticide users (growers, mosquito abaters, forest pest controllers, and pest-control operators) don't pay the bill, they simply pass it on to the consumer, who doubles as taxpayer. But the story doesn't end with money needlessly spent; there are also ecological and social impacts, which add immensely to the cost of the prevailing chemical control strategy.

What I have just described for California pretty much characterizes pest control for the United States in general, and for that matter, other of the world's modern agri-technologies. Chemical pest control, like so many of our modern practices, is a technology gone wild under the merchandising imperative. And as with our other excesses, this rampant technology must be brought under rein if irreparable damage is to be avoided. I am convinced that we pest-control researchers (particularly entomologists) have the capacity to turn things around through integrated control, and perhaps coincidentally establish a model of technological responsibility for other disciplines.

Critical Thinking

1. Why can chemical pesticides not provide long-term control of insect pests?
2. Why is the use of natural enemies a good approach to pest control?
3. What is integrated pest management?
4. Why do advocates of integrated pest management not try to eradicate insect pests?

Living Downstream: An Ecologist Looks at Cancer and the Environment

Sandra Steingraber

Biologist Sandra Steingraber is the author of *Post-Diagnosis* (Firebrand Books, 1995), a volume of poetry based on personal experiences during her battle with bladder cancer. That struggle, the subsequent loss of a close, young friend to cancer, and the knowledge that cancer has affected many members of her immediate family, inspired Steingraber to carefully study the notes produced by Rachel Carson during her research for the writing of *Silent Spring* (Houghton Mifflin, 1962). Steingraber then explored the numerous sources of environmental pollution that she was personally exposed to while growing up in central Illinois. Fields were frequently doused with pesticides and industrial plants located on the banks of the Illinois River emitted various toxins into the air and water. This research resulted in the writing of *Living Downstream: An Ecologist Looks at Cancer and the Environment* (Addison-Wesley, 1997), from which the following selection is taken.

In this widely acclaimed book, Steingraber blends poetic anecdotes and vivid descriptions of reckless industrial and agricultural pollution with a wealth of data from scientific and medical literature. The result is a compelling analysis of what is known and unknown about the relationship between environmental factors and cancer. Steingraber bemoans the imbalance between funding devoted to studies of genetic predispositions to cancer and the relative paucity of funding devoted to studies of potential environmental contributions to cancer incidence. She argues persuasively that while we can do little to change our genetic inheritance, there is much that can be done to reduce human exposure to environmental carcinogens.

Living Downstream became the focus of a heated controversy in December 1997. It was revealed that the *New England Journal of Medicine* was guilty of an ethical lapse by publishing a scathing review of the book without identifying the reviewer as a senior official at W. R. Grace & Company. The book contains a detailed discussion of W. R. Grace's role in the pollution of the water in Woburn, Massachusetts; a town where a cluster of fatal childhood leukemia cases has occurred. Steingraber was vindicated when she was appointed to the National Action Plan on Breast Cancer by the U.S. Department of Health and Human Services. She has continued her work in the book *Having Faith: An Ecologist's Journey to Motherhood* (Perseus, 2001).

Key Concept: the need for further research on environmental factors in cancer incidence

I had bladder cancer as a young adult. If I tell people this fact, they usually shake their heads. If I go on to mention that cancer runs in my family, they usually start to nod. *She is from one of those cancer families*, I can almost hear them thinking. Sometimes, I just leave it at that. But, if I am up for blank stares, I add that I am adopted and go on to describe a study of cancer among adoptees that found correlations within their adoptive families but not within their biological ones. ("Deaths of adoptive parents from cancer before the age of 50 increased the rate of mortality from cancer fivefold among the adoptees. . . . Deaths of biological parents from cancer had no detectable effect on the rate of mortality from cancer among the adoptees.") At this point, most people become very quiet.

These silences remind me how unfamiliar many of us are with the notion that families share environments as well as chromosomes or with the concept that our genes work in communion with substances streaming in from the larger, ecological world. What runs in families does not necessarily run in blood. And our genes are less an inherited set of teacups enclosed in a cellular china cabinet than they are plates used in a busy diner. Cracks, chips, and scrapes accumulate. Accidents Happen.

*M*y Aunt Jean died of bladder cancer. Raymond and Violet both died of colon cancer. LeRoy is currently under treatment. These are my father's relatives. About Uncle Ray I remember very little, except that he, along with my dad, was one of the less loud of the concrete-pouring, brick-laying Steingraber brothers.

Aunt Jean laughed a lot and once asked me to draw a pig so she could tape it to her refrigerator door. Red-haired Aunt Vi cooked magnificent dinners, was partial to wearing pink, and was married to a man truly untempted by silence. Together, she once remarked, the two of them sure knew how to enjoy themselves. Her widowed husband, my Uncle Ed, is now being treated aggressively for prostate cancer. Nonetheless, at last report, he was busy building a shrine to his wife out in the backyard. When it comes to expressions of grief, my father's side of the family tends toward large-scale construction projects.

The man who was to be my brother-in-law was stricken with intestinal cancer at the age of twenty-one. He cleaned out chemical drums for a living. Three years before Jeff's diagnosis, I was diagnosed with bladder cancer, and three years before my diagnosis, my mother learned she had metastatic breast cancer. That she is still alive today is a topic of considerable wonder among her doctors. Mom is matter-of-fact about this, although she will, if prompted, shyly point out that she has outlived her oncologist and three of her other doctors, two of whom died of cancer.

My mother was first diagnosed in 1974, a year that is considered an anomaly in the annals of breast cancer. Graphs displaying U.S. breast cancer incidence rates across the decades show a gently rising line that suddenly zooms skyward, falls back, then continues its slow ascent. The story behind the blip of '74 has been deemed a textbook lesson in statistical artifacts.

In this year, First Lady Betty Ford and Second Lady Happy Rockefeller both underwent mastectomies. The words *breast cancer* entered public conversation. Women who might otherwise have delayed routine checkups or who were hesitant to seek medical opinion about a lump were propelled into doctors' offices. The result was that a lot of women were diagnosed with breast cancer within a short period of time, my mother among them.

When I, at age fifteen, inquired why my mother was in the hospital, the answer was "Because she has what Mrs. Ford has." When my mother, at age forty-four, questioned whether a radical mastectomy was necessary, she was told, "If it's good enough for Happy, it's good enough for you." . . .

Ecological Roots

In 1983, I took the train home to Illinois for the holidays—and an appointment at the hospital.

The scheduling of cancer checkups is always an elaborate decision. The calendar date must sound auspicious. Monday or Tuesday appointments are best; otherwise, one risks waiting through the weekend for the results of a laggard lab test or delayed radiology report. It's also best if these appointments fall within a hectic, deadline-filled month so that frenetic activity can preclude fretfulness. During the years I was a graduate student, this meant the ends of semesters, which explains why some half dozen Christmas carols now remind me of outpatient waiting rooms. This particular appointment was destined to turn out fine. What I remember most clearly is my journey there by train.

Something about the landscape changes abruptly between northern and central Illinois. I am not sure what it is exactly, but it happens right around the little towns of Wilmington and Dwight. The horizon recedes, and the sky becomes larger. Distances increase, as though all objects are moving slowly away from each other. Lines become more sharply drawn. These changes always make me restless and, when driving, drive faster. But since I am in a train, I close the book I am reading and begin impatiently straightening the pages of a newspaper strewn over the adjacent seat.

That is when my eye catches the headline of a back-page article: SCIENTISTS IDENTIFY GENE RESPONSIBLE FOR HUMAN BLADDER CANCER. Pulling the newspaper onto my lap, I stare out the window and become very still. It is only early evening, but the fields are already dark, a patchwork of lights quilted over and across them. They have always soothed me. I look for signs of snow. There are none. Finally, I read the article.

Researchers at the Massachusetts Institute of Technology, it seems, had extracted DNA from the cells of a human bladder tumor and used it to transform normal mouse cells into cancerous ones. Through this process, they located the segment of DNA responsible for the transformation. And by comparing this segment to its unmutated form in noncancerous human cells, they were able to pinpoint the exact alteration that had caused a respectable gene to go bad.

In this case, the mutation turned out to be a substitution of one unit of genetic material for another in a single rung of the DNA ladder. Namely, at some point during DNA replication, a double-ringed base called guanine was swapped for the single-ringed thymine. Like a typographical error in which one letter replaces another—*snow* instead of *show*, *block* instead of *black*—the message sent out by this gene was utterly changed. Instead of instructing the cell to manufacture the amino acid glycine, the altered gene now specified for valine. (Nine years later, other researchers would determine that this substitution alters the structure of proteins involved in signal transduction—the crucial line of communication between the cell membrane and the nucleus that helps coordinate cell division.)

Guanine instead of thymine. Valine instead of glycine. I look away again—this time at my face superimposed over the landscape by the window's mirror. If, in fact, this mutation was involved in my cancer, when did it happen? Where was I? Why had it escaped repair? I had been betrayed. But by what?

Thirteen years later, I possess a bulging file of scientific articles documenting an array of genetic changes involved in bladder cancer. Besides the oncogene just described, two tumor suppressor genes, p15 and p16, have also been discovered to play a role. Their deletion is a common event in transitional cell carcinoma, the kind of cancer I had. Mutations of the famous p53 tumor suppressor gene, with guest-star appearances in so many different cancers, have been detected in more than half of invasive bladder tumors. Also associated with transitional cell carcinomas are surplus numbers of growth factor receptors. Their overexpression has been linked to the kinds of gross genetic injuries that appear near the end of the malignant process.

The nature of the transaction between these various genes and certain bladder carcinogens has likewise been worked out in the years since a newspaper article introduced me to the then new concept of oncogenes. Consider, for example, that redoubtable class of bladder carcinogens called aromatic amines—present as contaminants in cigarette smoke; added to rubber during vulcanization; formulated as dyes for cloth, leather, and paper; used in printing and color photography; and featured in the manufacture of certain pharmaceuticals and pesticides. Aniline, benzidine, naphthylamine, and *o*-toluidine are all members of this group. The first reports of excessive bladder cancers among workers in the aniline dye industry were published in 1895. . . . More than a century later, we now know that anilines and other aromatic amines ply their wickedness by forming DNA adducts in the cells of the tissues lining the bladder, where they arrive as contaminants of urine.

We also now know that aromatic amines are gradually detoxified by the body through a process called acetylation. Like all such processes, it is carried out by a special group of detoxifying enzymes whose actions are controlled and modified by a number of genes. People who are slow acetylators have low levels of these enzymes and are at greater risk of bladder cancer from exposure to aromatic amines. Members of this population can be readily identified because they bear significantly higher burdens of adducts than fast acetylators at the same exposure levels. These genetically suspectible individuals hardly constitute a tiny minority: more than half of Americans and Europeans are estimated to be slow acetylators.

Very likely, I am one. You may be one, too.

We know a lot about bladder cancer. Bladder carcinogens were among the earliest human carcinogens ever identified, and one of the first human oncogenes ever decoded was isolated from some unlucky fellow's bladder tumor. More than most malignancies, bladder cancer has provided researchers with a picture of the sequential genetic changes that unfold from initiation through promotion to progression, from precursor lesions to increasingly more aggressive tumors.

Sadly, all this knowledge about genetic mutations, inherited risk factors, and enzymatic mechanisms has not translated into an effective campaign to prevent the disease. The fact remains that the overall incidence rate of bladder cancer increased 10 percent between 1973 and 1991. Increases are especially dramatic among African Americans: among black men, bladder cancer incidence has risen 28 percent since 1973, and among black women, 34 percent.

Somewhat less than half of all bladder cancers among men and one-third of all cases among women are thought to be attributable to cigarette smoking, which is the single largest known risk factor for this disease. . . . [T]he question still remains: What is causing bladder cancer in the rest of us, the majority of bladder cancer patients for whom tobacco is not a factor?

I also possess another bulging file of scientific articles. These concern the ongoing presence of known and suspected bladder carcinogens in rivers, groundwater, dump sites, and indoor air. For example, industries reporting to the Toxics Release Inventory disclosed environmental releases of the aromatic amine *o*-toluidine that totaled 14,625 pounds in 1992 alone. Detected also in effluent from refineries and other manufacturing plants, *o*-toluidine exists as residues in the dyes of commercial textiles, which may, according to the *Seventh Annual Report on Carcinogens*, expose members of the general public who are consumers of these goods: "The presence of *o*-toluidine, even as a trace contaminant, would be a cause for concern." A 1996 study investigated a sixfold excess of bladder cancer among workers exposed years before to *o*-toluidine and aniline in the rubber chemicals department of a manufacturing plant in upstate New York. Levels of these contaminants are now well within their legal workplace limits, and yet blood and urine collected from current employees were found to contain substantial numbers of DNA adducts and detectable levels of *o*-toluidine and aniline. Another recent investigation revealed an eightfold excess of bladder cancer among workers employed in a Connecticut pharmaceuticals plant that manufactured a variety of aromatic amines. This study was reported as having national implications because the main suspect, dichlorobenzidine, has been widely used throughout the United States.

What my various file folders do *not* contain is a considered evaluation of all known and suspected bladder carcinogens—their sources, their possible interactions with each other, and our various routes of exposure to them. As we have seen, trihalomethanes—those unwanted by-products of water chlorination—have been linked to bladder cancer, as has the dry-cleaning solvent and sometime-contaminant of drinking-water pipes, tetrachloroethylene. I possess individual reports on each of these topics. What I do not have is a comprehensive description of how all these substances behave in combination. What are the risks of multiple trace exposures? What happens when we drink trihalomethanes, absorb aromatic amines, and inhale tetrachloroethylene? Furthermore, what is the ecological fate of these substances once they are released into the environment? What happens when dyed cloth, colored paper, and leather goods are laundered, landfilled, or incinerated? And why—almost a century after some of them were so identified—do powerful bladder carcinogens such as amine dyes continue to be manufactured, imported, used, and released into the environment in the first place? However improved the record of effort to regulate them, why have safer substitutes not replaced them all? These questions remain, to my knowledge, largely unaddressed by the cancer research community.

Several obstacles, I believe, prevent us from addressing cancer's environmental roots. An obsession with genes and heredity is one.

Cancer research currently directs considerable attention to the study of inherited cancers. Most immediately, this approach facilitates the development of genetic testing, which attempts to predict an individual's risk of succumbing to cancer, based on the presence or absence of certain genetic alterations. These efforts may also reveal which genes are common targets of acquired mutation in the general population. (Hereditary mutations are present at the time of conception, and they are carried in the DNA of all body cells; acquired mutations, which accumulate over an individual's lifetime, are passed only to the direct descendents of the cells in which they arise.)

Hereditary cancers, however, are the rare exception. Collectively, fewer than 10 percent of all malignancies are thought to involve inherited mutations. Between 1 and 5 percent of colon cancers, for example, are of the hereditary variety, and only about 15 percent exhibit any sort of familial component. The remaining 85 percent of colon cancers are officially classified as "sporadic," which, confesses one prominent researcher, "is a fancy medical term for 'we don't know what the hell causes it.' " Breast cancer also shows little connection to heredity (probably between 5 and 10 percent). Finding "cancer genes" is not going to prevent the vast majority of cancers that develop.

Moreover, even when rare, inherited mutations play a role in the development of a particular cancer, environmental influences are inescapably involved as well. Genetic risks are not exclusive of environmental risks. Indeed, the direct consequence of some of these damaging mutations is that people become even more sensitive to environmental carcinogens. In the case of hereditary colon cancer, for example, what is passed down the generations is a faulty DNA repair gene. Its human heirs are thereby rendered less capable of coping with environmental assaults on their genes or repairing the spontaneous mistakes that occur during normal cell division. These individuals thus become more likely to accumulate the series of *acquired* mutations needed for the formation of a colon tumor.

Cancer incidence rates are not rising because we are suddenly sprouting new cancer genes. Rare, heritable genes that predispose their hosts to cancer by creating special susceptibilities to the effects of carcinogens have undoubtedly been with us for a long time. The ill effects of some of these genes might well be diminished by lowering the burden of environmental carcinogens to which we are all exposed. In a world free of aromatic amines, for example, being born a slow acetylator would be a trivial issue, not a matter of grave consequence. The inheritance of a defective carcinogen-detoxifying gene would matter less in a culture that did not tolerate carcinogens in air, food, and water. By contrast, we cannot change our ancestors. Shining the spotlight on inheritance focuses us on the one piece of the puzzle we can do absolutely nothing about. . . .

During the last year of her life, Rachel Carson discussed before a U.S. Senate subcommittee her emerging ideas about the relationship between environmental contamination and human rights. The problems addressed in *Silent Spring*, she asserted, were merely one piece of a larger story—namely, the threat to human health created by reckless pollution of the living world. Abetting this hidden menace was a failure to inform common citizens about the senseless and frightening dangers they were being asked, without their consent, to endure. In *Silent Spring*, Carson had predicted that full knowledge of this situation would lead us to reject the counsel of those who claim there is simply no choice but to go on filling the world with poisons. Now she urged recognition of an individual's right to know about poisons introduced into one's environment by others and the right to protection against them. These ideas are Carson's final legacy.

The process of exploration that results from asserting our right to know about carcinogens in our environment is a different journey for every person who undertakes it. For all of us, however, I believe it necessarily entails a three-part inquiry. Like the Dickens character Ebenezer Scrooge, we must first look back into our past, then reassess our present situation, and finally summon the courage to imagine an alternative future.

We begin retrospectively for two reasons. First, we carry in our bodies many carcinogens that are no longer produced and used domestically but which linger in the environment and in human tissue. Appreciating how, even today, we remain in contact with banned chemicals such as PCBs and DDT requires a historical understanding. Second, because cancer is a multicausal disease that unfolds over a period of decades, exposures during young adulthood, adolescence, childhood—and even prior to birth—are relevant to our present cancer risks. We need to find out what pesticides were sprayed in our neighborhoods and what sorts of household chemicals were stored under our parents' kitchen sink. Reminiscing with neighbors, family members, and elders in the community where one grew up can be an eye-opening first step.

This part of the journey is, in essence, a search for our ecological roots. Just as awareness of our genealogical roots offers us a sense of heritage and cultural identity, our ecological roots provide a particular appreciation of who we are biologically. It means asking questions about the physical environment we have grown up within and whose molecules are woven together with the strands of DNA inherited from our genetic ancestors. After all, except for the original blueprint of our chromosomes, all the material that is us—from bone to blood to breast tissue—has come to us from the environment. . . .

In full possession of our ecological roots, we can begin to survey our present situation. This requires a human rights approach. Such an approach recognizes that the current system of regulating the use, release, and disposal of known and suspected carcinogens—rather than preventing their generation in the first place—is intolerable. So is the decision to allow untested chemicals free access to our bodies, until which time they are finally assessed for carcinogenic properties. Both practices show reckless disregard for human life.

A human rights approach would also recognize that we do not all bear equal risks when carcinogens are allowed to circulate within our environment. Workers who manufacture carcinogens are exposed to higher levels, as are those who live near the chemical graveyards that serve as their final resting place. Moreover, people are not uniformly vulnerable to effects of environmental carcinogens. Individuals with genetic predispositions, infants whose detoxifying mechanisms are not yet fully developed, and those with significant prior exposures may all be affected more profoundly. Cancer may be a lottery, but we do not each of us hold equal chances of "winning." When carcinogens are deliberately or accidentally introduced into the environment, some number of vulnerable persons are consigned to death. The impossibility of tabulating an exact body count does not alter this fact. A human rights approach to cancer strives, nonetheless, to make these deaths visible.

Suppose we assume for a moment that the most conservative estimate concerning the proportion of cancer deaths due to environmental causes is absolutely accurate. This estimate, put forth by those who dismiss environmental carcinogens as negligible, is 2 percent. Though others have placed this number far higher, let's assume for the sake of argument that this lowest value is absolutely correct. Two percent means that 10,940 people in the United States die each year from environmentally caused cancers. This is more than the number of women who die each year from hereditary breast cancer—an issue that has launched multi-million-dollar research initiatives. This is more than the number of children and teenagers killed each year by firearms—an issue that is considered a matter of national shame. It is more than three times the number of nonsmokers estimated to die each year of lung cancer caused by exposure to secondhand smoke—a problem so serious it warranted sweeping changes in laws governing air quality in public spaces. It is the annual equivalent of wiping out a small city. It is thirty funerals every day.

None of these 10,940 Americans will die quick, painless deaths. They will be amputated, irradiated, and dosed with chemotherapy. They will expire privately in hospitals and hospices and be buried quietly. Photographs of their bodies will not appear in newspapers. We will not know who most of them are. Their anonymity, however, does not moderate violence. These deaths are a form of homicide.

All activities with potential public health consequences should be guided by the *principle of the least toxic alternative*, which presumes that toxic substances will not be used as long as there is another way of accomplishing the task. This means choosing the least harmful way of solving problems—whether it be ridding fields of weeds, school cafeterias of cockroaches, dogs of fleas, woolens of stains, or drinking water of pathogens. Biologist Mary O'Brien advocates a system of alternatives assessment in which facilities regularly evaluate the availability of alternatives to the use and release of toxic chemicals. Any departure from zero should be preceded by a finding of necessity. These efforts, in turn, should be coordinated with active attempts to develop and make available affordable, nontoxic alternatives for currently toxic processes and with systems of support for those making the transition—whether farmer, corner drycleaner, hospital, or machine shop. Receiving the highest priority for transformation should be all processes that generate dioxin or require the use or release of any known human carcinogen such as benzene and vinyl chloride.

The principle of the least toxic alternative would move us away from protracted, unwinnable debates over how to quantify the cancer risks from each individual carcinogen released into the environment and where to set legal maximum limits for their presence in air, food, water, workplace, and consumer goods. As O'Brien observed, "Our society proceeds on the assumption that toxic substances *will* be used and the only question is how much. Under the current system, toxic chemicals are used, discharged, incinerated, and buried without ever requiring a finding that these activities are necessary." The principle of the least toxic alternative looks toward the day when the availability of safer choices makes the deliberate and routine release of chemical carcinogens into the environment as unthinkable as the practice of slavery.

Critical Thinking

1. Why is it hard to tell whether—and how much—a chemical is carcinogenic in humans?
2. What is wrong with the present system of regulating the use, release, and disposal of known and suspected carcinogens?
3. Should there be more effort to study the links between cancer and environmental factors?

Our Stolen Future

Theo Colborn, Dianne Dumanoski, and John Peterson Myers

Until recently most efforts to assess the potential toxicity of synthetic industrial chemicals focused almost exclusively on their role as carcinogens. This was because of legitimate public concern about rising cancer rates and the belief among scientists that cancer was most likely caused by exposure to low levels of synthetic chemicals. Although some scientists urged public health officials to give serious consideration to other possible health effects of environmental contaminants, they were generally ignored. This was due to limited funding for research and the common belief that the toxic effects of synthetic chemicals required larger exposure than what the public ordinarily experienced.

In the late 1980s Theo Colborn, a research scientist for the World Wildlife Fund who was then working on a study of pollution in the Great Lakes, began the process of linking together the results of a growing series of isolated studies. Researchers in the Great Lakes region, as well as in Florida, on the West Coast of the United States, and in Northern Europe, had observed widespread evidence of serious and frequently lethal physiological problems. These problems included abnormal reproductive development, unusual sexual behavior, and neurological impairment, and were exhibited by a diverse group of animal species. As a result of Colborn's insights, communications among researchers, and further studies, a hypothesis was developed that all of these wildlife problems were manifestations of abnormal estrogenic activity. The causative agents were identified as more than 50 synthetic chemical compounds that have been shown in laboratory studies to either mimic the action, or disrupt the normal function, of the powerful hormones responsible for sexual development and many other biological functions.

The following selection is taken from *Our Stolen Future* (Dutton, 1996) in which Colborn and her coauthors, journalists Dianne Dumanoski and John Peterson Myers, Senior Advisor to The United Nations Foundation and a Senior Fellow at Commonweal, summarize their research on the effects of exposure to environmental hormone mimics. The title of the book reflects their warning that unless action is taken to reduce such exposure, the fertility, intelligence, and even the survival of future generations may be threatened. In fact, such action has begun. In response to the mounting evidence that environmental estrogens may be a serious health threat, the United States Congress passed legislation requiring that all pesticides be screened for estrogenic activity and that the Environmental Protection Agency develop procedures for detecting environmental estrogenic contaminants in drinking water supplies.

Key Concept: disruptive developmental effects of environmental hormone mimics

Altered Destinies

"Our fate is connected with the animals," Rachel Carson wrote more than three decades ago in *Silent Spring*, a now classic indictment of synthetic pesticides and human hubris that helped launch the modern environmental movement. This has long been a guiding belief among environmentalists, wildlife biologists, and others who recognize two fundamental realities—our shared evolutionary inheritance and our shared environment. What is happening to the animals in Florida, English rivers, the Baltic, the high Arctic, the Great Lakes, and Lake Baikal in Siberia has immediate relevance to humans. The damage seen in lab animals and in wildlife has ominously foreshadowed symptoms that appear to be increasing in the human population.

. . . [B]asic physiological processes such as those governed by the endocrine system have persisted relatively unchanged through hundreds of millions of years of evolution. Evolutionary narratives tend to highlight the innovations of natural selection, ignoring the stubborn conservative streak that has marked the history of life on Earth. At the same time that evolution experimented greatly with form, shaping the vessels in various and wondrous ways, it has strayed surprisingly little from an ancient recipe for life's biochemical brew. In examining our place in the evolutionary lineage, humans tend to focus inordinately on those characteristics that make us unique. But these differences are small, indeed, when compared to how much we share not only with other primates such as chimpanzees and gorillas, but with mice, alligators, turtles, and other vertebrates. Though turtles and humans bear little physical similarity, our kinship is unmistakable. The estrogen circulating in the painted turtle seen basking on logs during lazy summer afternoons is exactly the same as the estrogen rushing through the human bloodstream.

Humans and animals share a common environment as well as a common evolutionary legacy. Living in a man-made landscape, we easily forget that our well-being is rooted in natural systems. Yet all human enterprise rests on the foundation of natural systems that provide a myriad of invisible life-support services. Our connections to these natural systems may be less direct and obvious than those of an eagle or an otter, but we are no less deeply implicated in life's web. No one has stated this fundamental ecological principle more simply than the early twentieth-century American environmental philosopher, John Muir. "When we try to pick out anything by itself, we find that it is bound by a thousand invisible cords to everything in the universe."

Our regrettable experience with persistent chemicals over the past half century has demonstrated the reality of this deep and complex interconnection. Whether we live in Tokyo, New York, or a remote Inuit village in the Arctic thousands of miles from farm fields or sources of industrial pollution, all of us have accumulated a store of persistent synthetic chemicals in our body fat. Through this web of inescapable connection, these chemicals have found their way to each and every one of us just as they have found their way to the birds, seals, alligators, panthers, whales, and polar bears. With this shared biology and shared contamination, there is little reason to expect that humans will in the long term have a separate fate.

Yet, some skeptics question whether animal studies provide a useful tool for forecasting threats to humans. The refrain that "mice are not little people" has been heard frequently in the ongoing debate about whether animal testing accurately predicts whether a chemical poses a cancer risk to humans. Critics have also attacked testing procedures for using unrealistically high doses, arguing, for example, that mice tested to discover if DDT caused cancer were fed more than eight hundred times the average amount humans would take in eating a typical diet.

Whatever the merits of these criticisms regarding cancer testing, they have little relevance to the use of animals to predict the effects of hormone-disrupting chemicals. Because scientists have only an incomplete understanding of the basic mechanisms that induce cancer, extrapolating from one species to another has admitted uncertainties. In contrast, scientists have a good grasp of the mechanisms and actions of hormones. They understand how chemical messages are sent and received and how some synthetic chemicals disrupt this communication. They know that hormones guide development in basically the same way in all mammals, and if there was any doubt, the DES [the synthetic hormone drug diethylstilbestrol] experience has verified the similarity of disruption across many species, including humans. Time after time, the abnormalities first seen in laboratory experiments with DES later showed up in the children of women who had taken this drug during pregnancy.

The relevance of the DES experience to the threat from environmental endocrine disruptors has also been questioned because of the very high doses given to pregnant women and to laboratory animals in experiments. While most of the early experiments did indeed use high doses, recent studies using much lower doses have produced no less alarming results. In fact, in some cases a high dose may paradoxically cause *less* damage than a lower dose. In exploring the effects of much lower doses of DES, Fred vom Saal has found that the response increases with dose for a time and then, with even higher doses, begins to diminish.

Vom Saal's dose response curve looks like an inverted U. Its shape is profoundly important to the interaction between the endocrine system and synthetic contaminants. Neither linear nor always moving in the same direction, the inverted U seems characteristic of hormone systems and it means that they do not conform to the assumptions that underlie classical toxicology—that a biological response always increases with dose. It means that testing with very high doses will miss some effects that would show up if the animals were given lower doses. The inverted U is another example of how the action of endocrine disruptors challenges prevailing notions about toxic chemicals. Extrapolation from high-dose tests to lower doses may in some cases seriously underestimate risks rather than exaggerate them.

Because the endocrine disruption question has surfaced so recently, the scientific case on the extent of the threat is still far from complete. Nevertheless, if one looks broadly at a wide array of existing studies from various branches of science and medicine, the weight of the evidence indicates that humans are in jeopardy and are perhaps already affected in major ways. Taken together, the pieces of this scientific patchwork quilt have, despite admitted gaps, a cumulative power that is compelling and urgent.

This was the lesson from the historic meeting on endocrine disruption that took place in July 1991 at the Wingspread Conference Center in Racine, Wisconsin. Over the years, dozens of scientists have explored isolated pieces of the puzzle of hormone disruption, but the larger picture did not emerge until Theo Colborn and Pete Myers finally brought twenty-one of the key researchers together. At this unique gathering, specialists from diverse disciplines ranging from anthropology to zoology shared what they knew about the role of hormones in normal development and about the devastating impacts of hormone-disrupting chemicals on wildlife, laboratory animals, and humans.

For the first time, Ana Soto, Frederick vom Saal, Michael Fry, Howard Bern, John McLachlan, Earl Gray, Richard Peterson, Peter Reijnders, Pat Whitten, Melissa Hines, and others explored the exciting connections between their work and the ominous implications that arose from this exercise. As the evidence was laid out, the parallels proved remarkable and deeply disturbing. The conclusion seemed inescapable: the hormone disruptors threatening the survival of animal populations are also jeopardizing the human future.

At the end of the session, the scientists issued the Wingspread Statement, an urgent warning that humans in many parts of the world are being exposed to chemicals that have disrupted development in wildlife and laboratory animals, and that unless these chemicals are controlled, we face the danger of widespread disruption in human embryonic development and the prospect of damage that will last a lifetime.

The pressing question is whether humans are already suffering damage from half a century of exposure to endocrine-disrupting synthetic chemicals. Have these man-made chemicals already altered individual destinies by scrambling the chemical messages that guide development?

Many of those familiar with the scientific case believe the answer is yes. Given human exposure to dioxinlike chemicals, for example, it is probable that some humans, especially the most sensitive individuals, are suffering some effects. But whether hormone-disrupting chemicals are now having a broad impact across the human population is difficult to assess and even harder to prove. This is inescapable in light of the nature of the contamination, the transgenerational effects, the often long lag time before damage becomes evident, and the invisible nature of much of this damage. Those trying to document whether perceived increases in specific problems reflect genuine trends in human health find themselves thwarted by a dearth of reliable medical data.

Few disease registries exist for anything except cancers. A number of pediatricians from various parts of the United States have expressed their concern about an increasing frequency of genital abnormalities in children such as undescended testicles, extremely small penises, and hypospadias, a defect in which the urethra that carries urine does not extend to the end of the penis, but it is virtually impossible to document these anecdotal reports. Unfortunately, the problems caused by endocrine disruption may have to reach crisis proportion before we have a clear sign that something serious is happening.

In the face of these difficulties, the animal studies provide a touchstone for identifying and investigating what might be happening in humans. They can alert us to the probable kinds of disruptions and help focus research efforts. They can also provide early warnings about the hazards of current levels of contamination. Because of the diversity of life, some animals are bound to be more readily exposed to contaminants than humans. Transgenerational effects, such as changes in behavior and diminished fertility, are also likely to show up faster in wildlife because most animals mature and reproduce more quickly than humans. Experimental work with animals adds another equally invaluable dimension. As the history of DES demonstrates, laboratory experiments with rats and mice accurately forecasted damage that later showed up in humans. The tragedy is that we ignored the warnings.

Critical Thinking

1. Is cancer the only health effect of environmental contaminants that should concern us?
2. What is an environmental hormone mimic (or disruptor)?
3. What effects are environmental hormone mimics *known* to have on animals?
4. What effects are environmental hormone mimics *known* to have on humans?

Part 5: Environment and Society

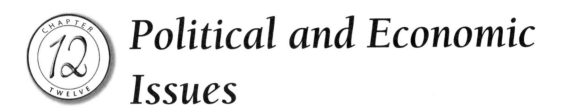 *Political and Economic Issues*

Learning Outcomes

After reading this unit, you should be able to:

1. Explain how people do not always act in regard to the environment as they think they should.
2. Explain why monetary considerations are not the only ones that pertain to the environment.
3. Describe what environmental justice is and explain why it matters.
4. Explain how "free" ecosystem services add benefit to the human economy.
5. Describe what the "replacement value" of an ecosystem service is.

At the Shrine of Our Lady of Fatima or Why Political Questions Are Not All Economic, as seen in Arizona Law Review

Mark Sagoff

The fact that at a local, national, or global level, insufficient public and private resources exist to address every conceivable environmental problem is not seriously contested by anyone. For this reason alone it is necessary to establish a system of priorities to guide those responsible for setting ecological and environmental protection policies. How to achieve this task is not obvious. Who should decide? On what basis should the decisions be made? Numerous surveys have shown that the ranking of the significance of environmental threats—such as hazardous wastes, air pollutants, pesticides, and ozone-depleting chemicals—by the general public and by environmental scientists result in very different sets of priorities.

Mark Sagoff is director of the Institute for Philosophy and Public Policy at the University of Maryland and President of the International Society of Environmental Ethics. He has published numerous articles describing his views on the proper place of economics and risk-benefit analyses in decision making about environmental issues in philosophy of law and environmental journals. His most recent books are *The Economy of the Earth*, 2nd Ed. (Cambridge University Press, 2008) and *Price, Principle, and the Environment* (Cambridge University Press, 2004)."

As indicated in the following selection from "At the Shrine of Our Lady of Fatima or Why Political Questions Are Not All Economic," *Arizona Law Review* (vol. 23, 1981), Sagoff believes that environmental problems are complex and require more comprehensive assessment than can be made on the basis of economic analyses alone. He maintains that it is impossible to place economic values on aesthetic satisfaction or physical well-being. More profoundly, he challenges the assumption that the true value of all things can be judged by what people are willing to pay for them in the marketplace.

Key Concept: the limitations of economics in making environmental policy

Lewiston, New York, a well-to-do community near Buffalo, is the site of the Lake Ontario Ordinance Works, where years ago the federal government disposed of the residues of the Manhattan Project. These radioactive wastes are buried but are not forgotten by the residents who say that when the wind is southerly, radon gas blows through the town. Several parents at a recent Lewiston conference I attended described their terror on learning that cases of leukemia had been found among area children. They feared for their own lives as well. On the otherside of the table, officials from New York State and from local corporations replied that these fears were ungrounded. People who smoke, they said, take greater risks than people who live close to waste disposal sites.

One speaker talked in terms of "rational methodologies of decisionmaking." This aggravated the parents' rage and frustration.

The speaker suggested that the townspeople, were they to make their decision in a free market and if they knew the scientific facts, would choose to live near the hazardous waste facility. He told me later they were irrational—"neurotic"—because they refused to recognize or to act upon their own interests. The residents of Lewiston were unimpressed with his analysis of their "willingness to pay" to avoid this risk or that. They did not see what risk-benefit analysis had to do with the issues they raised.

If you take the Military Highway (as I did) from Buffalo to Lewiston, you will pass through a formidable wasteland. Landfills stretch in all directions and enormous trucks—tiny in that landscape—incessantly deposit sludge which great bulldozers then push into the ground. These machines are the only signs of life, for in the miasma that hangs in the air, no birds, not even scavengers, are seen. Along colossal power lines which criss-cross this dismal land, the dynamos at Niagra send electric power south, where factories have fled, leaving their remains to decay. To drive along this road is to feel, oddly, the mystery and awe one experiences in the presence of so much power and decadence. . . .

At the Shrine of Our Lady of Fatima, on a plateau north of the Military Highway, a larger than life sculpture of Mary looks into the chemical air. The original of this shrine stands in central Portugal where in May, 1917, three children said they saw a Lady, brighter than the sun, raised on a cloud in an evergreen tree. Five months later, on a wet and chilly October day, the Lady again appeared, this time before a large crowd. Some who were skeptical did not see the miracle. Others in the crowd reported, however, that "the sun appeared and seemed to tremble, rotate violently and fall, dancing over the heads of the throng. . . ."

The Shrine was empty when I visited it. The cult of Our Lady of Fatima, I imagine, has only a few devotees. The cult of Pareto Optimality, however, has many. Where some people see only environmental devastation, its devotees perceive efficiency, utility, and the maximization of wealth. They see the satisfaction of wants. They envision the good life. As I looked over the smudged and ruined terrain I tried to share that vision. I hoped that Our Lady of Fatima, worker of miracles, might serve, at least for the moment, as the Patroness of cost-benefit analysis. I thought of all the wants and needs that are satisfied in a landscape of honeymoon cottages, commercial strips, and dumps for hazardous waste. I saw the miracle of efficiency. The prospect, however, looked only darker in that light.

Political and Economic Decisionmaking

This essay concerns the economic decisions we make about the environment. It also concerns our political decisions about the environment. Some people have suggested that ideally these should be the same, that all environmental problems are problems in distribution. According to this view, there is an environmental problem only when some resource is not allocated in equitable and efficient ways.

This approach to environmental policy is pitched entirely at the level of the consumer. It is his or her values that count, and the measure of these values is the individual's willingness to pay. The problem of justice or fairness in society becomes, then, the problem of distributing goods and services so that more people get more of what they want to buy: a condo on the beach, a snowmobile for the mountains, a tank full of gas, a day of labor. The only values we have, according to this view, are those that a market can price.

How much do you value open space, a stand of trees, an "unspoiled" landscape? Fifty dollars? A hundred? A thousand? This is one way to measure value. You could compare the amount consumers would pay for a townhouse or coal or a landfill to the amount they would pay to preserve an area in its "natural" state. If users would pay more for the land with the house, the coal mine, or the landfill, then without—less construction and other costs of development—than the efficient thing to do is to improve the land and thus increase its value. That is why we have so many tract developments, pizza stands, and gas stations. How much did you spend last year to preserve open space? How much for pizza and gas? "In principle, the ultimate measure of environmental quality," as one basic text assures us, "is the value people place on these. . . services or their *willingness to pay.*"[1]

Willingness to pay: what is wrong with that? The rub is this: not all of us think of ourselves simply as *consumers*. Many of us regard ourselves *as citizens* as well. We act as consumers to get what we want *for ourselves*. We act as citizens to achieve what we think is right or best *for the community*. The question arises, then, whether what we want for ourselves individually as consumers is consistent with the goals we would set for ourselves collectively as citizens. Would I vote for the sort of things I shop for? Are my preferences as a consumer consistent with my judgments as a citizen?

They are not. I am schizophrenic. Last year, I fixed a couple of tickets and was happy to do so since I saved fifty dollars. Yet, at election time, I helped to vote the corrupt judge out of office. I speed on the highway; yet I want the police to enforce laws against speeding. I used to buy mixers in returnable bottles—but who can bother to return them? I buy only disposables now, but to soothe my conscience, I urge my state senator to outlaw one-way containers. I love my car; I hate the bus. Yet I vote for candidates who promise to tax gasoline to pay for public transportation. And of course I applaud the Endangered Species Act, although I have no earthly use for the Colorado squawfish or the Indiana bat. I support almost any political cause that I think will defeat my consumer interests. This is because I have contempt for—although I act upon—those interests. I have an "Ecology Now" sticker on a car that leaks oil everywhere it's parked.

The distinction between consumer and citizen preferences has long vexed the theory of public finance. Should the public economy serve the same goals as the household economy? May it serve, instead, goals emerging from our association as citizens? The question asks if we may collectively strive for and achieve only those items we individually compete for and consume. Should we aspire, instead, to public goals we may legislate as a nation?. . .

Cost-benefit Analysis vs. Regulation

On February 19, 1981, President Reagan published Executive Order 12,291 requiring all administrative agencies and departments to support every new major regulation with a cost-benefit analysis establishing that the benefits of the regulation to society outweigh its costs. The order directs the Office of Management and Budget (OMB) to review every such regulation on the basis of the adequacy of the cost-benefit analysis supporting it. This is a departure from tradition. Historically, regulations have been reviewed not by OMB but by the courts on the basis of the relation of the regulation to authorizing legislation, not to cost-benefit analysis.

A month earlier, in January, 1981, the Supreme Court heard lawyers for the American Textile Manufacturers Institute argue against a proposed Occupational Safety and Health Administration (OSHA) regulation which would have severely restricted the acceptable levels of cotton dust in textile plants. The lawyers for industry argued that the benefits of the regulation would not equal the costs. The lawyers for the government contended that the law required the tough standard. OSHA, acting consistently with Executive Order 12,291, asked the Court not to decide the cotton dust case in order to give the agency time to complete the cost-benefit analysis required by the textile industry. The Court declined to accept OSHA's request and handed down its opinion in *American Textile Manufacturer v. Donovan* on June 17, 1981.

The Supreme Court, in a 5–3 decision, found that the actions of regulatory agencies which conform to the OSHA law need not be supported by cost-benefit analysis. In addition, the Court asserted that Congress, in writing a statute, rather than the agencies in applying it, has the primary responsibility for balancing benefits and costs. The Court said:

> When Congress passed the Occupational Health and Safety Act in 1970, it chose to place pre-eminent value on assuring employees a safe and healthful working environment, limited only by the feasibility of achieving such an environment. We must measure the validity of the Secretary's actions against the requirements of that Act.

The opinion upheld the finding of the District of Columbia Court of Appeals that "Congress itself struck the balance between costs and benefits in the mandate to the agency."

The Appeals Court opinion in *American Textile Manufacturers v. Donovan* supports the principle that legislatures are not necessarily bound to a particular conception of regulatory policy. Agencies that apply the law therefore may not need to justify on cost-benefit grounds the standards they set. These standards may conflict with the goal of efficiency and still express our political will as a nation. That is, they may reflect not the personal choices of self-interested individuals, but the collective

judgments we make on historical, cultural, aesthetic, moral, and ideological grounds.

The appeal of the Reagan Administration to cost-benefit analysis, however, may arise more from political than economic considerations. The intention seen in the most favorable light, may not be to replace political or ideological goals with economic ones, but to make economic goals more apparent in regulation. This is not to say that Congress should function to reveal a collective willingness-to-pay just as markets reveal an individual willingness-to-pay. It is to suggest that Congress should do more to balance economic with ideological, aesthetic, and moral goals. To think that environmental or worker safety policy can be based exclusively on aspiration for a "natural" and "safe" world is as foolish as to hold that environmental law can be reduced to cost-benefit accounting. The more we move to one extreme, as I found in Lewiston, the more likely we are to hear from the other.

Substituting Efficiency for Safety

The labor unions won an important political victory when Congress passed the Occupational Safety and Health Act of 1970. That Act, among other things, severely restricts worker exposure to toxic substances. It instructs the Secretary of Labor to set "the standard which most adequately assures, to the extent feasible. . .that no employee will suffer material impairment of health or functional capacity even if such employee has regular exposure to the hazard. . .for the period of his working life."

Pursuant to this law, the Secretary of Labor in 1977 reduced from ten to one part per million (ppm) the permissable ambient exposure level for benzene, a carcinogen for which no safe threshold is known. The American Petroleum Institute thereupon challenged the new standard in court. It argued, with much evidence in its favor, that the benefits (to workers) of the one ppm standard did not equal the costs (to industry). The standard therefore did not appear to be a rational response to a market failure in that it did not strike an efficient balance between the interests of workers in safety and the interests of industry and consumers in keeping prices down.

The Secretary of Labor defended the tough safety standard on the ground that the law demanded it. An efficient standard might have required safety until it cost industry more to prevent a risk than it cost workers to accept it. Had Congress adopted this vision of public policy—one which can be found in many economics texts—it would have treated workers not as ends-in-themselves but as means for the production of overall utility. This, as the Secretary saw it, was what Congress refused to do.

The United States Court of Appeals for the Fifth Circuit agreed with the American Petroleum Institute and invalidated the one ppm benzene standard. On July 2, 1980, the Supreme Court affirmed the decision in *American Petroleum Institute v. Marshall* and remanded the benzene standard back to OSHA for revision. The narrowly based Supreme Court decision was divided over the role economic considerations should play in judicial review. Justice Marshall, joined in dissent by three other justices, argued that the Court had undone on the basis of its own theory of regulatory policy an act of Congress inconsistent with that theory. He concluded that the plurality decision of the Court "requires the American worker to return to the political arena to win a victory that he won before in 1970."

The decision of the Supreme Court is important not because of its consequences, which are likely to be minimal, but because of the fascinating questions it raises. Shall the courts uphold only those political decisions that can be defended on economic grounds? Shall we allow democracy only to the extent that it can be construed either as a rational response to a market failure or as an attempt to redistribute wealth? Should the courts say that a regulation is not "feasible" or "reasonable"—terms that occur in the OSHA law—unless it is supported by a cost-benefit analysis?

The problem is this: An efficiency criterion, as it is used to evaluate public policy, assumes that the goals of our society are contained in the preferences individuals reveal or would reveal in markets. Such an approach may appear attractive, even just, because it treats everyone as equal, at least theoretically, by according to each person's preferences the same respect and concern. To treat a person with respect, however, is also to listen and to respond intelligently to his or her views and opinions. This is not the same thing as to ask how much he or she is willing to pay for them. The cost-benefit analyst does not ask economists how much they are willing to pay for what they believe, that is, that the workplace and the environment should be made efficient. Why, then, does the analyst ask workers, environmentalists, and others how much they are willing to pay for what they believe is right? Are economists the only ones who can back their ideas with reasons while the rest of us can only pay a price? The cost-benefit approach treats people as of equal worth because it treats them as of no worth, but only as places or channels at which willingness to pay is found.

Liberty: Ancient and Modern

When efficiency is the criterion of public safety and health, one tends to conceive of social relations on the model of a market, ignoring competing visions of what we as a society should be like. Yet it is obvious that there are competing conceptions of what we should be as a society. There are some who believe on principle that worker safety and environmental quality ought to be protected only insofar as the benefits of protection balance the costs. On the other hand, people argue—also on principle—that neither worker safety nor environmental quality should be treated merely as a commodity to be traded at the margin for other commodities, but rather each should be valued for its own sake. The conflict between these two principles is logical or moral, to be resolved by argument or debate. The question whether cost-benefit analysis should play a decisive role in policy-making is not to be decided by cost-benefit analysis. A contradiction between principles—between contending visions of the good society—cannot be settled by asking how much partisans are willing to pay for their beliefs.

The role of the *legislator*, the political role, may be more important to the individual than the role of *consumer*. The person, in other words, is not to be treated merely as a bundle of preferences to be juggled in cost-benefit analyses. The individual is to be respected as an advocate of ideas which are to be judged according to the reasons for them. If health and environmental statutes reflect a vision of society as something other than a market by requiring protections beyond what are efficient, then this may express not legislative ineptitude but legislative responsiveness to public values. To deny this vision because it is economically inefficient is simply to replace it with another vision. It is to insist that the ideas of the citizen be sacrificed to the psychology of the consumer.

We hear on all sides that government is routinized, mechanical, entrenched, and bureaucratized; the jargon alone is enough to dissuade the most mettlesome meddler. Who can make a difference? It is plain that for many of us the idea of a national political community has an abstract and suppositious quality. We have only our private conceptions of the good, if no way exists to arrive at a public one. This is only to note the continuation, in our time, of the trend Benjamin Constant described in the essay *De La Liberte des Anciens Comparee a Celle des Modernes*. Constant observes that the modern world, as opposed to the ancient, emphasizes civil over political liberties, the rights of privacy and property over those of community and participation. "Lost in the multitude," Constant writes, "the individual rarely perceives the influence that he exercises," and, therefore, must be content with "the peaceful enjoyment of private independence."[2] The individual asks only to be protected by laws placed by the consumer; the tradition of Rousseau has been supplanted by that of Locke and Mill.

Nowhere are the rights of the moderns, particularly the rights of privacy and property, less helpful than in the area of the natural environment. Here the values we wish to protect—cultural, historical, aesthetic, and moral—are public values. They depend not so much upon what each person wants individually as upon what he or she thinks is right for the community. We refuse to regard worker health and safety as commodities; we regulate hazards as a matter of right. Likewise, we refuse to treat environmental resources simply as public goods in the economist's sense. Instead, we prevent significant deterioration of air quality not only as a matter of individual self-interest but also as a matter of collective self-respect. How shall we balance efficiency against moral, cultural, and aesthetic values in policy for the workplace and the environment? No better way has been devised to do this than by legislative debate ending in a vote. This is very different from a cost-benefit analysis terminating in a bottom line.

Values Are Not Subjective

It is the characteristic of cost-benefit analysis that it treats all value judgments other than those made on its behalf as nothing but statements of preference, attitude, or emotion, insofar as they are value judgments. The cost-benefit analyst regards as true the judgment that we should maximize efficiency or wealth. The analyst believes that this view can be backed by reasons, but does not regard it as a preference or want for which he or she must be willing to pay. The cost-benefit analyst tends to treat all other normative views and recommendations as if they were nothing but subjective reports of mental states. The analyst supposes in all such cases that "this is right" and "this is what we ought to do" are equivalent to "I want this" and "this is what I prefer." Value judgments are beyond criticism if, indeed, they are nothing but expressions of personal preference; they are incorrigible since every person is in the best position to know what he or she wants. All valuation, according to this approach, happens *in foro interno*; debate *in foro publico* has no point. With this approach, the reasons that people give for their views do not count; what does count is how much they are willing to pay to satisfy their wants. Those who are willing to pay the most, for all intents and purposes, have the right view; theirs is the more informed opinion, the better aesthetic judgment, and the deeper moral insight. . . .

Economists. . . argue that their role as policy-makers is legitimate because they are neutral among competing values in the client society. The political economist, according to James Buchanan, "is or should be ethically neutral: the indicated results are influenced by his own value scale only insofar as this reflects his membership in a larger group." The economist might be most confident of the impartiality of his or her policy recommendations if he or she could derive them formally or mathematically from individual preferences. If theoretical difficulties make such a social welfare function impossible, however, the next best thing, to preserve neutrality, is to let markets function to transform individual preference orderings into a collective ordering of social states. The analyst is able then to base policy on preferences that exist in society and are not necessarily his own.

Economists have used this impartial approach to offer solutions to many significant social problems, for example, the controversy over abortion. An economist argues that "there is an optimal number of abortions, just as there is an optimal level of pollution, or purity. . . .Those who oppose abortion could eliminate it entirely, if their intensity of feeling were so strong as to lead to payments that were greater at the margin than the price anyone would pay to have an abortion."[3] Likewise, economists, in order to determine whether the war in Vietnam was

justified, have estimated the willingness to pay of those who demonstrated against it. Following the same line of reasoning, it should be possible to decide whether Creationism should be taught in the public schools, whether black and white people should be segregated, whether the death penalty should be enforced, and whether the square root of six is three. All of these questions arguably depend upon how much people are willing to pay for their subjective preferences or wants. This is the beauty of cost-benefit analysis: no matter how relevant or irrelevant, wise or stupid, informed or uninformed, responsible or silly, defensible or indefensible wants may be, the analyst is able to derive a policy from them—a policy which is legitimate because, in theory, it treats all of these preferences as equally valid and good.

Preference or Principle?

In contrast, consider a Kantian conception of value. The individual, for Kant, is a judge of values, not a mere haver of wants, and the individual judges not for himself or herself merely, but as a member of a relevant community or group. The central idea in a Kantian approach to ethics is that some values are more reasonable than others and therefore have a better claim upon the assent of members of the community as such. The world of obligation, like the world of mathematics or the world of empirical fact, is public not private, and objective standards of argument and criticism apply. Kant recognized that values, like beliefs, are subjective states of mind which have an objective content as well. Therefore, both values and beliefs are either correct or mistaken. A value judgment is like an empirical or theoretical judgment in that it claims to be *true* not merely to be *felt*.

We have, then, two approaches to public policy before us. The first, the approach associated with normative versions of welfare economics, asserts that the only policy recommendation that can or need be defended on objective grounds is efficiency or wealth-maximization. The Kantian approach, on the other hand, assumes that many policy recommendations may be justified or refuted on objective grounds. It would concede that the approach of welfare economics applies adequately to some questions, for example, those which ordinary consumer markets typically settle. How many yo-yos should be produced as compared to how many frisbees? Shall pens have black ink or blue? Matters such as these are so trivial it is plain that markets should handle them. It does not follow, however, that we should adopt a market or quasi-market approach to every public question.

A market or quasi-market approach to arithmetic, for example, is plainly inadequate. No matter how much people are willing to pay, three will never be the square root of six. Similarly, segregation is a national curse and the fact that we are willing to pay for it does not make it better, but only us worse. The case for abortion must stand on the merits; it cannot be priced at the margin.

Our failures to make the right decisions in these matters are failures in arithmetic, failures in wisdom, failures in taste, failures in morality—but not market failures. There are no relevant markets which have failed.

What separates these questions from those for which markets are appropriate is that they involve matters of knowledge, wisdom, morality, and taste that admit of better or worse, right or wrong, true or false, and not mere economic optimality. Surely environmental questions—the protection of wilderness, habitats, water, land, and air as well as policy toward environmental safety and health—involve moral and aesthetic principles and not just economic ones. This is consistent, of course, with cost-effectiveness and with a sensible recognition of economic constraints.

The neutrality of the economist is legitimate if private preferences or subjective wants are the only values in question. A person should be left free to choose the color of his or her necktie or necklace, but we cannot justify a theory of public policy. . .on that basis. . . .

The neutrality of economics is not a basis for its legitimacy. We recognize it as an indifference toward value—an indifference so deep, so studied, and so assured that at first one hesitates to call it by its right name.

The Citizen as Joseph K.

The residents of Lewiston at the conference I attended demanded to know the truth about the dangers that confronted them and the reasons for those dangers. They wanted to be convinced that the sacrifice asked of them was legitimate even if it served interests other than their own. One official from a large chemical company dumping wastes in the area told them in reply that corporations were people and that people could talk to people about their feelings, interests, and needs. This sent a shiver through the audience. Like Joseph K. in *The Trial*,[4] the residents of Lewiston asked for an explanation, justice, and truth, and they were told that their wants would be taken care of. They demanded to know the reasons for what was continually happening to them. They were given a personalized response instead.

This response, that corporations are "just people serving people," is consistent with a particular view of power. This is the view that identifies power with the ability to get what one wants as an individual, that is, to satisfy one's personal preferences. When people in official positions in corporations or in the government put aside their personal interests, it would follow that they put aside their power as well. Their neutrality then justifies them in directing the resources of society in ways they determine to be best. This managerial role serves not their own interests but those of their clients. Cost-benefit analysis may be seen as a pervasive form of this paternalism. Behind this paternalism, as William Simon observes of the lawyer-client relationship, lies a theory of value that tends to personalize power. "It resists understanding power as a product of class, property, or institutions and collapses power into the personal needs and dispositions of the individuals who command and obey.". . .

In the 1980s, the citizens of Lewiston, surrounded by dynamos, high tension lines, and nuclear wastes, are powerless. They do not know how to criticize power, resist power, or justify power—for to do so depends on making distinctions between good and evil, right and wrong, innocence and guilt, justice and injustice, truth and lies. These distinctions cannot be made out and have no significance within an emotive or psychological theory of value. To adopt this theory is to imagine society as a market in which individuals trade voluntarily and without coercion. No individual, no belief, no faith has authority over them. To have power to act as a nation we must be able to act, at least at times, on a public philosophy, conviction, or faith. We cannot abandon the moral function of public law. The antinomianism of cost-benefit analysis is not enough.

Notes

1. A. Freeman, R. Haveman, A. Kneese, *The Economics of Environmental Policy* (1973).
2. *Oeuvres Politiques de Benjamin Constant*, 269 (C. Louandre, ed. 1874), *quoted in* S. Wolin, Politics and Vision 281 (1960).
3. H. Macaulay & B. Yandle, Environmental Use and the Market 120–21 (1978).
4. F. Kafka, The Trial (rev. ed. trans. 1957).

Critical Thinking

1. Mark Sagoff describes his behavior with regard to environmental issues as "schizophrenic." What does he mean? Is *your* behavior "schizophrenic" in the same way?
2. Why can we not evaluate environmental problems in economic terms alone?
3. What are the limitations of cost-benefit or risk-benefit analysis?

Environmental Justice for All

Robert D. Bullard

The environmental movement has often been described as reflecting the idealist aspirations of white middle- and upper-income people. Indeed, poor people and minority groups were not well represented among those who gathered for teach-ins and other events organized to celebrate the first Earth Day in April 1970. The planners and leaders of that event were all white and relatively affluent. The speeches they made and the goals they set rarely took into account the special needs of the poor or the possible relationship between racial oppression and the burdens of pollution.

The involvement of poor and minority people in grassroots environmental organizing has been growing dramatically for the past quarter century. The movement for environmental justice was triggered in 1982 by demonstrations protesting the decision to locate a poorly planned PCB (poly-chlorinated biphenyls) disposal site adjacent to impoverished African American and Native American communities in Warren County, North Carolina. It has since grown to encompass local, regional, and national groups organized to protest what they claim is systematic discrimination in the setting of environmental goals and the siting of polluting industries and waste disposal facilities in their backyards. At the 1992 UN Conference on Environment and Development in Rio de Janeiro, a set of "Principles of Environmental Justice" was widely circulated and discussed. In 1993 the U.S. Environmental Protection Agency opened an Office of Environmental Equity (now the Office of Environmental Justice; www.epa.gov/compliance/environmentaljustice/index.html) with plans for cleaning up sites in several poor communities. On February 11, 1994, President Bill Clinton made environmental equity a national priority by issuing a sweeping Executive Order on Environmental Justice.

Sociology Professor Robert D. Bullard has emerged as one of the most influential leaders of the environmental justice movement. He is Ware Professor of Sociology and Director of the Environmental Justice Resource Center at Clark Atlanta University. He is the author of numerous articles, scholarly papers, and books (notably *Dumping in Dixie: Race, Class, and Environmental Quality* [Westview Press, 1990]) that address the inequitable treatment of African Americans and other minorities in environmental planning and decision making, and he has taken an active political role in working with both political leaders and grassroots organizations to combat environmental discrimination.

In the following selection, Bullard describes the history of the environmental justice movement, argues that even though environmental racism has been recognized for a quarter of a century, it remains a problem, and calls for government to maintain its commitment to protecting the environment for all Americans.

Key Concept: environmental discrimination against racial minorities and the poor

Almost every day the media discovers an African American community fighting some form of environmental threat from land fills, garbage dumps, incinerators, lead smelters, petrochemical plants, refineries, highways, bus depots, and the list goes on. For years, residents watched helplessly as their communities became dumping grounds.

But citizens didn't remain silent for long. Local activists have been organizing under the mantle of environmental justice since as far back as 1968. In 1979, a landmark environmental discrimination lawsuit filed in Houston, followed by similar litigation efforts in the 1980s, rallied activists to stand up to corporations and demand government intervention. More than a decade ago, environmental activists from across the country came together to pool their efforts and organize.

In 1991, a new breed of environmental activists gathered in Washington D.C., to bring national attention to pollution problems threatening low-income and minority communities. Leaders introduced the concepts of environmental justice, protesting that Black, poor and working-class communities often received less environmental protection than White or more affluent communities. The first National People of Color Environmental Leadership

Summit effectively broadened what "the environment" was understood to mean. It expanded the definition to include where we live, work, play, worship and go to school, as well as the physical and natural world. In the process, the environmental justice movement changed the way environmentalism is practiced in the United States and, ultimately, worldwide.

Because many issues identified at the inaugural summit remain unaddressed, the second National People of Color Environmental Leadership Summit was convened in Washington, D.C., this past October [2002]. The second summit was planned for 500 delegates; but more than 1,400 people attended the four-day gathering.

"We are pleased that the Summit II was able to attract a record number of grassroots activists, academicians, students, researchers, planners, policy analysts and government officials. We proved to the world that our movement is alive and well, and growing," says Beverly Wright, director of the Deep South Center for Environmental Justice at Xavier University in New Orleans and chair of the summit.

The meeting produced two dozen policy papers that show powerful environmental and health disparities between people of color and Whites.

"This is not rocket science. We live these statistics every day, and the sad thing is many of us have to die to prove the point," says Donele Wilkins, executive director of Detroiters Working for Environmental Justice.

At the 2002 Essence Music Festival in New Orleans, the National Black Environmental Justice Network (NBEJN) launched its Healthy and Safe Communities Campaign, which targets childhood lead poisoning, asthma and cancer in the Black community. "We are dead serious about educating and mobilizing Black people to eliminate these diseases that cause so much pain, suffering and death in the Black community," says Damu Smith, executive director of NBEJN.

Birth of a Movement

More than three decades ago, the concept of environmental justice had not registered on the radar screens of many environmental or civil rights groups. But environmental justice fits squarely under the civil rights umbrella. It should not be forgotten that Dr. Martin Luther King Jr. went to Memphis on an environmental and economic justice mission in 1968, seeking support for striking garbage workers who were underpaid and whose basic duties exposed them to dangerous environmentally hazardous conditions. King was killed before completing his mission, but others continued it.

The first lawsuit to challenge environmental discrimination using civil rights law, *Bean v. Southwestern Waste Management, Inc.*, was filed in Houston in 1979. Black homeowners in a suburban middle-income Houston neighborhood and their attorney, Linda McKeever Bullard (my wife) filed a class action lawsuit challenging a waste facility siting.

From the early 1920's through 1978, more than 80 percent of Houston's garbage landfills and incinerators were located in mostly Black neighborhoods—even though Blacks made up only 25 percent of the city's population. The residents were not able to halt the landfill, but they were able to impact the city and state waste facility siting regulations.

The *Bean* lawsuit was filed three years before the environmental justice movement was catapulted into the national limelight after rural and mostly Black Warren County, N.C., was selected as the final resting place for toxic waste. Oil laced with highly toxic PCBs (polychlorinated biphenyls) was illegally dumped along roadways in 14 North Carolina counties in 1978 and cleaned up in 1982. The decision to dispose of the contaminated soil in a landfill in the county sparked protests and more than 500 arrests—marking the first time any Americans had been jailed protesting the placement of waste facility.

The Warren County protesters put "environmental racism" on the map. Environmental racism refers to any environmental policy, practice, or directive that negatively affects (whether intentionally or not) individuals, groups, or communities based on race or color. Although the Warren County protesters were also unsuccessful in blocking the PCB landfill, they galvanized Black church leaders, civil rights organizers, youth and grassroots activists around environmental issues in the Black community.

The events in Warren County prompted District of Columbia Delegate Walter Fauntroy to request a General Accounting Office (GAO) investigation of hazardous waste facilities in the EPA's Region IV—which includes Alabama, Florida, Georgia, Kentucky, Mississippi, South Carolina, North Carolina and Tennessee. The 1983 GAO study found that three of four off-site hazardous waste landfills in the region were located in predominantly Black communities, even though Blacks make up just 20 percent of the region's population.

The events in Warren County led the United Church of Christ (UCC) Commission for Racial Justice to publish its landmark 1987 "Toxic Wastes and Race in the U.S." report. The UCC study documented that three of every five Blacks live in communities with abandoned toxic waste sites.

The publication of my book *Dumping in Dixie: Race, Class and Environmental Quality* in 1990 offered the nation a first-hand glimpse of environmental racism struggles all across the South—a region whose "look-the-other-way" environmental policies allowed it to become the most environmentally befouled part of the country.

Sumter County, Ala., typifies the environmental racism pattern chronicled in *Dumping in Dixie*. The county is 71.8 percent Black, and 35.9 percent of the jurisdiction's residents live below the poverty line. The county is home to the nation's largest hazardous waste landfill, in Emelle, Ala., which is more than 90 percent Black. In

1978, the hazardous waste landfill was lured to Sumter County during a period when no Blacks held public office or served on governing bodies, including the state legislature, county commission, or industrial development board. The landfill, which had been dubbed the "Cadillac of Dumps," accepts hazardous wastes—as much as nearly 800,000 tons per year—from the 48 contiguous states and several foreign countries.

It's about Winning, Not Whining

Environmental justice networks and grassroots community groups are making their voices heard loud and clear. In 1996, after five years of organizing, Citizens Against Toxic Exposure convinced the EPA to relocate 358 Pensacola, Fla., families from a dioxin dump, marking the first time a Black community was relocated under the federal government's giant Superfund program.

After eight years in a struggle that began in 1989, Citizens Against Nuclear Trash (CANT) defeated the plans by Louisiana Energy Services (LES) to build the nation's first privately owned uranium enrichment plant in the mostly Black rural communities of Forest Grove and Center Springs, La. On May 1, 1997, a three-judge panel of the Nuclear Regulatory Commission Atomic Safety and Licensing Board ruled that "racial bias played a role in the selection process." The court decision was upheld on appeal April 4, 1998.

In September 1998, after more than 18 months of intense grassroots organizing and legal maneuvering, St. James Citizens for Jobs and the Environment forced the Japanese-owned Shintech Inc. to scrap its plan to build a giant polyvinyl chloride (PVC) plant in Convent, La.—a community that is more than 80 percent Black. The Shintech plant would have added 600,000 pounds of air pollutants annually.

In April 2001, a group of 1,500 Sweet Valley/Cobb Town neighborhood plaintiffs in Anniston, Ala., reached a $42.8 million out-of-court settlement with Monsanto. The group filed a class action lawsuit against Monsanto for contaminating the Black community with PCBs. Monsanto manufactured PCBs from 1927 through 1972 for use as insulation in electrical equipment. The Environmental Protection Agency (EPA) banned PCB production in the late 1970s amid questions of health risks.

In June 2002, victory finally came to the Norco, La., community, whose residents are sandwiched between a Shell Oil plant and the Shell/Motiva refinery. Concerned Citizens of Norco and their allies forced Shell to agree to a buyout that allowed residents to relocate. Shell also is considering a $200 million investment in environmental improvements to its facility.

"I am surrounded by 27 petrochemical companies and oil refineries. My house is located only three meters away from the 15-acre Shell chemical plant," says long-time Norco resident Margie Richard. "We are not treated as citizens with equal rights according to U.S. law and international human rights laws."

Getting government to respond to environmental justice problems has not been easy. In 1992, after meeting with environmental justice leaders, the EPA administrator William Reilly (under the first Bush administration) established the Office of Environmental Equity (the name was changed to the Office of Environmental Justice under the Clinton administration) and produced "Environmental Equity: Reducing Risks for All Communities," one of the first comprehensive government reports to examine environmental hazards and social equity.

In response to growing public concern and mounting scientific evidence, President Clinton on Feb. 11, 1994 issued Executive Order 12898, Federal Actions to Address Environmental Justice in Minority Populations and Low-Income Populations. The order attempts to address environmental justice within existing federal laws and regulations.

But the current Bush administration's actions may erode years of modest progress. The new air pollution rules proposed by the administration are alarming. The controversial rule changes, which have been a top priority of the White House, would make it easier for utilities and refinery operators to change operations and expand production without installing new controls to capture the additional pollution. Industry has argued that the old EPA regulations have hindered operations and prevented efficiency improvements.

Three decades after the first Clean Air Act was passed, the nation should be getting tougher on plants originally exempted from those standards, not more lenient. Communities that are already overburdened with dirty air need stronger, not weaker, air quality standards. Reducing emissions is a matter of public health. Rolling back the Clean Air Act and allowing polluters to spew their toxic fumes into the air would spell bad news for asthma sufferers, poor people and people of color who are concentrated in the most polluted urban areas.

If this nation is to achieve environmental justice, the environment in urban ghettos, barrios, rural "poverty pockets" and on reservations must be given the same protection as is provided to affluent suburbs. All communities—Black or White, rich or poor, urban or suburban—deserve to be protected from the ravages of pollution

Critical Thinking

1. What is "environmental justice"?
2. Does the environmental movement neglect issues of concern to the poor and minorities?
3. How do people with money keep polluting industries out of their neighborhoods?

Putting a Value on Nature's "Free" Services

Janet N. Abramowitz

Traditional economics views nature as a "free good." That is, forests generate oxygen and wood, clouds bring rain, bees pollinate crops, the sun warms the living room, and all without charge to the humans who benefit. At the same time, nature has provided ways to dispose of wastes without explicitly charging for the service. The "free" waste disposal has turned out to have hidden costs in the form of the health effects of air and water pollution (among other things), but it has been up to individuals and governments to bear the costs associated with those effects. The costs were real, but they were not borne by the businesses and other organizations responsible for them. They have thus come to be known as "external" costs.

Environmental economists have long recognized the problem of external costs, and governmental regulators have devised a number of ways to make those responsible accept the bill in the form of requirements for pollution control and fines for exceeding permitted emissions. Yet external costs remain outside standard economic calculations. They do not affect corporate profit-and-loss statements, and they play no part in calculations of Gross National Product (GNP). In fact, a nation with huge expenditures on health care due to pollution will have a large GNP and look economically healthy. There have been efforts to replace GNP with economic measures that include human health and well-being and the condition of a nation's environment (Indexes of Sustainable Economic Welfare, or ISEW; www.foe.co.uk/community/tools/isew/), but they have not been widely accepted.

The "ecosystem services" approach recognizes that undisturbed ecosystems do many things that benefit human society. In 1997, Robert Costanza and his colleagues published an influential paper in the journal *Nature*, which attempted to list ecosystem services and estimate what it would cost to replace those services if they were somehow lost (as by building a sewage treatment plant to replace a swamp). The total annual bill for the entire biosphere came to $33 trillion (the middle of a $16-54 trillion range), compared to a global GNP of $25 trillion, and Costanza, et al., noted that this was surely an underestimate.

When she wrote the following selection, Janet N. Abramowitz was a senior researcher at the Worldwatch Institute, a private, nonprofit organization devoted to the analysis of global economic and environmental issues, where she covered biodiversity, natural resources management, human development, and social equity. Here she argues that nature's services are responsible for the vast bulk of the value in the world's economy and that attaching economic value to those services may encourage their protection.

Key Concept: the enormous value of nature

During the last half of 1997, massive fires swept through the forests of Sumatra, Borneo, and Irian Jaya, which together form a stretch of the Indonesian archipelago as wide as all of Europe. By November, almost 2 million hectares had burned, leaving the region shrouded in haze and more than 20 million of its people breathing hazardous air. Tens of thousands of people had been treated for respiratory ailments. Hundreds had died from illness, accidents and starvation. The fires, though by then out of control, had been set deliberately and systematically—not by small farmers, and not by El Niño, but by commercial outfits operating with implicit government approval. Strange as this immolation of some of the world's most valuable natural assets may seem, it was not unique. The same year, a large part of the Amazon Basin in Brazil was blanketed by smoke for similar reasons. The fires in the Amazon have been set annually, but in 1997 they destroyed over 50 percent more forest than the year before, which in turn had recorded five times as many fires (some 19,115 fires during a single six-week period) as in 1995.

For the timber and plantation barons of Indonesia, as for the cattle ranchers and frontier farmers of Amazonia, setting fires to clear forests has become standard practice. To them, the natural rainforests are an obstruction that must be sold or burned to make way for their profitable pulp and palm oil plantations. Yet, these are the same forests that for many others serve as both homes and livelihoods. For the hundreds of millions who live in Indonesia and in the neighboring nations of Malaysia, Singapore, Brunei, southern Thailand and the Philippines, it is becoming painfully apparent that without healthy forests, it is difficult to remain healthy people. . . .

As the smoke billowed dramatically from Southeast Asia, a much less visible—but similarly costly—ecological loss was taking place in a very different kind of location. While the Indonesian haze was being photographed from satellites, this other loss might not be noticed by a person standing within an arm's length of the evidence—yet, in its implications for the human future, it is a close cousin of the Asian catastrophe. In the United States, more than 50 percent of all honeybee colonies have disappeared in the last 50 years, with half of that loss occurring in just the last 5 years. Similar losses have been observed in Europe. Thirteen of the 19 native bumblebee species in the United Kingdom are now extinct. These bees are just two of the many kinds of pollinators and their decline is costing farmers, fruit growers, and beekeepers hundreds of millions of dollars in losses each year.

What the ravaged Indonesian forests and disappearing bees have in common is that they are both examples of "free services" that are provided by nature and consumed by the human economy—services that have immense economic value, but that go largely unrecognized and uncounted until they have been lost. Many of those services are indispensable to the people who exploit them, yet are not counted as real benefits, or as a part of GNP.

Though widely taken for granted, the "free" services provided by the natural world form the invisible foundation that supports all societies and economies. We rely on the oceans to provide abundant fish, on forests for wood and new medicines, on insects and other creatures to pollinate our crops, on birds and frogs to keep pests in check, and on forests and rivers to supply clean water. We take it for granted that when we need timber we can cut trees, or that when we need water we can find a spring or drill a well. We assume that clean air will blow the smog out of our cities, that the climate will be stable and predictable, and that the mounting quantity of waste we generate will continue to disappear, if we can just get it out of sight. Nature's services have always been there, free for the taking, and our expectations—and economies—are based on the premise that they always will be. A timber magnate or farmer may have to pay a price for the land, but assumes that what happens naturally on the land—the growing of trees, or pollinating of crops by wild bees, or filtering of fresh water—usually happens for free. We are like young children who think that food comes from the refrigerator, and who do not yet understand that what now seems free is not.

Ironically, by undervaluing natural services, economies unwittingly provide incentives to misuse and destroy the very systems that produce those services; rather than protecting their assets, they squander them. Nature, in turn becomes increasingly less able to supply the prolific range of services that the earth's expanding population and economy demand. It is no exaggeration to suggest that the continued erosion of natural systems threatens not only the continuing viability of today's human enterprise, but ultimately the prospects for our continued existence.

Underpinning the steady stream of services nature provides to us, there is a more fundamental service these systems provide—a kind of self-regulating process by which ecosystems and the biosphere are kept relatively stable and resilient. The ability to withstand disturbances like fires, floods, diseases, and droughts, and to rebound from the shocks these events inflict, is essential to keeping the life-support system operating. As systems are simplified by monoculture or cut up by roads, and the webs that link systems become disconnected, they become more brittle and vulnerable to catastrophic, irreversible decline. We are being confronted by ample evidence, now—from the breakdown of the ozone layer to the increasingly severity of fires, floods and droughts, to the diminished productivity of fruit and seed sets in wild and agricultural plants—that the biosphere is becoming less resilient.

Unfortunately, much of the human economy is based on practices that convert natural systems into something simpler, either for ease of management (it's easier to harvest straight rows of trees that are all the same age than to harvest carefully from complex forests) or to maximize the production of a desired commodity (like corn). But simplified systems lack the resilience that allows them to survive short-term shocks such as outbreaks of diseases or pests, or forest fires, or even longer-term stresses such as that of global warming. One reason is that the conditions within these simplified systems are not hospitable to all of the numerous organisms and processes needed to keep such systems running. A tree plantation or fish farm may provide some of the products we need, but it cannot supply the array of services that natural diverse systems do—and must do—in order to survive over a range of conditions. To keep our own economies sustainable, then, we need to use natural systems in ways that capitalize on, rather than destroy, their regenerative capacity. For humans to be healthy and resilient, nature must be too.

Resiliency is destroyed by fragmentation, as well as by simplification. Fires in healthy rainforests are very rare. By nature, they are too wet to burn. But as they are opened up and fragmented by roads and logging and pasture, they become drier and more prone to fire. When fire strikes forests that are not adapted to fire (as is the case in the rainforests of both Brazil and Indonesia), it is exceptionally destructive and tends to kill a majority of the trees. The fires in Southeast Asia's peat swamp rainforests bring further disruption, by releasing long-sequestered carbon into the atmosphere. . . .

What Forests Do

Around the world, the degradation, fragmentation, and simplification—or "conversion"—of ecosystems is progressing rapidly. Today, only 1 to 5 percent of the original forest cover of the United States and Europe remains. One-third of Asia's forest has been lost since 1960, and half of what remains is threatened by the same industrial forest activities responsible for the Indonesian fires. In the Amazon, 13 percent of the natural cover has already been cleared, mostly for cattle pasture. In many countries, including some of the largest, more than half of the land has been converted from natural habitat to other uses that are less resilient. In countries that stayed relatively undisturbed until the 1980s, significant portions of remaining ecosystems have been lost in the last decade. These trends have been accelerating everywhere. As the natural ecosystems disappear, so do many of the goods and services they provide.

That may seem to contradict the premise that people want those goods and services and would not deliberately destroy them. But there's a logical explanation: governments and business owners typically perceive that the way they can make the most profit from an ecosystem is to maximize its production of a single commodity, such as timber from a forest. For the community (or society) as a whole, however, that is often the least profitable or sustainable use. The economic values of other uses, and the number of people who benefit, added up, can be enormous. A forest, if not cut down to make space for a one-commodity plantation, can produce a rich variety of nontimber forest products (NTFPs) on one hand, while providing essential watershed protection and climate regulation, on the other. These uses not only have more immediate economic value but can also be sustained over a longer term and benefit more people.

In 1992, alternative management strategies were reviewed for the mangrove forests of Bintuni Bay in Indonesia. When nontimber uses such as fish, locally used products, and erosion control were included in the calculations, the researchers found that the most economically profitable strategy was to keep the forest standing with only a modest amount of timber cutting—yielding $4,800 per hectare. If the forest was managed only for timber-cutting, it would yield only $3,600 per hectare. Over the longer term, it was calculated that keeping the forest intact would ensure continued local uses of the area worth $10 million a year (providing 70 percent of local income) and protect fisheries worth $25 million a year—values that would be lost if the forest were cut.

The variety and value of goods produced and collected from forests, and their importance to local livelihoods and national economies, is an economic reality worldwide. For instance, rattan—a vine that grows naturally in tropical forests—is widely used to make furniture. Global trade in rattan is worth $2.7 billion in exports each year, and in Asia it employs a half-million people. In Thailand, the value of rattan exports is equal to 80 percent of the legal timber exports. In India, such "minor" products account for three fourths of the net export earnings from forest produce, and provide more than half of the formal employment in the forestry sector. And in Indonesia, hundreds of thousands of people make their livelihoods collecting and processing NTFPs for export, a trade worth at least $25 million a year. Many of these forests were destroyed in the fire.

Even so, non-timber commodities are only part of what is lost when a forest is converted to a one-commodity industry. There is a nexus between the two catastrophes of the Indonesian fires and the North American and European bee declines, for example, since forests provide habitat for bees and other pollinators. They also provide habitat for birds that control disease-carrying and agricultural pests. Their canopies break the force of the winds and reduce rainfall's impact on the ground, which lessens soil erosion. Their roots hold soil in place, further stemming erosion. In purely monetary terms, a forest's capacity to protect a watershed alone can exceed the value of its timber. Forests also act as effective water-pumping and recycling machinery, helping to stabilize local climate. And, through photosynthesis, they generate enough of the planet's oxygen, while absorbing and storing so much of its carbon (in living trees and plants), that they are essential to the stability of climate worldwide.

Beyond these general functions, there are services that are specific to particular kinds of forests. Mangrove forests and coastal wetlands, notably, play critical roles in linking land and sea. They buffer coasts from storms and erosion, cycle nutrients, serve as nurseries for coastal and marine fisheries, and supply critical resources to local communities. For flood control alone, the value of mangroves has been calculated at $300,000 per kilometer of coastline in Malaysia—the cost of the rock walls that would be needed to replace them. Protecting coasts from storms will be especially important as climate change makes storms more violent and unpredictable. One force driving the accelerated loss of these mangroves in the last two decades has been the explosive growth of intensive commercial aquaculture, especially for shrimp export. Another has been the excess diversion of inland rivers and streams, which reduces downstream flow and allows the coastal waters to become too salty to support the coastal forests.

The planet's water moves in a continuous cycle, falling as precipitation and moving slowly across the landscape to streams and rivers and ultimately to the sea, being absorbed and recycled by plants along the way. Yet, human actions have changed even that most fundamental force of nature by removing natural plant cover, draining swamps and wetlands, separating rivers from their floodplains, and paving over land. The slow natural movement of water across the landscape is also vital for refilling nature's underground reservoirs, or aquifers, from which we draw much of our water. In many places, water now races across the landscape much too quickly, causing flooding and droughts, while failing to adequately recharge aquifers.

The value of a forested watershed comes from its capacity to absorb and cleanse water, recycle excess nutrients, hold soil in place, and prevent flooding. When plant cover is removed or disturbed, water and wind not only race across the land, but carry valuable topsoil with them. According to David Pimentel, an agricultural ecologist at Cornell University, exposed soil is eroded at several thousand times the natural rate. Under normal conditions, each hectare of land loses somewhere between 0.004 and 0.05 tons of soil to erosion each year—far less than what is replaced by natural soil building processes. On lands that have been logged or converted to crops and grazing, however, erosion typically takes away 17 tons in a year in the United States or Europe, and 30 to 40 tons in Asia, Africa, or South America. On severely degraded land, the hemorrhage can rise to 100 tons in a year. The eroded soil carries nutrients, sediments, and chemicals valuable to the system it leaves, but often harmful to the ultimate destination.

One way to estimate the economic value of an ostensibly free service like that of a forested watershed is to estimate what it would cost society if that service had to be replaced. New York City, for example, has always relied on the natural filtering capacity of its rural watersheds to cleanse the water that serves 10 million people each day. In 1996, experts estimated that it would cost $7 billion to build water treatment facilities adequate to meet the city's future needs. Instead, the city chose a strategy that will cost it only one-tenth that amount: simply helping upstream counties to protect the watersheds around its drinking water reservoirs.

Even an estimate like that tends to greatly understate the real value, however, because it covers the replacement cost of only one of the many services the ecosystem provides. A watershed, for example, also contributes to the regulation of the local climate. After forest cover is removed, an area can become hotter and drier, because water is no longer cycled and recycled by plants (it has been estimated that a single rainforest tree pumps 2.5 million gallons of water into the atmosphere during its lifetime.) Ancient Greece and turn-of-the-century Ethiopia, for example, were moister, wooded regions before extensive deforestation, cultivation, and the soil erosion that followed transformed them into the hot, rocky countries they are today. The global spread of desertification offers brutal evidence of the toll of lost ecosystem services.

The cumulative effects of local land use changes have global implications. One of the planet's first ecosystem services was the production of oxygen over billions of years of photosynthetic activity, which allowed oxygen-breathing organisms—such as ourselves—to evolve. Humans have begun to unbalance the global climate regulation system, however, by generating too much carbon dioxide and reducing the capacity of ecosystems to absorb it. Burning forests and peat deposits only makes the problem worse. The fires in Asia sent about as much carbon into the atmosphere last year as did all of the factories, power plants, and vehicles in the United Kingdom. For carbon sequestration alone, economists have been able to estimate the value of intact forests at anywhere from several hundred to several thousand dollars per hectare. As the climate changes the value of being able to regulate local and global climates will only increase.

What Bees Do

If we are often blind to the value of the free products we take from nature, it is even easier to overlook the value of those products we don't harvest directly—but without which our economies could not function. Among these less conspicuous assets are the innumerable creatures that keep potentially harmful organisms in check, build and maintain soils, and decompose dead matter so it can be used to build new life, as well as those that pollinate crops. These various birds, insects, worms, and microorganisms demonstrate that small things can have hugely disproportionate value. Unfortunately, their services are in increasingly short supply because pesticides, pollutants, disease, hunting, and habitat fragmentation or destruction have drastically reduced their numbers and ability to function. As Stephen Buchmann and Gary Paul Nabhan put it in a recent book on pollinators, "nature's most productive workers [are] slowly being put out of business."

Pollinators, for example, are of enormous value to agriculture and the functioning of natural ecosystems. Without them, plants cannot produce the seeds that ensure their survival—and ours. Unlike animals, plants cannot roam around looking for mates. To accomplish sexual reproduction and ensure genetic mixing, plants have evolved strategies for moving genetic material from one plant to the next, sometimes over great distances. Some rely on wind or water to carry pollen to a receptive female, and some can self-pollinate. The most highly evolved are those that use flowers, scents, oils, pollens, and nectars to attract and reward animals to do the job. In fact, more than 90 percent of the world's quarter-million flowering plant species are animal-pollinated. When animals pick up the flower's reward, they also pick up its pollen on various body parts—faces, legs, torsos. Laden with sticky yellow cargo, they can appear comical as they veer through the air—but their evolutionary adaptations are uncannily potent.

Developing a mutually beneficial relationship with a pollinator is a highly effective way for a plant to ensure reproductive success, especially when individuals are isolated from each other. Spending energy producing nectars and extra pollen is a small price to pay to guarantee reproduction. Performing this matchmaking service are between 120,000 and 200,000 animal species, including bees, beetles, butterflies, moths, ants, and flies, along with more than 1,000 species of vertebrates such as birds, bats, possums, lemurs, and even geckos. New evidence shows that many more of these pollinator species than previously believed are threatened with extinction.

Eighty percent of the world's 1,330 cultivated crop species (including fruits, vegetables, beans and legumes, coffee and tea, cocoa, and spices) are pollinated by wild and semi-wild pollinators. One-third of U.S. agricultural output is from insect-pollinated plants (the remainder is from wind-pollinated grain plants such as wheat, rice, and corn). In dollars, honeybee pollination services are 60 to 100 times more valuable than the honey they produce. The value of wild blueberry bees is so great, with each bee pollinating 15 to 19 liters (about 40 pints) of blueberries in its life, that they are viewed by farmers as "flying $50 bills."

Without pollinator services, crops would yield less, and wild plants would produce few seeds—with large economic and ecological consequences. In Europe, the contribution of honey bee pollination to agriculture was estimated to be worth $100 billion in 1989. In the Piedmont region of Italy, poor pollination of apple and apricot orchards cost growers $124 million in 1996. The most pervasive threats to pollinators include habitat fragmentation and disturbance, loss of nesting and over-wintering sites, intense exposure of pollinators to pesticides and of nectar plants to herbicides, breakdown of "nectar corridors" that provide food sources to pollinators during migration, new diseases, competition from exotic species, and excessive hunting. The rapid spread of two parasitic mites in the United States and Europe has wiped out substantial numbers of honeybee colonies. A "forgotten pollinators" campaign was recently launched by the Arizona Sonoran Desert Museum and others, to raise awareness of the importance and plight of these service providers.

Ironically, many modern agricultural practices actually limit the productivity of crops by reducing pollination. According to one estimate, for example, the high levels of pesticides used on cotton reduce annual yields by 20 percent (worth $400 million) in the United States alone by killing bees and other insect pollinators. One-fifth of all honeybee losses involve pesticide exposure, and honeybee poisonings may cost agriculture hundreds of millions of dollars each year. Wild pollinators are particularly vulnerable to chemical poisoning because their colonies cannot be picked up and moved in advance of spraying the way domesticated hives can. Herbicides can kill the plants that pollinators need to sustain themselves during the "off-season" when they are not at work pollinating crops. Plowing to the edges of fields to maximize planting area can reduce yields by disturbing pollinator nesting sites. Just one hectare of unplowed land, for example, provides nesting habitat for enough wild alkali bees to pollinate 100 hectares of alfalfa.

Domesticated honeybees cannot be expected to fill the gap left when wild pollinators are lost. Of the world's major crops, only 15 percent are pollinated by domesticated and feral honeybees, while at least 80 percent are serviced by wild pollinators. Honeybees do not "fit" every type of flower that needs pollination. And because honeybees visit so many different plant species, they are not very "efficient"—that is, there is no guarantee that the pollen will be carried to a potential mate of the same species and not deposited on a different species.

Many plants have developed interdependencies with particular species of pollinators. In peninsular Malaysia, the bat *Eonycteris spelea* is thought to be the exclusive pollinator of the durian, a large spiny fruit that is highly valued in Southeast Asia. The bats' primary food supply is a coastal mangrove that flowers continuously throughout the year. The bats routinely fly tens of kilometers from their roost sites to the mangrove stands, pollinating durian trees along the way. However, mangrove stands in Malaysia and elsewhere are under siege, as are the inland forests. Without both, the bats are unlikely to survive. . . .

Many of the disturbances that have harmed pollinators are also hurting creatures that provide other beneficial services, such as biological control of pests and disease. Much of the wild and semi-wild habitat inhabited by beneficial predators such as birds has been wiped out. The "pest control services" that nature provides are incalculable, and do not have the fundamental flaws of chemical pesticides (which kill beneficial insects along with the pests and harm people). Individual bat colonies in Texas can eat 250 tons of insects each night. Without birds, leaf-eating insects are more abundant and can slow the growth of trees or damage crops. Biologists Paul and Anne Ehrlich speculate that without birds, insects would have become so dominant that humans might never have been able to achieve the agricultural revolution that set the stage for the rise of civilization.

It is not too late to provide essential protections to the providers of such essential services—by using no-till farming to reduce soil erosion and allow nature's underground economy to flourish, by cutting back on the use of toxic agricultural chemicals, and by protecting migratory routes and nectar corridors to ensure the survival of wild pollinators and pest control agents.

Buffer areas of native vegetation and trees can have numerous beneficial effects. They can serve as havens for resident and migratory insects and animals that pollinate crops and control pests. They can also help to reduce wind erosion, and to absorb nutrient pollution that leaks from agricultural fields. Such zones have been eliminated from many agricultural areas that are modernized to accommodate new equipment or larger field sizes. The "sacred groves" in South Asian and African villages—natural areas intentionally left undeveloped—still provide such havens. Where such buffers have been removed, they can be reestablished; they can be added not only around farmers' fields, but along highways and river banks, links between parks, and in people's backyards.

People can also encourage pollinators by providing nesting sites, such as hollow logs, or by ensuring that pollinators have the native plants they need during the "off-season" when they are not working on the agricultural crops. Changing some prevalent cultural or industrial practices, too, can help. There is the practice, for example, of growing tidy rows of cocoa trees. These may make for a handsome plantation. But midges, the only known pollinator of cultivated cacao (the source of chocolate), prefer an abundance of leaf litter and trees in a more natural array. Plantations that encourage midges can have ten times the yield of those that don't.

Scientists have begun to ratchet up their study of wild pollinators and to domesticate more of them. The bumblebee, for example, was domesticated ten years ago and is now a pollinator of valuable greenhouse grown crops.

The Other Service Economy

Natural services have been so undervalued because, for so long, we have viewed the natural world as an inexhaustible resource and sink. Human impact has been seen as insignificant or beneficial. The tools used to gauge the economic health and progress of a nation have tended to reinforce and encourage these attitudes. The gross domestic product (GDP), for example, supposedly measures the value of the goods and services produced in a nation. But the most valuable goods and services—the ones provided by nature, on which all else rests—are measured poorly or not at all. The unhealthy dynamic is compounded by the fact that activities that pollute or deplete natural capital are counted as contributions to economic wellbeing. As ecologist Norman Myers puts it, "Our tools of economic analysis are far from able to apprehend, let alone comprehend, the entire range of values implicit in forests."

When economies and societies use misleading signals about what is valuable, people are encouraged to make decisions that run counter to their own long-range interests—and those of society and future generations. Economic calculations grossly underestimate the current and future value of nature. While a fraction of nature's goods are counted when they enter the marketplace, many of them are not. And nature's services—the life-support systems—are not counted at all. When the goods are considered free and therefore valued at zero, the market sends signals that they are only economically valuable when converted into something else. For example, the profit from deforesting land is counted as a plus on a nation's ledger sheet, because the trees have been converted to saleable lumber or pulp, but the depletions of the timber stock, watershed, and fisheries are not subtracted.

Last year, an international team of researchers led by Robert Costanza of the University of Maryland's Institute for Ecological Economics, published a landmark study on the importance of nature's services in supporting human economies. The study provides, for the first time, a quantification of the current economic value of the world's ecosystem services and natural capital. The researchers synthesized the findings of over 100 studies to compute the average per hectare value for each of the 17 services that world's ecosystems provide. They concluded that the current economic value of the world's ecosystem services is in the neighborhood of $33 trillion per year, exceeding the global GNP of $25 trillion.

Placing a monetary value on nature in this way has been criticized by those who believe that it commoditizes and cheapens nature's infinite value. But in practice, we all regularly assign value to nature through the choices we make. The problem is that in normal practice, many of us don't assign such value to nature until it is converted to something man-made—forests to timber, or swimming fish to a restaurant meal. With a zero value, it's easy to see why nature has almost always been the loser in standard economic equations. As the authors of the Costanza study note, ". . . the decisions we make about ecosystems imply valuations (although not necessarily expressed in monetary terms). We can choose to make these valuations explicit or not . . . but as long as we are forced to make choices, we are going through the process of valuation." The study is also raising a powerful new challenge to those traditional economists who are accustomed to keeping environmental costs and benefits "external" to their calculations.

While some skeptics will doubtless argue that the global valuation reported by Costanza and his colleagues overestimates the current value of nature's services, if anything it is actually a very conservative estimate. As the authors point out, values for some biomes (such as mountains, arctic tundra, deserts, urban parks) were not included. Further, they note that as ecosystem services become scarcer, their economic value will only increase.

Clearly, failure to value nature's services is not the only reason why these services are misused. Too often, illogical and inequitable resource use continues—even in the face of evidence that it is ecologically, economically, and socially unsustainable—because powerful interests are able to shape policies by legal or illegal means. Frequently, some individuals or entities get the financial benefits from a resource while the losses are distributed across society. Economists call this "socializing costs." Stated simply, the people who get the benefits are not the ones who pay the costs. Thus, there is little economic incentive for those exploiting a resource to use it judiciously or in a manner that maximizes public good. Where laws are lax or are ignored, and where people do not have an opportunity for meaningful participation in decision-making, such abuses will continue.

The liquidation of 90 percent of the Philippines' forest during the 1970s and 1980s under the Ferdinand Marcos dictatorship, for example, made a few hundred families over $42 billion richer. But 18 million forest dwellers became much poorer. The nation as a whole went from being the world's second largest log exporter to a net importer. Likewise, in Indonesia today, the "benefits" from burning the forest will enrich a relatively few well-connected individuals and companies but tens of millions of others are bearing the costs. Even in wealthy nations, such as Canada, the forest industry wields heavy influence over how the forests are managed, and for whose benefit.

We have already seen that the loss of ecosystem services can have severe economic, social, and ecological costs even though we can only measure a fraction of them. The loss of timber and lives in the Indonesian fires, and the lower production of fruits and vegetables from inadequate pollination, are but the tip of the iceberg. The other consequences for nature are often unforeseen and unpredictable. The loss of individual species and habitat, and the degradation and simplification of ecosystems, impair nature's ability to provide the services we need. Many of these changes are irreversible, and much of what is lost is simply irreplaceable.

By reducing the number of species and the size and integrity of ecosystems, we are also reducing nature's capacity to evolve and create new life. Almost half of the forests that once covered the Earth are now gone, and much of what remains is in fragmented patches. In just a few centuries we have gone from living off nature's interest to spending down the capital that has accumulated over millions of years of evolution. At the same time we are diminishing the capacity of nature to create new capital. Humans are only one part of the evolutionary product. Yet we have taken on a major role in shaping its future production course and potential. We are pulling out the threads of nature's safety net even as we depend on it to support the world's expanding human population and economy.

In that expanding economy, consumers now need to recognize that it is possible to reduce and reverse the destructive impact of our activities by consuming less and by placing fewer demands on those services we have so mistakenly regarded as free. We can, for example, reduce the high levels of waste and overconsumption of timber and paper. We can also increase the efficiency of water and energy use. In agricultural fields we can leave hedgerows and unplowed areas that serve as nesting and feeding sites for pollinators. We can sharply reduce reliance on agricultural chemicals, and improve the timing of their application to avoid killing pollinators.

Maintaining nature's services requires looking beyond the needs of the present generation, with the goal of ensuring sustainability for many generations to come. We have no honest choice but to act under the assumption that future generations will need at least the same level of nature's services as we have today. We can neither practically nor ethically decide what future generations will need and what they can survive without.

Critical Thinking

1. In what sense are ecosystem services "free"?
2. In what ways do "free" ecosystem services add value to the human economy?
3. What is meant by the "replacement value" of an ecosystem service?
4. Discuss how assigning value to ecosystem services may help with managing ecosystems (see Selection 11).

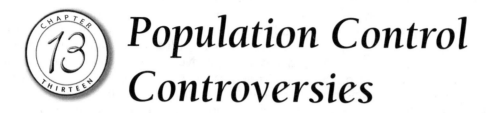

Population Control Controversies

Selection 33

DAVE FOREMAN, from "The Human Population Explosion and the Future of Life," The Rewilding Institute (March 11, 2008).

Selection 34

BETSY HARTMANN, from *Reproductive Rights and Wrongs: The Global Politics of Population Control* (South End Press, 1995)

Selection 35

JOEL E. COHEN, from *How Many People Can the Earth Support?* (W. W. Norton, 1995)

Learning Outcomes

After reading this unit, you should be able to:

1. Describe how growing populations threaten the environment.
2. Explain why some people object to discussing measures to stabilize or reduce population size.
3. Explain what carrying capacity is.
4. Explain how human choices can affect the Earth's carrying capacity for human beings.

The Human Population Explosion and the Future of Life[1]

Dave Foreman

The dire consequences of ever-continued population growth were first brought to public attention in 1798, when British economist Thomas Malthus warned that because population grows exponentially (the graph over time looks like a hockey stick) while food supply grows linearly, population must inevitably outstrip its food supply (not to mention other essential resources) and experience famine. In more modern times, Paul R. Ehrlich's *The Population Bomb* (Sierra Club-Ballantine, 1968) made the fundamental point that the root cause of environmental degradation is worldwide population growth and described the inevitable consequences of unchecked population growth. It conveyed a dramatic message to its wide readership about the severity of the "population explosion." Both men have been ridiculed because the crisis has not yet happened, but a careful look at history shows that we have been fortunate. In the 1800s, many new lands were opened to farming, and in the 1900s the "Green Revolution" increased productivity per hectare dramatically. It is certain that we cannot count on such good fortune to continue forever.

    Dave Foreman is the director and senior fellow of the Rewilding Institute (www.rewilding.org), best known for promoting the restoration of the North American wilderness. His books include *Confessions of an Eco-Warrior* (Harmony, 1991) and *Rewilding North America* (Island Press, 2004). In this essay, he argues that though the impacts of population growth are exacerbated by the worldwide trend toward increased standards of living and use of technology (both of which increase demand on resources and energy use, increase pollution, and threaten wildlife and ecosystems), by far the greater threat is the sheer number of humans on the Earth, coupled with continued growth in population. He believes it is essential that world population be stabilized as soon as possible.

Key Concept: population growth, resource use, and technology combine to threaten our future

Shortly after the end of World War Two, visionary conservationists and scientists such as Fairfield Osborn began to warn that continued human population growth would cause all kinds of problems including heightened plundering of wild Nature. It was not until the late 1960s, however, that population growth moved to the front burner of the conservation stove as shown by the Sierra Club's publication of a book called *The Population Bomb* by a young biologist named Paul Ehrlich. During the next decade those who were worried studied, wrote, and warned about human population growth and its consequences.

Ehrlich and physicist John Holdren (currently in the highly prestigious position of president of the American Association for the Advancement of Science) suggested a formula for understanding the consequences of human growth: I = PAT. This formula, once recognizable but now widely forgotten, means that human impact is a production of population, affluence (consumption), and some measure of technology. At the time, P (population size) was seen as the underlying and key factor for determining the magnitude of human impact. During the last two decades, however, it has become fashionable to discount P and stress A (affluence or consumption).

The level of consumption is a key multiplier of population's impact and individuals worldwide have vastly different levels of consumption of goods and services. Nevertheless, some "environmentalists" and social engineers (right and left) now argue that population size or even continued growth is relatively unimportant; they say it is the level of consumption of certain groups that is key for calculating how much damage an individual or population causes. Such activists argue that reducing consumption is much more important than stabilizing population. Others of us still see population as the big rock. Consumption vs. population may be an intractable debate since it is grounded in worldview as much as in evidence. In general, those who are biologically (or scientifically) oriented are more likely to see population as paramount in I = PAT, while those socially and economically directed tend to stress consumption. I would argue that biologists deal with a more fundamental and real world than do culturalists.

Let me offer just two examples to show how total population is the key. China's remarkable and frightening economic explosion in the last few years has now thrust it into the lead of nations cranking out greenhouse gases. However, were it not for the draconian population policies of China since the 1960s, the population of China would be closer to two billion instead of a billion and a half. How much more greenhouse pollution would China be pumping out had it not taken extreme measures to reduce the birth rate?

Had the United States instituted reasonable immigration policies in the 1960s before immigration became such a volatile issue, our population would likely have stabilized around 250 million. Because we didn't deal wisely with immigration, our population is now over 300 million and rapidly headed toward 400 million. How much less sprawl, greenhouse gas production, resource consumption, etc., would the U.S. boast today, had we stabilized our population then?

We can sort out the kinds of impacts population growth or PAT cause into five stacks:

Land degradation
Resource depletion
Famine
Social, political disruption
Ecological/Evolutionary wounds

Overall, discussion about the problems of population growth has looked at the first four sets. As a conservationist and Nature lover, however, I believe the impact of the population explosion on the rest of Nature is paramount. I also agree with Virginia Tech Professor Eileen Crist that population stabilization and resource depletion activists have made a strategic error by focusing on the first four. As she points out, we cannot predict when we will run out of a resource such as oil or when famine will strike. Indeed, attempting to do so has harmed

our credibility. On the other hand, we can accurately and strikingly show how the spread and population growth of humans is harming Nature.

In *Rewilding North America* [Island Press, 2004], I sorted out how humans harm Nature into Seven Ecological Wounds: direct killing, habitat loss and modification, habitat fragmentation, exotic species invasion, loss and disruption of ecological and evolutionary processes, biocide pollution, and the greenhouse gas effect. These different wounds are cumulative and synergistic.

Taken simplistically, the I = PAT equation lumps together the impacts of all kinds of people. This is how we get comparisons such as the one claiming that one typical American causes thirty-five times the impact of one typical Bangladeshi. To more accurately understand ecological wounds, though, we need to shake out how global and local populations have differing effects. For example, an average American is responsible for much more greenhouse gas production, but a poor Bangladeshi farmer may be much more dangerous to the survival of tigers. Someone with a million-dollar starter castle certainly consumes more than someone living in an old shack on a national forest inholding. But the relatively poorer rural dweller might contribute more to the extinction of the Mexican wolf in the wild. So, we need to look at both IG = PAT and IL = PAT.

Seventy years ago, Aldo Leopold saw the protection of rare species as the most important task of conservation. Since the mid-1970s, we've known that mass extinction caused by humans is the leading problem, conservation or otherwise, facing us. In addition to wiping out many of our fellow Earthlings, mass extinction destroys raw material for evolution and disrupts evolutionary processes. Therefore the shocking, explosive growth of human population has not just short-term but also very long-term catastrophic consequences. Indeed, should the vast, sprawling human population break off major limbs of the tree of life instead of just plucking twigs and leaves, our impact will last forever in terms of Earthly life.

If we are to bring back population stabilization as a bedrock goal of conservation, I think it is essential to show how high human populations and continued growth cause and exacerbate the Seven Ecological Wounds. What I sketch here is just a beginning, only a few examples of what conservationists need to document on how the human population explosion drives the biodiversity crisis. For each of the wounds, I'll give an example of how both global and local populations drive them. (It can be more complicated when global and local people work together to cause wounds, but I will save that issue for a later time.)

Direct Killing

I remember how as a kid I read the *Weekly Reader* telling us how better exploitation of the oceans would feed growing populations. Well, we did that. The consequences are collapsing fisheries around the world, die-off of coral reefs, and the functional extinction of once abundant and/or highly interactive species such as cod, sharks, and tuna. As hungry rural populations swell and spread, they vacuum the rainforest and other semi-wild ecosystems of larger animals for food. Experienced tropical researchers have told how even a small village with primitive technology can clean out the larger animals in a nearby protected area. As their population grows, hunters go ever farther afield with snares, nets, and old guns.

Habitat Destruction and Modification

Other factors, such as lust for larger homes and moving to the Sunbelt, grow the cancerous sprawl of suburban and exurban bedroom communities in the United States. But the absolute growth of numbers of Americans also contributes to the destruction of wildlife habitat by new home developments. In the United States, immigration plays the big role in increasing our population. Even if people come here poor, their goal is to increase their standard of living—which they do, thereby consuming more than had they stayed home. Regardless of overconsumption, if the U.S. had fifty or sixty million fewer residents, there would be fewer home developments in the California coastal chaparral, the Sonoran Desert around Phoenix, woods and forests surrounding Atlanta, and so on.

In India, rapid growth of very poor peasants and tribal peoples is putting irresistible pressure on tiger reserves. Plans are afoot to allow people to move into once-well-protected areas that provide the last secure habitat for tigers and other species. Indian conservationists predict that tigers will vanish where this is allowed to happen.

Fragmentation

No matter if it is starter-castle suburbs and wide freeways in the United States or new slash-and-burn crop patches and logging roads in poorer countries, humans create barriers and fracture zones in wildlife habitats that isolate animal and plant populations into smaller and smaller areas and that will prevent migration north (south below the equator) or to higher elevations with radical climate change. Increased numbers of people are a key cause of our spread into once unpopulated or lightly populated regions where wild critters could range as widely as they needed before the human invasion.

Ecological/Evolutionary Processes

High population densities and the further spread of humans lead to disruption of vital ecological and evolutionary processes, such as wildfire, river flooding and drying, predation, and pollination. Growing numbers of people crowd fertile river bottoms creating pressure

for upstream flood-control dams, which, of course, stop normal hydrological processes. In the United States and Canada, the spread of homes to forested areas leads to suppression of naturally occurring wildfires, which harms forest health and creates conditions for larger, unnatural, uncontrollable conflagrations in the future. Humans, whether Denver suburbanites or Indian peasants, are intolerant of big cats, wolves, and other carnivores, leading to their extermination and the loss of essential topdown regulation of prey species.

Exotic Species

The spread of people spreads invasive species. Increased global trade between growing (in numbers and affluence) populations spreads invasive species. The impacts people have on the ground when they move into a new area create the conditions most beneficial for invasive, exotic, weedy species that outcompete native species. Among the most destructive organisms spread by growing populations are plant diseases, which play absolute havoc with native plants and forests.

Biocide Pollution

People use biocides and cause pollution. On a global scale, wealthy populations cause much more pollution than do poor populations. But locally, all people produce pollution of varying kinds and growing population density leads inevitably to more local pollution, which acts as biocides to many species.

Greenhouse Gases

It can be hard to tell what the limiting resource will be when a population of any species grows to approach a point of overshooting the carrying capacity of its habitat. Unlike other species and other communities of humans before the industrial revolution, we are now a global species. The entire Earth is our habitat. For the global human community, the resource in most short supply has turned out to be the ability of the atmosphere and the seas to handle industrial farts of carbon dioxide and methane, among other gases. By industry, I do not just include factories, smelters, motor vehicles, and such, but also the industrial exploitation of forests and other lands. Wealthy people throughout the world share the greatest guilt for producing greenhouse gases, but poorer people with their fires in forests, grasslands, and woodlands also contribute.

Here affluence (A) and technology (T) play large roles in how much impact individuals may have. However, the influence of sheer numbers of people cannot be brushed aside on the greenhouse gas question. What drives the clearing and burning of the Amazon? Too many people and too high a rate of growth in Brazil, plus the burgeoning numbers of hungry people in the world who need food the "virgin lands" of Brazil supposedly can produce—resulting in a speedy increase in Brazil's greenhouse gas production and loss of forests that could continue to sequester carbon from the atmosphere and sea. China passed the United States as the number one greenhouse gas emitting country in the world because of its massive population charging after greater affluence and technology. But if China had only half a billion people instead of close to a billion and a half, it would not have passed the U.S. If China's population growth had not been harshly curtailed, China would be producing much more greenhouse pollution and may have by this time already pushed past a deeply consequential tipping point. The one-child policy in China may be giving us a few more years to deal with the staggering greenhouse problem.

As I ponder what factors cause each of these wounds on global and local scales, I do not deny that affluence and technology play big parts. My point, however, is that we cannot allow that obvious reality to overshadow the probably even-greater role of high population numbers and population growth rates in driving ecological wounds, from direct killing of threatened species to production of carbon dioxide.

Should conservationists find the wisdom and courage to come back to calling for population stabilization, we must stress how the population explosion causes the ecological wounds that result in mass extinction and destruction of the biosphere. There is our expert province. Because species extinction and destruction of wilderness has consistently been overshadowed by the other consequences of the population explosion, pointing this out as a new concern in a thoughtful, convincing way could help return the world community to a more rational approach about population growth. . . .

Critical Thinking

1. In what ways does a large and growing human population harm the environment?
2. Why do you think, as Dave Foreman says, society has largely forgotten the environmental impact of human population size?

Reproductive Rights and Wrongs

Betsy Hartmann

Biologists Paul R. Ehrlich and Garret Hardin have made cases for aggressive population control as a primary, essential strategy to prevent what they see as an otherwise inevitable future of global food and resource shortages and environmental devastation. Other experts, such as Barry Commoner, Frances M. Lappé, Joseph Collins, and Cary Fowler, have focused on inappropriate technology and political systems that promote inequitable distribution of resources as the causes of food shortages and environmental problems in the present and immediate future. This latter perspective has been elaborated on in recent years by women's organizations and representatives of developing nations.

Betsy Hartmann, an author and researcher, is currently director of the Population and Development Program and associate Professor of Development Studies at Hampshire College in Amherst, Massachusetts. She has written that "the threat to livelihoods, democracy and the environment posed by the fertility of poor women hardly compares to that posed by the consumption patterns of the rich or the ravages of militaries." Hartmann has lived in villages in India and Bangladesh and has made significant contributions to the theories and strategies supported by the international women's health and reproductive rights movement.

In the following selection from her book *Reproductive Rights and Wrongs: The Global Politics of Population Control,* rev. ed. (South End Press, 1995), Hartmann explains the opposition of women's organizations to the policies of the "population establishment." The selection includes the statement of an alternative program supported by the international Committee on Women, Population and the Environment (www.cwpe.org/).

In the years since this book was published, she has continued to develop her arguments. See, for instance, "The 'New' Population Control Craze: Retro, Racist, Wrong Way to Go," *On the Issues* (Fall 2009) (www.ontheissuesmagazine.com/2009fall/2009fall_hartmann.php).

Key Concept: feminist rejection of the establishment's population control policies

In my wildest dreams I wish "population" could be dropped all together from the development lexicon and replaced by concern for real people, real environments, not the fixed images of dark babies as bombs, women as wombs, statistical manipulations as absolute truth. Human welfare, yes: education, employment, health care, social and economic justice, reproductive rights, and the long-awaited peace dividend. Environmental protection, yes: curbs on pollution and waste, demilitarization, support for alternative technologies, farming systems and values which strengthen a sense of democratic community, not crass materialism.

Population Control, No!

This was the slogan of the Women's International Tribunal and Meeting on Reproductive Rights held in Amsterdam in 1984, where 400 women from around the world came together in a powerful condemnation of both population control and antiabortion forces. That meeting solidified my own activism in the international women's health movement. It was held at a time when the movement, particularly its reproductive rights wing, was coming into its own. On the local level feminist initiatives were defining birth control in very different ways than the population establishment. They were responding to women's needs not only through the provision of more comprehensive reproductive health services and sexuality education, but through grassroots organizing around women's economic and political rights.

On the national level, in countries such as Brazil, the Philippines, and India, coalitions of women's groups were exerting pressure on the state to reform health policy, while internationally, the movement was expanding.

The centralization of power in Northern countries was eroding, as the movement became more democratic and geographically diverse.

At the same time feminists within the population establishment were engaged in their own reform efforts, mainly through articulating a quality of care agenda. As the antiabortion movement gained strength, finding a powerful political ally in U.S. President Ronald Reagan, other members of the population establishment recognized the value of a strategic alliance with the international women's health movement. In the mid-1980s, dialogues began between representatives of the two groups, often organized by the New York-based International Women's Health Coalition (IWHC) which acted as a political broker. IWHC maintained that it was possible to "balance the scales" between population control and women's health. The first stones had been laid in the foundation of the Cairo consensus.

Who has gained most from the deal is still an open question. Without the participation of the women's movement, the Cairo consensus would no doubt be less attentive to women's rights and more Malthusian than it already is. But have women's organizations given up too much in the process? In particular, should they accept working within a framework which still blames population growth disproportionately for economic and environmental problems, sets targets for population stabilization and scarcely addresses the much more salient issues of unequal terms of trade, debt and structural adjustment, income distribution, and arms control?

This question now divides the movement. A number of groups have agreed to work within a population policy framework, although they have tried in the process to expand it. According to the Women's Declaration on Population Policies prepared in advance of Cairo:

> Population policies . . . need to address a wide range of conditions that affect the reproductive health and rights of women and men. These include unequal distribution of material and social resources among individuals and groups, based on gender, age, race, religion, social class, rural-urban residence, nationality and other social criteria; changing patterns of sexual and family relationships; political and economic policies that restrict girls' and women's access to health services and methods of fertility regulation; and ideologies, laws and practices that deny women's basic rights.

Despite this opening, the Declaration essentially sets forth a reproductive health agenda, with calls for more "women decision-makers" and financial resources for meeting program requirements. There is no fundamental challenge to the population paradigm, and government and international agencies, despite mention of their past shortcomings, are perceived as basically benign.

There are two key reasons women's organizations have agreed to work within the population framework. Some reproductive rights activists genuinely believe in the urgency of slowing population growth. Marge Berer, for example, writes that the women's movement should "acknowledge that the world cannot sustain an unlimited number of people, just as women's bodies cannot sustain unlimited pregnancies." Likening the planet to a woman's body, however, is a comparison fraught with peril.

For others it is more of a strategic choice—they would argue that accepting the legitimacy of a population framework allows women greater access to decision-makers, or in some cases, the opportunity to be decision-makers themselves. It reflects the movement's growing sophistication and professionalization, with some kind of compromise as the inevitable price of success.

But does influencing and interacting with the establishment necessarily depend on articulating women's concerns within a population framework? No, write Judith Richter and Loes Keysers, who set forth an alternative agenda of engagement. They note that more liberal people in the establishment may welcome support from the women's movement in their struggle against coercive population control, on the one hand, and the Vatican and fundamentalists, on the other.

> Women's advocates are thus in a relative position of strength. Could the splits in the population field not be used differently? Why not—instead of embracing a population agenda—enroll the population soft liners and family planners, for example, into a people's alliance for reproductive rights and health? This would allow feminists to set the terms of reference. It would create space for a shift of paradigm, rather than a shift from hard to soft population control.

The Declaration of People's Perspectives on "Population," issued by women from 23 countries meeting in 1993 in Comilla, Bangladesh, strongly rejects the population framework:

> Women's basic needs of food, education, health, work, social and political participation, a life free of violence and oppression should be addressed on their own merit. Meeting women's needs should be de-linked from population policy including those expressed as apparent humanitarian concerns for women.

The Comilla Declaration also makes clear its support for women's access to safe contraception and legal abortion as part of general health care. "Our resistance to population control policies must never be confused with the opposition of the religious and political right to the same policies."

Nevertheless, women who insist on remaining outside of the population framework are often accused of playing into the hands of the Vatican. There are also debates on which is the greater enemy of reproductive rights: religious fundamentalism or population control. Clearly, this depends on one's specific situation—whether one is fighting restrictive abortion legislation in Latin America and Eastern Europe, for example, or sterilization abuse in India and China. In many places, including the United States, both are enemies, both must be confronted simultaneously.

So far the movement has managed to maintain an uneasy unity and some sense of common identity—its strength is its members' ability to discuss and debate openly, to accept difference and heterogeneity, to insist on democratic processes and leadership. But the road ahead will be difficult. Besides the debate over population, there are many other unresolved issues: If it is politically pragmatic for some members to work within mainstream institutions, how can they remain accountable to the outside movement? Power doesn't necessarily corrupt, but it often separates and isolates. Does participation in official processes, such as U.N. conferences, siphon too much energy away from grassroots organizing? How can the movement, especially at the international level, maintain any grassroots authenticity? In an era of scarce resources, will the agenda conform too much to the funders' priorities? Where will funding come from?

I myself would argue for a strategy of principled pragmatism. While it makes political sense to dialogue and interact with the establishment periodically, as well as to have sympathetic women in positions of power within it, I believe the international women's health movement should not accept—and does not need to accept—the population framework in its efforts to reform health and family planning policy. When the movement does accept the framework, it loses its critical edge, dulls its tools of analysis, and ends up endorsing narrow technocratic agendas, rather than a broader politics of social and economic transformation. It divorces itself further from the poor women it is supposed to represent and places too much faith in official rhetoric. After Cairo, when the consensus snake begins to shed its skin, its progressive trappings will probably be the first to go. Already, Bangladesh and Indonesia are being put forward as the models which other countries should follow.

Stripped of all the economic arguments, political justifications, and soft-sell marketing, population control at heart is a philosophy without a heart, in which human beings become objects to be manipulated. It is a philosophy of domination, for its architects must necessarily view people of different sex, race, and class as inferior, less human than themselves, or otherwise they could not justify the double standards they employ.

Population control profoundly distorts our world view, and negatively affects people in the most intimate areas of their lives. Instead of promoting ethics, empathy, and true reproductive choice, it encourages us to condone coercion. And even on the most practical level, it is no solution to the serious economic, political, and environmental problems we face at the end of the century.

In saying no to population control from a prochoice, feminist perspective, one often feels like a voice in the wilderness, especially in the United States where

Malthusianism is a popular religion, a veritable article of faith. But inhabiting the political wilderness is preferable to accepting conventional wisdom that is unwise. And if one listens closely, one can hear many other voices raised in protest and one can join them until collectively, they are too loud and clear to ignore.

The Committee on Women, Population and the Environment is an alliance of women activists, community organizers, health practitioners, and scholars of diverse races, cultures, and countries of origin working for women's empowerment and reproductive freedom, and against poverty, inequality, racism and environmental degradation. Issued in 1992, their statement, "Women, Population and the Environment: Call for a New Approach" continues to gather individual and organizational endorsements from around the world.

Call For a New Approach

We are troubled by recent statements and analyses that single out population size and growth as a primary cause of global environmental degradation.

We believe the major causes of global environmental degradation are:

- Economic systems that exploit and misuse nature and people in the drive for short-term and short-sighted gains and profits.

- The rapid urbanization and poverty resulting from migration from rural areas and from inadequate planning and resource allocation in towns and cities.

- The displacement of small farmers and indigenous peoples by agribusiness, timber, mining, and energy corporations, often with encouragement and assistance from international financial institutions, and with the complicity of national governments.

- The disproportionate consumption patterns of the affluent the world over. Currently, the industrialized nations, with 22 percent of the world's population, consume 70 percent of the world's resources. Within the United States, deepening economic inequalities mean that the poor are consuming less, and the rich more.

- Technologies designed to exploit but not to restore natural resources.

- Warmaking and arms production which divest resources from human needs, poison the natural environment and perpetuate the militarization of culture, encouraging violence against women.

Environmental degradation derives thus from complex, interrelated causes. Demographic variables can have an impact on the environment, but reducing population growth will not solve the above problems. In many countries, population growth rates have declined yet environmental conditions continue to deteriorate.

Moreover, blaming global environmental degradation on population growth helps to lay the groundwork for the re-emergence and intensification of top-down, demographically driven population policies and programs which are deeply disrespectful of women, particularly women of color and their children.

In Southern countries, as well as in the United States and other Northern countries, family planning programs have often been the main vehicles for dissemination of modern contraceptive technologies. However, because so many of their activities have been oriented toward population control rather than women's reproductive health needs, they have too often involved sterilization abuse; denied women full information on contraceptive risks and side effects; neglected proper medical screening, follow-up care, and informed consent; and ignored the need for safe abortion and barrier and male methods of contraception. Population programs have frequently fostered a climate where coercion is permissible and racism acceptable.

Demographic data from around the globe affirm that improvements in women's social, economic, and health status and in general living standards, are often keys to declines in population growth rates. We call on the world to recognize women's basic right to control their own bodies and to have access to the power, resources, and reproductive health services to ensure that they can do so.

National governments, international agencies, and other social institutions must take seriously their obligation to provide the essential prerequisites for women's development and freedom. These include:

1. Resources such as fair and equitable wages, land rights, appropriate technology, education and access to credit.
2. An end to structural adjustment programs, imposed by the IMF [International Monetary Fund], the World Bank, and repressive governments, which sacrifice human dignity and basic needs for food, health, and education to debt repayment and 'free market', male-dominated models of unsustainable development.
3. Full participation in the decisions which affect our own lives, our families, our communities, and our environment, and incorporation of women's knowledge systems and expertise to enrich these decisions.
4. Affordable, culturally appropriate, and comprehensive health care and health education for women of all ages and their families.
5. Access to safe, voluntary contraception and abortion as part of broader reproductive health services which also provide pre-and post-natal care, infertility services, and prevention and treatment of sexually transmitted diseases including HIV and AIDS.
6. Family support services that include child-care, parental leave and elder care.
7. Reproductive health services and social programs that sensitize men to their parental responsibilities and to the need to stop gender inequalities and violence against women and children.
8. Speedy ratification and enforcement of the UN Convention on the Elimination of All Forms of Discrimination Against Women as well as other UN conventions on human rights.

People who want to see improvements in the relationship between the human population and natural environment should work for the full range of women's rights; global demilitarization; redistribution of resources and wealth between and within nations; reduction of consumption rates of polluting products and processes and of non-renewable resources; reduction of chemical dependency in agriculture; and environmentally responsible technology. They should support local, national, and international initiatives for democracy, social justice, and human rights.

Critical Thinking

1. Betsy Hartmann calls population "a philosophy without a heart." Explain what she means.
2. Which is more important to future human welfare, stabilizing or reducing population or improving women's freedom, well-being, and health?
3. What, according to Betsy Hartmann, are the chief causes of population growth?

Human Carrying Capacity

Joel E. Cohen

Joel E. Cohen is the professor of populations and head of the Laboratory of Populations at The Rockefeller University and Columbia University, New York. His book *How Many People Can the Earth Support?* (W. W. Norton, 1995), from which the following selection is taken, has been called an "admirable tour de force on human population" by the distinguished paleontologist and curator of the American Museum of Natural History, Niles Eldredge. On the controversial question referred to in the book's title, Cohen takes the middle ground between the position of deep ecologists, such as David Foreman, who assert that the earth's population has already far exceeded its sustainable limit and optimistic economists who echo Julian Simon's view that the larger the population, the better. Cohen's readers will not find a definitive response to the dispute about the earth's carrying capacity, but they should become convinced that it is a highly complex and subjective issue. Cohen maintains that any analysis that results in definitive answers to questions about the relationships among population, resources, and the environment should be assessed with a highly critical eye. One must be careful when contemplating value-laden assumptions. Nevertheless, Cohen warns against passivity in the face of irrefutable evidence that uncontrolled population growth and environmentally inappropriate technological developments pose serious threats to the future of human civilization.

Key Concept: the earth's carrying capacity as a complex but vital issue

The question of how many people the world can support is unanswerable in a finite sense. What do we want?

Are there global limits, absolute limits beyond which we cannot go without catastrophe or overwhelming costs? There are, most certainly.

—George Woodwell 1985

Case Study: Easter Island

The constraints on the Earth's human carrying capacity are just as real as the wide range of choices within those boundaries. The history of Easter Island provides a case study of human choices and natural constraints in a small world. While exotic in location and culture, Easter Island is of general interest as one example of the many civilizations that undercut their own ecological foundations.

The island is one of the most isolated bits of land on the Earth. The inhabited land nearest to Easter Island is Pitcairn Island, 2,250 kilometers northwest; the nearest continental place, Concepción, Chile, is 3,747 kilometers southeast. The island is roughly triangular in plan, with sides of 16, 18 and 22 kilometers and an area of 166.2 square kilometers (a bit larger than Staten Island, a borough of the city of New York). About 2.5 million

years old, the volcanic island rose from the sea floor 2,000 meters below sea level. A plateau occupies the middle of the island and a peak rises nearly 1,000 meters above sea level.

Radiocarbon dating suggests that people, almost certainly Polynesians, occupied the island by A.D. 690 at the latest; scattered earlier radiocarbon dates from the fourth and fifth centuries are uncertain. The first arrivals found an island covered by a rainforest of huge palms. The islanders were probably isolated from outside human contact until the island was spotted by Dutch sailors in 1722.

During this millennium or millennium and a half of isolation, a fantastic civilization arose. Its most striking material remains are 800 to 1,000 giant statues, or *moai*, two to ten meters high, carved in volcanic tuff and scattered over the island. Many are probably still buried by rubble and soil. The largest currently known is 20 meters (65 feet) long and weighs about 270 tonnes. It was left unfinished.

According to pollen cores recently taken from volcanic craters on the island, a tree used for rope was originally dominant on the island. At different times, depending on the site, between the eighth and the tenth centuries, forest pollen began to decline. Forest pollen reached its lowest level around A.D. 1400, suggesting that the last forests were destroyed by then. The deforestation coincided with soil erosion, visible in soil profiles. The Polynesian rat introduced for food by the original settlers consumed the seeds of forest trees, preventing regeneration. Freshwater supplies on the island diminished. In the 1400s or 1500s, large, stemmed obsidian flakes used as daggers and spearheads appeared for the first time; previously obsidian had been used only for tools.

While early visitors in 1722 and 1770 do not mention fallen *moai*, Captain Cook in 1774 reported that many statues had fallen next to their platforms and that the statues were not being maintained. Something drastic, probably some variant of intergroup warfare, probably happened between 1722, when the Dutch thought the statue cult was still alive, and 1774, when Cook thought it finished. A visitor in 1786 observed that the island no longer had a chief.

The population history of the island is full of uncertainties. The prehistory is based on the dating of sites of agricultural and human occupation. The best current estimate is that the population began with a boatload of settlers in the first half millennium after Christ, perhaps around A.D. 400. The population remained low until about A.D. 1100. Growth then accelerated and the population doubled every century until around 1400. Slower growth continued until at most 6,000 to 8,000 people occupied the island around 1600. The maximum population may have reached 10,000 people in A.D. 1680. A decline then set in. Jean François de Galaup, Comte de La Pérouse, who visited the island in 1786, estimated a population of 2,000, and this estimate is now accepted as roughly correct. Smallpox swept the island in the 1860s, introduced by returning survivors among the islanders who had been enslaved and taken to Peru. The population numbered 111 by 1877. In 1888, the island was attached to Chile. Since then, Chileans have added to the population. The present population of 2,100 includes 800 children.

The plausibility of these numbers can be checked from the annual rates of population growth or decline that they imply. An increase from 50 people in A.D. 400 to 10,000 people in A.D. 1680 requires an annual increase of 0.41 percent. If the number of original settlers were 100 instead of 50, the implied population growth rate would be 0.36 percent; if 25 instead of 50, 0.47 percent. The long-term growth rate of around 0.4 percent per year is within the historical experience of developing countries before the post–World War II public health evolution. If the population declined from 10,000 in 1680 to 111 in 1877 (including removals by Peruvian slave traders), the annual rate of decline was 2.3 percent. If the population maximum in 1680 was only 6,000 instead of 10,000, then the annual rate of decline was 2.0 percent. A population decline from 10,000 in 1680 to the 2,000 reported by La Pérouse in 1786 requires an annual decline of 1.5 percent.

In round numbers, Easter Island's human population seems to have increased by about 0.4 percent per year for about 13 centuries, then to have declined by about 2 percent per year for about two centuries before resuming a rise in the twentieth century.

Paul Bahn, a British archeologist and writer, and John Flenley, an ecologist and geographer in New Zealand, synthesized the archeological and historical data in an interpretive model.

> Forest clearance for the growing of crops would have led to population increase, but also to soil erosion and decline of soil fertility. Progressively more land would have had to be cleared. Trees and shrubs would also be cut down for canoe building, firewood, house construction, and for the timbers and ropes needed in the movement and erection of statues. Palm fruits would be eaten, thus reducing regeneration of the palm. Rats, introduced for food, could have fed on the palm fruits, multiplied rapidly and completely prevented palm regeneration. The overexploitation of prolific sea bird resources would have eliminated these from all but the offshore islets. Rats could have helped in this process by eating eggs. The abundant food provided by fishing, sea birds and rats would have encouraged rapid initial human population growth. Unrestrained human population increase would later put pressure on availability of land, leading to disputes and eventually warfare. Non-availability of timber and rope would make it pointless to carve further statues. A disillusionment with the efficacy of the statue religion in providing the wants of the people could lead to the abandonment of this cult. Inadequate canoes would restrict fishing to inshore waters, leading to further decline in protein supplies. The result could have been general famine, warfare and the collapse of the whole economy, leading to a marked population decline.
>
> Of course, most of this is hypothesis. Nevertheless, there is evidence, as we have seen, that many features of this model did in fact occur. There certainly was deforestation, famine, warfare, collapse of civilization and population decline.

Supposing you accept this summary of the island's history, would you accept the following conclusion of Bahn and Flenley? "We consider that Easter Island was a microcosm which provides a model for the whole planet." Easter Island shares important features with the whole planet and differs in others. Draw your own conclusion.

Living in the Land of Lost Illusions: Human Carrying Capacity as an Indicator

A number or range of numbers, presented as a constraint independent of human choices, is an inadequate answer to the question "How many people can the Earth support?" While trying to answer this question, I learned to question the question.

If an absolute numerical upper limit to human numbers on the Earth exists, it lies beyond the bounds that human beings would willingly tolerate. Human physical requirements for bare minimal subsistence are very modest, closer to the level of Auschwitz than to the modest comforts of the Arctic Inuit or the Kalahari bushmen. For most people of the world, expectations of well-being have risen so far beyond subsistence that human choices will prevent human numbers from coming anywhere near absolute upper limits. If human choices somehow failed to prevent population size from approaching absolute upper limits, then gradually worsening conditions for human and other life on the Earth would first prompt and eventually enforce human choices to stop such an approach. As different people have different expectations of well-being, some people would be moved to change their behavior sooner than others. Social scientists focus on the choices and minimize the constraints; natural scientists do the reverse. In reality, neither choices nor constraints can be neglected.

An ideal tool for estimating how many people the Earth can support would be a model, simple enough to be intelligible, complicated enough to be potentially realistic and empirically tested enough to be credible. The model would require users to specify choices concerning technology, domestic and international political institutions, domestic and international economic arrangements (including recycling), domestic and international demographic arrangements, physical, chemical and biological environments, fashions, tastes, moral values, a desired typical level of material well-being and a distribution of well-being among individuals and areas. Users would specify how much they wanted each characteristic to vary as time passes and what risk they would tolerate that each characteristic might go out of the desired range of variability. Users would state how long they wanted their choices to remain in effect. They would specify the state of the world they wished to leave at the end of the specified period. The model would first check all these choices for internal consistency, detect any contradictions and ask users to resolve them or to specify a balance among contradictory choices. The model would then attempt to reconcile the choices with the constraints imposed by food, water, energy, land, soil, space, diseases, waste disposal, nonfuel minerals, forests, biological diversity, biologically accessible nitrogen, phosphorus, climatic change and other natural constraints. The model would generate a complete set of possibilities, including human population sizes, consistent with the choices and the constraints. . . .

The speed registered on the speedometer of a car, the current total fertility rate of a population and the gross national or domestic product of an economy are, every one, indirect and incomplete summaries of more complicated realities: they are summary indicators, approximate but useful. Likewise, estimates of the human carrying capacity of the Earth are indicators. They indicate the population that can be supported under various assumptions about the present or future. Estimates of the Earth's human carrying capacity are conditional on current choices and on natural constraints, all of which may change as time passes. This view of estimates of human carrying capacity as conditional and changing differs sharply from a common view that there is one right number (perhaps imperfectly known) for all time.

Human carrying capacity is more difficult to estimate than some of the standard demographic indicators, like expectation of life or the total fertility rate, because human carrying capacity depends on populations and activities around the world. The expectation of life of a country can be determined entirely from the mortality experienced by the people within the country. But that same country's human carrying capacity depends not only on its soils and natural resources and population and culture and economy, but also on the prices of its products in world markets and on the resources and products other countries can and are willing to trade. When the world consisted of largely autonomous localities, it may have made sense to think of the Earth's human carrying capacity as the sum of local human carrying capacities; but no longer.

Beyond Equilibrium

Think of a man engaged in four activities: lying on his back on the floor with his arms and legs relaxed; standing erect but at ease; walking at a comfortable pace; and running. When lying on his back, the man is at a passive equilibrium. If you push him gently on one side, he may rock a bit but will roll back to his original position. If you push him hard enough, he may switch from a passive supine equilibrium to a passive prone equilibrium. Whether he is supine or prone, he can remain in his present equilibrium without effort.

Standing is a much more complicated matter. Opposing muscles are constantly adjusting their tension to maintain upright posture, under the guidance of the body's sensory and nervous systems for maintaining balance. The man may not appear to be working, but his oxygen consumption increases and he will fall to the floor if he relaxes completely. If pushed hard, he may not be able to stay standing. His apparent equilibrium is dynamically maintained by constant control.

Walking is controlled falling. The man's center of gravity moves forward from his area of support and he puts one foot forward with just the right timing and placement to catch himself. He then pivots over the forward foot and continues to fall forward with just-in-time support from alternating legs. Anyone who has watched a child learn to walk, or an adult learn to walk with crutches, appreciates the complex sequential coordination, muscular strength and balance required to walk. The equilibrium of walking is not a stationary state at all, but a sustained motion.

Finally, running is more than an acceleration of walking because the runner may have both legs off the ground at once. The effort required, the speed of the motion and the vulnerability to collapse increase compared to walking. On a rocky mountain ridge or a crowded city street, running is impossible; simplification and control of the environment are required to sustain the equilibrium of steady running.

The purpose of this foray into kinesiology is to find some old and new analogies for the situation of humans on the Earth. If the population size of the human species was ever in a passive equilibrium regulated by the environment, it must surely have been before people gained control of fire. By using fire, early peoples massively reshaped their environment to their own advantage, with the effect of increasing their own population size. For example, they periodically burned grasslands to encourage plants desirable to themselves and to the game they hunted. After the mastery of fire, people moved from a supine equilibrium to a controlled balance analogous to standing. With the invention of shifting cultivation some ten or so millennia ago, and then settled cultivation, the human species initiated a form of controlled forward falling analogous to walking. Humans invented cities and farms and learned to coordinate them. Where agriculture failed, civilizations collapsed. In the last four centuries, surpluses of food released huge numbers of people from being tied to the land and enabled them to make machines and technologies that further loosened their ties to the land. Machines for handling energy, materials and information released people from old work, imposed new work and transformed much of the natural world. More than ever before, the land that still supported people became a partly human creation. For humans now, the notion of a static, passive equilibrium is inappropriate, useless. So is the notion of a static "human carrying capacity" imposed by the natural world on a passive human species. There is no choice but to try to control the direction, speed, risks, duration and purposes of our falling forward.

Critical Thinking

1. What is "carrying capacity"?
2. In what sense does the Earth not have a single carrying capacity for human beings?

Environmental Ethics and Worldviews

Selection 36

PAUL W. TAYLOR, from "The Ethics of Respect for Nature,"
Environmental Ethics (Fall 1981)

Selection 37

VANDANA SHIVA, from "Women's Indigenous Knowledge and
Biodiversity Conservation," in Maria Mies and Vandana Shiva,
Ecofeminism (Zed Books, 1993)

Selection 38

JARED DIAMOND, from "A Tale of Two Farms," *Collapse: How
Societies Choose to Fail or Succeed* (Viking, 2005)

Learning Outcomes

After reading this unit, you should be able to:

1. Explain why sustainable development is an important goal.
2. Choose between anthropocentric and biocentric environmental ethics
   and explain the grounds for the choice.
3. Discuss whether women have special knowledge relating to the
   conservation of biodiversity.
4. Explain how the choices of individuals and societies can affect the
   societies' long-term success or failure.

The Ethics of Respect for Nature

Paul W. Taylor

Historically, Western philosophy has paid little attention to the environment. The anthropocentric (human-centered) perspective that has dominated philosophical thought accords no intrinsic value to either nonhuman animate or inanimate objects. Thus, from this perspective, land that was unimproved by human labor was considered valueless, and the value of animals and plants was considered exclusively in terms of their satisfaction of human needs or interests.

During the twentieth century, however, the ecological perspective that views the need for human beings to exist in harmony with nature has begun to supersede the notion that nature is a hostile environment that needs to be subdued and dominated. Although anthropocentrism is still a dominant perspective, the biocentric, or ecocentric, views expressed in the writings of such environmentalists as John Muir and Aldo Leopold have grown in popularity. Leopold's proposal of a land ethic based upon the notion of a community that includes nonhuman life and inanimate objects as well as human beings has stimulated philosophers to consider the possibility of developing a general system of ethics based on an ecological perspective.

Of the recent attempts to create and justify a system of environmental ethics, the work of philosophy professor emeritus Paul W. Taylor of the City University of New York may be the most complete. The following selection is from "The Ethics of Respect for Nature," *Environmental Ethics* (Fall 1981). He expanded upon this article for his book *Respect for Nature: A Theory of Environmental Ethics* (Princeton University Press, 1986). In the following selection, Taylor explains the meaning of the "inherent worth" of all living organisms and how this concept coupled with a rejection of human superiority can form the basis for a justifiable system of environmental ethics.

Key Concept: a life-centered system of ethics

Human-Centered and Life-Centered Systems of Environmental Ethics

In this paper I show how the taking of a certain ultimate moral attitude toward nature, which I call "respect for nature," has a central place in the foundations of a life-centered system of environmental ethics. I hold that a set of moral norms (both standards of character and rules of conduct) governing human treatment of the natural world is a rationally grounded set if and only if, first, commitment to those norms is a practical entailment of adopting the attitude of respect for nature as an ultimate moral attitude, and second, the adopting of that attitude on the part of all rational agents can itself be justified. When the basic characteristics of the attitude of respect for nature are made clear, it will be seen that a life-centered system of environmental ethics need not be holistic or organicist in its conception of the kinds of entities that are deemed the appropriate objects of moral concern and consideration. Nor does such

a system require that the concepts of ecological homeostasis, equilibrium, and integrity provide us with normative principles from which could be derived (with the addition of factual knowledge) our obligations with regard to natural ecosystems. The "balance of nature" is not itself a moral norm, however important may be the role it plays in our general outlook on the natural world that underlies the attitude of respect for nature. I argue that finally it is the good (well-being, welfare) of individual organisms, considered as entities having inherent worth, that determines our moral relations with the Earth's wild communities of life.

In designating the theory to be set forth as life-centered, I intend to contrast it with all anthropocentric views. According to the latter, human actions affecting the natural environment and its nonhuman inhabitants are right (or wrong) by either of two criteria: they have consequences which are favorable (or unfavorable) to human well-being, or they are consistent (or inconsistent) with the system of norms that protect and implement human rights. From this human-centered standpoint it is to humans and only to humans that all duties are ultimately owed. We may have responsibilities *with regard to* the natural ecosystems and biotic communities of our planet, but these responsibilities are in every case based on the contingent fact that our treatment of those ecosystems and communities of life can further the realization of human values and/or human rights. We have no obligation to promote or protect the good of nonhuman living things, independently of this contingent fact.

A life-centered system of environmental ethics is opposed to human-centered ones precisely on this point. From the perspective of a life-centered theory, we have prima facie moral obligations that are owed to wild plants and animals themselves as members of the Earth's biotic community. We are morally bound (other things being equal) to protect or promote their good for *their* sake. Our duties to respect the integrity of natural ecosystems, to preserve endangered species, and to avoid environmental pollution stem from the fact that these are ways in which we can help make it possible for wild species populations to achieve and maintain a healthy existence in a natural state. Such obligations are due those living things out of recognition of their inherent worth. They are entirely additional to and independent of the obligations we owe to our fellow humans. Although many of the actions that fulfill one set of obligations will also fulfill the other, two different grounds of obligation are involved. Their well-being, as well as human well-being, is something to be realized *as an end in itself.*

If we were to accept a life-centered theory of environmental ethics, a profound reordering of our moral universe would take place. We would begin to look at the whole of the Earth's biosphere in a new light. Our duties with respect to the "world" of nature would be seen as making prima facie claims upon us to be balanced against our duties with respect to the "world" of human civilization. We could no longer simply take the human point of view and consider the effects of our actions exclusively from the perspective of our own good.

The Good of a Being and the Concept of Inherent Worth

What would justify acceptance of a life-centered system of ethical principles? In order to answer this it is first necessary to make clear the fundamental moral attitude that underlies and makes intelligible the commitment to live by such a system. It is then necessary to examine the considerations that would justify any rational agent's adopting that moral attitude.

Two concepts are essential to the taking of a moral attitude of the sort in question. A being which does not "have" these concepts, that is, which is unable to grasp their meaning and conditions of applicability, cannot be said to have the attitude as part of its moral outlook. These concepts are, first, that of the good (well-being, welfare) of

a living thing, and second, the idea of an entity possessing inherent worth. I examine each concept in turn.

(1) Every organism, species population, and community of life has a good of its own which moral agents can intentionally further or damage by their actions. To say that an entity has a good of its own is simply to say that, without reference to any *other* entity, it can be benefited or harmed. One can act in its overall interest or contrary to its overall interest, and environmental conditions can be good for it (advantageous to it) or bad for it (disadvantageous to it). What is good for an entity is what "does it good" in the sense of enhancing or preserving its life and well-being. What is bad for an entity is something that is detrimental to its life and well-being.

We can think of the good of an individual nonhuman organism as consisting in the full development of its biological powers. Its good is realized to the extent that it is strong and healthy. It possesses whatever capacities it needs for successfully coping with its environment and so preserving its existence throughout the various stages of the normal life cycle of its species. The good of a population or community of such individuals consists in the population or community maintaining itself from generation to generation as a coherent system of genetically and ecologically related organisms whose average good is at an optimum level for the given environment. (Here *average good* means that the degree of realization of the good of *individual organisms* in the population or community is, on average, greater than would be the case under any other ecologically functioning order of interrelations among those species populations in the given ecosystem).

The idea of a being having a good of its own, as I understand it, does not entail that the being must have interests or take an interest in what affects its life for better or for worse. We can act in a being's interest or contrary to its interest without its being interested in what we are doing to it in the sense of wanting or not wanting us to do it. It may, indeed, be wholly unaware that favorable and unfavorable events are taking place in its life. I take it that trees, for example, have no knowledge or desires or feelings. Yet it is undoubtedly the case that trees can be harmed or benefited by our actions. We can crush their roots by running a bulldozer too close to them. We can see to it that they get adequate nourishment and moisture by fertilizing and watering the soil around them. Thus we can help or hinder them in the realization of their good. It is the good of trees themselves that is thereby affected. We can similarly act so as to further the good of an entire tree population of a certain species (say, all the redwood trees in a California valley) or the good of a whole community of plant life in a given wilderness area, just as we can do harm to such a population or community.

When construed in this way, the concept of a being's good is not coextensive with sentience or the capacity for feeling pain. William Frankena has argued for a general theory of environmental ethics in which the ground of a creature's being worthy of moral consideration is its sentience. I have offered some criticisms of this view elsewhere, but the full refutation of such a position, it seems to me, finally depends on the positive reasons for accepting a life-centered theory of the kind I am defending in this essay.

It should be noted further that I am leaving open the question of whether machines—in particular, those which are not only goal-directed, but also self-regulating—can properly be said to have a good of their own. Since I am concerned only with human treatment of wild organisms, species populations, and communities of life as they occur in our planet's natural ecosystems, it is to those entities along that the concept "having a good of its own" will here be applied. I am not denying that other living things, whose genetic origin and environmental condition have been produced, controlled, and manipulated by humans for human ends, do have a good of their own in the same sense as do wild plants and animals. It is not my purpose in this essay, however, to set out or defend the principles that should guide our conduct with regard to their good. It is only insofar as their production and use by humans have good or ill effects upon natural ecosystems and their wild inhabitants that the ethics of respect for nature comes into play.

(2) The second concept essential to the moral attitude of respect for nature is the idea of inherent worth. We take that attitude toward wild living things (individuals, species populations, or whole biotic communities) when and only when we regard them as entities possessing inherent worth. Indeed, it is only because they are conceived in this way that moral agents can think of themselves as having validly binding duties, obligations, and responsibilities that are *owed* to them as their *due*. I am not at this juncture arguing why they *should* be so regarded; I consider it at length below. But so regarding them is a presupposition of our taking the attitude of respect toward them and accordingly understanding ourselves as bearing certain moral relations to them. This can be shown as follows:

What does it mean to regard an entity that has a good of its own as possessing inherent worth? Two general principles are involved: the principle of moral consideration and the principle of intrinsic value.

According to the principle of moral consideration, wild living things are deserving of the concern and consideration of all moral agents simply in virtue of their being members of the Earth's community of life. From the moral point of view their good must be taken into account whenever it is affected for better or worse by the conduct of rational agents. This holds no matter what species the creature belongs to. The good of each is to be accorded some value and so acknowledged as having some weight in the deliberations of all rational agents. Of course, it may be necessary for such agents to act in ways contrary to the good of this or that particular organism or group or organisms in order to further the good of others, including the good of humans. But the principle of moral consideration prescribes that, with respect to each being an entity having its own good, every individual is deserving of consideration.

The principle of intrinsic value states that, regardless of what kind of entity it is in other respects, if it is a member of the Earth's community of life, the realization of its good is something *intrinsically* valuable. This means that its good is prima facie worthy of being preserved or promoted as an end in itself and for the sake of the entity whose good it is. Insofar as we regard any organism, species population, or life community as an entity having inherent worth, we believe that it must never be treated as if it were a mere object or thing whose entire value lies in being instrumental to the good of some other entity. The well-being of each is judged to have value in and of itself.

Combining these two principles, we can now define what it means for a living thing or group of living things to possess inherent worth. To say that it possesses inherent worth is to say that its good is deserving of the concern and consideration of all moral agents, and that the realization of its good has intrinsic value, to be pursued as an end in itself and for the sake of the entity whose good it is.

The duties owed to wild organisms, species populations, and communities of life in the Earth's natural ecosystems are grounded on their inherent worth. When rational, autonomous agents regard such entities as possessing inherent worth, they place intrinsic value on the realization of their good and so hold themselves responsible for performing actions that will have this effect and for refraining from actions having the contrary effect.

The Attitude of Respect for Nature

Why should moral agents regard wild living things in the natural world as possessing inherent worth? To answer this question we must first take into account the fact that, when rational, autonomous agents subscribe to the principles of moral consideration and intrinsic value and so conceive of wild living things as having that kind of worth, such agents are *adopting a certain ultimate moral attitude toward the natural world.* This is the attitude I call "respect for nature." It parallels the attitude of respect for persons in human ethics. When we adopt the attitude of respect for persons as the proper (fitting, appropriate) attitude to take toward all persons as persons, we consider the fulfillment of the basic interests of each individual to have intrinsic value. We thereby make a moral commitment to live a certain kind of life in relation to other persons. We place ourselves under the direction of a system of standards and rules that we consider validly binding on all moral agents as such.

Similarly, when we adopt the attitude of respect for nature as an ultimate moral attitude we make a commitment to live by certain normative principles. These principles constitute the rules of conduct and standards of character that are to govern our treatment of the natural world. This is, first, an *ultimate* commitment because it is not derived from any higher norm. The attitude of respect for nature is not grounded on some other, more general, or more fundamental attitude. It sets the total framework for our responsibilities toward the natural world. It can be justified, as I show below, but its justification cannot consist in referring to a more general attitude or a more basic normative principle.

Second, the commitment is a *moral* one because it is understood to be a disinterested matter of principle. It is this feature that distinguishes the attitude of respect for nature from the set of feelings and dispositions that comprise the love of nature. The latter seems from one's personal interest in and response to the natural world. Like the affectionate feelings we have toward certain individual human beings, one's love of nature is nothing more than the particular way one feels about the natural environment and its wild inhabitants. And just as our love for an individual person differs from our respect for all persons as such (whether we happen to love them or not), so love of nature differs from respect for nature. Respect for nature is an attitude we believe all moral agents ought to have simply as moral agents, regardless of whether or not they also love nature. Indeed, we have not truly taken the attitude of respect for nature ourselves unless we believe this. To put it in a Kantian way, to adopt the attitude of respect for nature is to take a stance that one wills it to be a universal law for all rational beings. It is to hold that stance categorically, as being validly applicable to every moral agent without exception, irrespective of whatever personal feelings toward nature such an agent might have or might lack.

Although the attitude of respect for nature is in this sense a disinterested and universalizable attitude, anyone who does adopt it has certain steady, more or less permanent dispositions. These dispositions, which are themselves to be considered disinterested and universalizable, comprise three interlocking sets: dispositions to seek certain ends, dispositions to carry on one's practical reasoning and deliberation in a certain way, and dispositions to have certain feelings. We may accordingly analyze the attitude of respect for nature into the following components. (a) The disposition to aim at, and to take steps to bring about, as final and disinterested ends, the promoting and protecting of the good of organisms, species populations, and life communities in natural ecosystems. (These ends are "final" in not being pursued as means to further ends. They are "disinterested" in being independent of the self-interest of the agent.) (b) The disposition to consider actions that tend to realize those ends to be prima facie obligatory *because* they have that tendency. (c) The disposition to experience positive and negative feelings toward states of affairs in the world *because* they are favorable or unfavorable to the good of organisms, species populations, and life communities in natural ecosystems.

The logical connection between the attitude of respect for nature and the duties of a life-centered system of environmental ethics can now be made clear. Insofar as one sincerely takes that attitude and so has the three sets of dispositions, one will at the same time be disposed to comply with certain rules of duty (such as nonmaleficence and noninterference) and with standards of character (such as fairness and benevolence) that determine the obligations and virtues of moral agents with regard to the Earth's wild living things. We can say that the actions one performs and the character traits one develops in fulfilling these moral requirements are the way one *expresses* or *embodies* the attitude in one's conduct and character. In his famous essay, "Justice as Fairness," John Rawls describes the rules of the duties of human morality (such as fidelity, gratitude, honesty, and justice) as "forms of conduct in which recognition of others as persons is manifested." I hold that the rules of duty governing our treatment of the natural world and its inhabitants are forms of conduct in which the attitude of respect for nature is manifested.

The Justifiability of the Attitude of Respect for Nature

I return to the question posed earlier, which has not yet been answered: why *should* moral agents regard wild living things as possessing inherent worth? I now argue that the only way we can answer this question is by showing how adopting the attitude of respect for nature is justified for all moral agents. Let us suppose that we were able to establish that there are good reasons for adopting the attitude, reasons which are intersubjectively valid for every rational agent. If there are such reasons, they would justify anyone's having the three sets of dispositions mentioned above as constituting what it means to have the attitude. Since these include the disposition to promote or protect the good of wild living things as a disinterested and ultimate end, as well as the disposition to perform actions for the reason that they tend to realize that end, we see that such dispositions commit a person to the principles of moral consideration and intrinsic value. To be disposed to further, as an end in itself, the good of any entity in nature just because it is that kind of entity, is to be disposed to give consideration to *every* such entity and to place intrinsic value on the realization of its good. Insofar as we subscribe to these two principles we regard living things as possessing inherent worth. Subscribing to the principles is what is *means* to so regard them. To justify the attitude of respect for nature, then, is to justify commitment to these principles and thereby to justify regarding wild creatures as possessing inherent worth.

We must keep in mind that inherent worth is not some mysterious sort of objective property belonging to living things that can be discovered by empirical observation or scientific investigation. To ascribe inherent worth to an entity is not to describe it by citing some feature discernible by sense perception or inferable by inductive reasoning. Nor is there a logically necessary connection between the concept of a being having a good of its own and the concept of inherent worth. We do not contradict ourselves by asserting that an entity that has a good of its own lacks inherent worth. In order to show that such an entity "has" inherent worth we must give good reasons for ascribing that kind of value to it (placing that kind of value upon it, conceiving of it to be valuable in that way). Although it is humans (persons, valuers) who must do the valuing, for the ethics of respect for nature, the value so ascribed is not a human value. That is to say, it is not a value derived from considerations regarding human well-being or human rights. It is a value that is ascribed to nonhuman animals and plants themselves, independently of their relationship to what humans judge to be conducive to their own good.

Whatever reasons, then, justify our taking the attitude of respect for nature as defined above are also reasons that show why we *should* regard the living things of the natural world as possessing inherent worth. We saw earlier that, since the attitude is an ultimate one, it cannot be derived from a more fundamental attitude nor shown to be a special case of a more general one. On what sort of grounds, then, can it be established?

The attitude we take toward living things in the natural world depends on the way we look at them, on what kind of beings we conceive them to be, and on how we understand the relations we bear to them. Underlying and supporting our attitude is a certain *belief system* that constitutes a particular world view or outlook on nature and the place of human life in it. To give good reasons for adopting the attitude of respect for nature, then, we must first articulate the belief system which underlies and supports that attitude. If it appears that the belief system is internally coherent and well-ordered, and if, as far as we can now tell, it is consistent with all known scientific truths relevant to our knowledge of the object of the attitude (which in this case includes the whole set of the Earth's natural ecosystems and their communities of life), then there remains the task of indicating why scientifically informed and rational thinkers with a developed capacity of reality awareness can find it acceptable as a way of conceiving of the natural world and our place in it. To the extent we can do this we provide at least a reasonable argument for accepting the belief system and the ultimate moral attitude it supports.

I do not hold that such a belief system can be *proven* to be true, either inductively or deductively. As we shall see, not all of its components can be stated in the form of empirically verifiable propositions. Nor is its internal order governed by purely logical relationships. But the system as a whole, I contend, constitutes a coherent, unified, and rationally acceptable "picture" or "map" of a total world. By examining each of its main components and seeing how they fit together, we obtain a scientifically informed and well-ordered conception of nature and the place of humans in it.

The belief system underlying the attitude of respect for nature I call (for want of a better name) "the biocentric outlook on nature." Since it is not wholly analyzable into empirically confirmable assertions, it should not be thought of as simply a compendium of the biological sciences concerning our planet's ecosystems. It might best be described as a philosophical world view, to distinguish it from a scientific theory or explanatory system. However, one of its major tenets is the great lesson we have learned from the science of ecology: the interdependence of all living things in an organically unified order whose balance and stability are necessary conditions for the realization of the good of its constituent biotic communities.

Before turning to an account of the main components of the biocentric outlook, it is convenient here to set forth the overall structure of my theory of environmental ethics as it has now emerged. The ethics of respect for nature is made up of three basic elements: a belief system, an ultimate moral attitude, and a set of rules of duty and standards of character. These elements are connected with each other in the following manner. The belief system provides a certain outlook on nature which supports and makes intelligible an autonomous agent's adopting, as an ultimate moral attitude, the attitude of respect for nature. It supports and makes intelligible the attitude in the sense that, when an autonomous agent understands its moral relations to the natural world in terms of this outlook, it recognizes the attitude of respect to be the only *suitable* or *fitting* attitude to take toward all wild forms of life in the Earth's biosphere. Living things are now viewed as *the appropriate objects of the attitude of respect* and are accordingly regarded as entities possessing inherent worth. One then places intrinsic value on the promotion and protection of their good. As a consequence of this, one makes a moral commitment to abide by a set of rules of duty and to fulfill (as far as one can by one's own efforts) certain standards of good character. Given one's adoption of the attitude of respect, one makes that moral commitment because one considers those rules and standards to be validly binding on all moral agents. They are seen as embodying forms of conduct and character structures in which the attitude of respect for nature is manifested.

This three-part complex which internally orders the ethics of respect for nature is symmetrical with a theory of human ethics grounded on respect for persons. Such a theory includes, first, a conception of oneself and others as persons, that is, as centers of autonomous choice. Second, there is the attitude of respect for persons as persons. When this is adopted as an ultimate moral attitude it involves the disposition to treat every person as having inherent worth or "human dignity." Every human being, just in virtue of her or his humanity, is understood to be worthy of moral consideration, and intrinsic value is placed on the autonomy and well-being of each. This is what Kant meant by conceiving of persons as ends in themselves. Third, there is an ethical system of duties which are acknowledged to be owed by everyone to everyone. These duties are forms of conduct in which public recognition is given to each individual's inherent worth as a person.

This structural framework for a theory of human ethics is meant to leave open the issue of consequentialism (utilitarianism) versus nonconsequentialism (deontology). That issue concerns the particular kind of system of rules defining the duties of moral agents toward persons. Similarly, I am leaving open in this paper the question of what particular kind of system of rules defines our duties with respect to the natural world.

The Biocentric Outlook on Nature

The biocentric outlook on nature has four main components. (1) Humans are thought of as members of the Earth's community of life, holding that membership on the same terms as apply to all the nonhuman members. (2) The Earth's natural ecosystems as a totality are seen as a complex web of interconnected elements, with the sound biological functioning of each being dependent on the sound biological functioning of the others. (This is the component referred to above as the great lesson that the science of ecology has taught us). (3) Each individual organism is conceived as a teleological center of life, pursuing its own good in its own way. (4) Whether we are concerned with standards of merit or with the concept of inherent worth, the claim that humans by their very nature are superior to other species in a groundless claim and, in the light of elements (1), (2), and (3) above, must be rejected as nothing more than an irrational bias in our own favor.

The conjunction of these four ideas constitutes the biocentric outlook on nature.

Critical Thinking

1. What is a "life-centered" environmental ethic?
2. What is an anthropocentric environmental ethic?
3. What does it mean to have respect for nature?

Women's Indigenous Knowledge and Biodiversity Conservation

Vandana Shiva

Ecologically focused feminists who call themselves *ecofemists* see important connections between the domination of women and the domination of nature under the patriarchal social and political framework that characterizes most of the world's human cultures. As with most groups that are defined by certain shared ideological perspectives, ecofeminists differ among themselves about the set of principles and beliefs that distinguish their perspective from that of other environmentalists. For example, not all of them agree that women have an intrinsic feminine perspective on the relationship between humans and nature that is distinct from that of males. Most ecofeminists, however, believe that their experiences as women in male-dominated societies provides them with a different way of knowing and thinking about environmental issues.

Vandana Shiva is a physicist, philosopher, and feminist. She directs the Research Foundation for Science, Technology, and Natural Resource Policy in Dehradun, India. Shiva is an environmental activist as well as a prolific writer and lecturer. Her ecological perspective is informed by the domination and exploitation of Third World peoples by Western industrial nations as well as by the oppression of women in patriarchal societies. Among her most recent books are *Stolen Harvest: The Hijacking of the Global Food Supply* (South End Press, 1999), *Water Wars; Privatization, Pollution, and Profit* (South End Press, 2002), *Globalization's New Wars: Seed, Water and Life Forms* (Women Unlimited, New Delhi, 2005), *Earth Democracy; Justice, Sustainability, and Peace* (South End Press, 2005), *Soil Not Oil: Environmental Justice in an Age of Climate Crisis* (South End Press, 2008), and *Staying Alive: Women, Ecology and Survival in India* (Women Unlimited, New Delhi, 2009).

The following selection is from "Women's Indigenous Knowledge and Biodiversity Conservation," which was published in Maria Mies and Vandana Shiva, *Ecofeminism* (Zed Books, 1993). In it, Shiva argues that women's traditional roles in society make them particularly resistant to the economically driven policies that threaten to destroy biodiversity.

Key Concept: women's knowledge as central to the preservation of biodiversity

Gender and diversity are linked in many ways. The construction of women as the 'second sex' is linked to the same inability to cope with difference as is the development paradigm that leads to the displacement and extinction of diversity in the biological world. The patriarchal world view sees man as the measure of all value, with no space for diversity, only for hierarchy. Woman, being different, is treated as unequal and inferior. Nature's diversity is seen as not intrinsically valuable in itself, its value is conferred only through economic exploitation for commercial gain. This criterion of commercial value thus reduces diversity to a problem, a deficiency. Destruction of diversity and the creation of monocultures becomes an imperative for capitalist patriarchy.

The marginalization of women and the destruction of biodiversity go hand in hand. Loss of diversity is the price paid in the patriarchal model of progress which pushes inexorably towards monocultures, uniformity and homogeneity. In this perverted logic of progress, even conservation suffers. Agricultural 'development' continues to work towards erasing diversity, while the same global interests that destroy biodiversity urge the Third World to conserve it. This separation of production and consumption, with 'production' based on uniformity and 'conservation' desperately attempting to preserve diversity militates against protecting biodiversity. It can be protected only by making diversity the basis, foundation and logic of the technology and economics of production.

The logic of diversity is best derived from biodiversity and from women's links to it. It helps look at dominant structures from below, from the ground of diversity, which reveal monocultures to be unproductive and the knowledge that produces them as primitive rather than sophisticated.

Diversity is, in many ways, the basis of women's politics and the politics of ecology; gender politics is largely a politics of difference. Eco-politics, too, is based on nature's variety and difference, as opposed to industrial commodities and processes which are uniform and homogeneous.

These two politics of diversity converge when women and biodiversity meet in fields and forest, in arid regions and wetlands.

Diversity as Women's Expertise

Diversity is the principle of women's work and knowledge. This is why they have been discounted in the patriarchal calculus. Yet it is also the matrix from which an alternative calculus of 'productivity' and 'skills' can be built that respects, not destroys, diversity.

The economies of many Third World communities depend on biological resources for their sustenance and well-being. In these societies, biodiversity is simultaneously a means of production, and an object of consumption. The survival and sustainability of livelihoods is ultimately connected to the conservation and sustainable use of biological resources in all their diversity. Tribal and peasant societies' biodiversity-based technologies, however, are seen as backward and primitive and are, therefore, displaced by 'progressive' technologies that destroy both diversity and people's livelihoods.

There is a general misconception that diversity-based production systems are low-productivity systems. However, the high productivity of uniform and homogenous systems is a contextual and theoretically constructed category, based on taking into account only one-dimensional yields and outputs. The alleged low productivity of the one against the alleged high productivity of the other is, therefore, not a neutral, scientific measure but biased towards commercial interests for whom maximizing the one-dimensional output is an economic imperative.

Crop uniformity, however, undermines the diversity of biological systems which form the production system as well as the livelihoods of people whose work is associated with diverse and multiple-use systems of forestry, agriculture and animal husbandry. For example, in the state of Kerala in India (its name derives from the coconut palm), coconut is cultivated in a multi-layered, high-intensity cropping system, along with betel and pepper vines, bananas, tapioca, drumstick, papaya, jackfruit, mango and vegetables. The annual labour requirement in a monoculture of coconut palm is 157 man-days per ha, while in a mixed cropping system, it is 960 man-days per ha. In the dry-land farming systems of the Deccan, the shift from mixed cropping millets, pulses and oilseeds to eucalyptus monocultures led to an annual loss of employment of 250 man-days per ha.

When labour is scarce and costly, labour displacing technologies are productive and efficient, but when labour is abundant, labour displacement is unproductive because it leads to poverty, dispossession and destruction of livelihoods. In Third World situations, sustainability has therefore to be achieved at two levels simultaneously: sustainability of natural resources and sustainability of livelihoods. Consequently, biodiversity conservation must be linked to conservation of livelihoods derived from biodiversity.

Women's work and knowledge is central to biodiversity conservation and utilization both because they work between 'sectors' and because they perform multiple tasks. Women, as farmers, have remained invisible despite their contribution. Economists tend to discount women's work as 'production' because it falls outside the so-called 'production boundary'. These omissions arise not because too few women work, but too many women do too much work of too many different kinds.

Statisticians and researchers suffer a conceptual inability to define women's work inside and outside the house—and farming is usually part of both. This recognition of what is and is not labour is exacerbated by the great volume and variety of work that women do. It is also related to the fact that although women work to sustain their families and communities, most of what they do is not measured in wages. Their work is also invisible because they are concentrated outside market-related or remunerated work, and they are normally engaged in multiple tasks.

Time allocation studies, which do not depend on an a priori definition of work, reflect more closely the multiplicity of tasks undertaken, and the seasonal, even daily movement in and out of the conventional labour force which characterize most rural women's livelihood strategy. Gender studies now being published, confirm that women in India are major producers of food in terms of value, volume and hours worked.

In the production and preparation of plant foods, women need skills and knowledge. To prepare seeds they need to know about seed preparation, germination requirements and soil choice. Seed preparation requires visual discrimination, fine motor co-ordination, sensitivity to humidity levels and weather conditions. To sow and strike seeds demands knowledge of seasons, climate, plant requirements, weather conditions, micro-climatic factors and soil-enrichment; sowing seeds requires physical dexterity and strength. To properly nurture plants calls for information about the nature of plant diseases, pruning, staking, water supplies, companion planting, predators, sequences, growing seasons and soil maintenance. Persistence and patience, physical strength and attention to plant needs are essential. Harvesting requires judgements in relation to weather, labour and grading; and knowledge of preserving, immediate use and propagation.

Women's knowledge has been the mainstay of the indigenous dairy industry. Dairying, as managed by women in rural India, embodies practices and logic rather different from those taught in dairy science at institutions of formal education in India, since the latter is essentially an import from Europe and North America. Women have been experts in the breeding and feeding of farm animals, including not only cows and buffaloes but also pigs, chickens, ducks and goats.

In forestry too, women's knowledge is crucial to the use of biomass for feed and fertilizer. Knowledge of the feed value of different fodder species, the fuel value of firewood types, and of food products and species is essential to agriculture-related forestry in which women are predominately active. In low input agriculture, fertility is transferred from forest and farm trees to the field by women's work either directly or via animals.

Women's work and knowledge in agriculture is uniquely found in the spaces 'in between' the interstices of 'sectors', the invisible ecological flows between sectors, and it is through these linkages that ecological stability, sustainability and productivity under resource-scarce conditions are maintained. The invisibility of women's work and knowledge arises from the gender bias which has a blind spot for realistic assessment of women's contributions. It is also rooted in the sectoral, fragmented and reductionist approach to development which treats forests, livestock and crops as independent of each other.

The focus of the 'green revolution' has been increasing grain yields of rice and wheat by techniques such as dwarfing, monocultures and multicropping. For an Indian woman farmer, rice is not only food, but also a source of cattle fodder and straw for thatch. High yield varieties (HYVs) can increase women's work; the shift from local varieties and indigenous crop-improvement strategies can also take away women's control over seeds and genetic resources. Women have been seed custodians since time immemorial, and their knowledge and skills should be the basis of all crop-improvement strategies.

Women: Custodians of Biodiversity

In most cultures women have been the custodians of biodiversity. They produce, reproduce, consume and con-serve biodiversity in agriculture. However, in common with all other aspects of women's work and knowledge, their role in the development and conservation of biodiversity has been rendered as non-work and non-knowledge. Their labour and expertise has been defined into nature, even though it is based on sophisticated cultural and scientific practises. But women's biodiversity conservation differs from the dominant patriarchal notion of biodiversity conservation.

Recent concern with biodiversity at the global level has grown as a result of the erosion of diversity due to the expansion of large-scale monoculture-based agricultural production and its associated vulnerability. Nevertheless, the fragmentation of farming systems linked to the spread of monocultures continues to be the guiding paradigm for biodiversity conservation. Each element of the farm eco-system is viewed in isolation, and conservation of diversity is seen as an arithmetical exercise of collecting varieties.

In contrast, in the traditional Indian setting, biodiversity is a relational category in which each element acquires its characteristics and value through its relationships with other elements. Biodiversity is ecologically and culturally embedded. Diversity is reproduced and conserved through the reproduction and conservation of culture, in festivals and rituals which not only celebrate the renewal of life, but also provide a platform for subtle tests for seed selection and propagation. The dominant world-view does not regard these tests as scientific because they do not emerge from the laboratory and the experimental plot, but are integral to the total world-view and lifestyle of people and are carried out, not by men in white coats, but by village woman. But because it is thus that the rich biological diversity in agriculture has been preserved they are systematically reliable.

When women conserve seed, they conserve diversity and therefore conserve balance and harmony. *Navdanya* or nine seeds are the symbol of this renewal of diversity and balance, not only of the plant world, but of the planet and of the social world. This complex relationship web gives meaning to biodiversity in Indian culture and has been the basis of its conservation over millennia. . . .

Biotechnology and the Destruction of Biodiversity

There are a number of crucial ways in which the Third World women's relationship to biodiversity differs from corporate men's relationship to biodiversity. Women produce through biodiversity, whereas corporate scientists produce through uniformity.

For women farmers, biodiversity has intrinsic value—for global seed and agribusiness corporations, biodiversity derives its value only as 'raw material' for the biotechnology industry. For women farmers the essence of the seed is the continuity of life. For multinational corporations, the value of the seed lies in the discontinuity of its life. Seed corporations deliberately breed seeds that cannot give rise to future generations so that farmers are transformed from seed custodians into seed consumers. Hybrid seeds are 'biologically patented' in that the offspring cannot be used as seeds as farmers must go back to corporations to buy seed every year. Where hybrids do not force the farmers back to the market, legal patents and 'intellectual property rights' are used to prevent farmers from saving seed. Seed patents basically imply that corporations treat seed as their 'creation.' Patents prevent others from 'making' the patented product, hence patented seed cannot be used for making seed. Royalties have to be paid to the company that gets the patent.

The claim of 'creation' of life by corporate scientists is totally unjustified, it is in fact an interruption in the life flow of creation. It is also unjustified because nature and Third World farmers have made the seed that corporations are attempting to own as their innovation and their private property. Patents on seeds are thus a twenty-first century form of piracy, through which the shared heritage and custody of Third World women peasants is robbed and depleted by multinational corporations, helped by global institutions like GATT [General Agreement on Tariffs and Trade].

Patents and biotechnology contribute to a two-way theft. From Third World producers they steal biodiversity. From consumers everywhere they steal safe and healthy food.

Genetic engineering is being offered as a 'green' technology worldwide. President Bush ruled in May 1992 that genetically engineered foods should be treated as 'natural' and hence safe. However, genetic engineering is neither natural nor safe.

A number of risks associated with genetically engineered foods have been listed recently by the Food and Drug Administration of the US:

- New toxicants may be added to genetically engineered food.
- Nutritional quality of engineered food may be diminished.
- New substances may significantly alter the composition of food.
- New proteins that cause allergic reactions may enter the food supply.
- Antibiotic resistant genes may diminish the effectiveness of some antibiotics to human and domestic animal diseases.
- The deletion of genes may have harmful side effects.
- Genetic engineering may produce 'counterfeit freshness'.
- Engineered food may pose risks to domestic animals.
- Genetically engineered food crops may harm wildlife and change habitats.

When we are being asked to trust genetically engineered foods, we are being asked to trust the same companies that gave us pesticides in our food. Monsanto, which is now selling itself as Green was telling us that 'without chemicals, millions more would go hungry'. Today, when Bhopal has changed the image of these poisons, we are being told by the Monsantos, Ciba-Geigys, Duponts, ICIs and Dows that they will now give us Green products. However, as Jack Kloppenberg has recently said, 'Having been recognized as wolves, the industrial semoticians want to redefine themselves as sheep, and green sheep at that.'

Critical Thinking

1. What is an "ecofeminist"?
2. In what sense, according to Vandana Shiva, is Third World-women's work in agriculture "invisible"?
3. Do women, as Vandana Shiva claims, have an intrinsically feminine perspective on the relationship between humans and nature that is necessarily distinct from that of males?

Collapse: How Societies Choose to Fail or Succeed

Jared Diamond

George Santayana once said "Those who cannot learn from history are doomed to repeat it." One might think that this is of little relevance to today's environmental problems, for population explosions, climate changes, and resource depletions all seem very new. But the only thing new about them is their global scale. Humans have been living on the Earth for a long time, and there have been environmental crises before. Some have lasted only briefly, as when the eruption of the Indonesian volcano Tambora in 1815 caused the "year without a summer" in 1816, also known as "eighteen-hundred-and-froze-to-death" (see http://en.wikipedia.org/wiki/Year_Without_a_Summer). Some lasted longer, as when prolonged drought may have caused the Anasazi culture of what is now the American Southwest to vanish.

The effects of the crises also varied. Some societies survived, and even thrived. Some suffered, declined, and even died. According to Jared Diamond, Professor of Geography at the UCLA Department of Geography and a noted environmental historian and popular science writer (his book *Guns, Germs, and Steel* [1997] won a Pulitzer Prize), among the reasons must be counted the choices people made, both when they saw the crises coming (if they did) and in response to the crises. The selection below is from the Prologue to his recent book, *Collapse: How Societies Choose to Fail or Succeed* (Viking, 2005). His point is that today's problems also call for choices. Some of those choices will be difficult. They will require going against decades and perhaps even centuries of "business as usual." But if we wish our children and grandchildren to have lives remotely resembling our own, they are choices that we must make.

Key Concept: the importance for tomorrow of human choices today

A Tale of Two Farms

Norse Greenland is just one of many past societies that collapsed or vanished, leaving behind monumental ruins such as those that Shelley imagined in his poem "Ozymandias." By collapse, I mean a drastic decrease in human population size and/or political/economic/social complexity, over a considerable area, for an extended time. The phenomenon of collapses is thus an extreme form of several milder types of decline, and it becomes arbitrary to decide how drastic the decline of a society must be before it qualifies to be labeled as a collapse. Some of those milder types of decline include the normal minor rises and falls of fortune, and minor political/economic/social restructurings, of any individual society; one society's conquest by a close neighbor, or its decline linked to the neighbor's rise, without change in the total population size or complexity of the whole region; and the replacement or overthrow of one governing elite by another. By those standards, most people would consider the following past societies to have been famous victims of full-fledged collapses rather than of just minor declines: the Anasazi and Cahokia within the boundaries of the modern U.S., the Maya cities in Central America, Moche and Tiwanaku societies in South America, Mycenean Greece and Minoan Crete in Europe, Great Zimbabwe in Africa, Angkor Wat and the Harappan Indus Valley cities in Asia, and Easter Island in the Pacific Ocean.

The monumental ruins left behind by those past societies hold a romantic fascination for all of us. We marvel at them when as children we first learn of them through pictures. When we grow up, many of us plan vacations in order to experience them at firsthand as tourists. We feel drawn to their often spectacular and haunting beauty, and also to the mysteries that they pose. The scales of the ruins testify to the former wealth and power of their builders—they boast "Look on my works, ye mighty, and despair!" in Shelley's words. Yet the builders vanished, abandoning the great structures that they had created at such effort. How could a society that was once so mighty end up collapsing? What were the fates of its individual citizens?—did they move away, and (if so) why, or did they die there in some unpleasant way? Lurking behind this romantic mystery is the nagging thought: might such a fate eventually befall our own wealthy society? Will tourists someday stare mystified at the rusting hulks of New York's skyscrapers, much as we stare today at the jungle-overgrown ruins of Maya cities?

It has long been suspected that many of those mysterious abandonments were at least partly triggered by ecological problems: people inadvertently destroying the environmental resources on which their societies depended. This suspicion of unintended ecological suicide—ecocide—has been confirmed by discoveries made in recent decades by archaeologists, climatologists, historians, paleontologists, and palynologists (pollen scientists). The processes through which past societies have undermined themselves by damaging their environments fall into eight categories, whose relative importance differs from case to case: deforestation and habitat destruction, soil problems (erosion, salinization, and soil fertility losses), water management problems, overhunting, overfishing, effects of introduced species on native species, human population growth, and increased per-capita impact of people.

Those past collapses tended to follow somewhat similar courses constituting variations on a theme. Population growth forced people to adopt intensified means of agricultural production (such as irrigation, double-cropping, or terracing), and to expand farming from the prime lands first chosen onto more marginal land, in order to feed the growing number of hungry mouths. Unsustainable practices led to environmental damage of one or more of the eight types just listed, resulting in agriculturally marginal lands having to be abandoned again. Consequences for society included food shortages, starvation, wars among too many people fighting for too few resources, and overthrows of governing elites by disillusioned masses. Eventually, population decreased through starvation, war, or disease, and society lost some of the political, economic, and cultural complexity that it had developed at its peak. Writers find it tempting to draw analogies between those trajectories of human societies and the trajectories of individual human lives—to talk of a society's birth, growth, peak, senescence, and death—and to assume that the long period of senescence that most of us traverse between our peak years and our deaths also applies to societies. But that metaphor proves erroneous for many past societies (and for the modern Soviet Union): they declined rapidly after reaching peak numbers and power, and those rapid declines must have come as a surprise and shock to their citizens.

In the worst cases of complete collapse, everybody in the society emigrated or died. Obviously, though, this grim trajectory is not one that all past societies followed unvaryingly to completion: different societies collapsed to different degrees and in somewhat different ways, while many societies didn't collapse at all.

The risk of such collapses today is now a matter of increasing concern; indeed, collapses have already materialized for Somalia, Rwanda, and some other Third World countries. Many people fear that ecocide has now come to overshadow nuclear war and emerging diseases as a threat to global civilization. The environmental problems facing us today include the same eight that undermined past societies, plus four new ones: human-caused climate change, buildup of toxic chemicals in the environment, energy shortages, and full human utilization of the Earth's photosynthetic capacity. Most of these 12 threats, it is claimed, will become globally critical within the next few decades: either we solve the problems by then, or the problems will undermine not just Somalia but also First World societies. Much more likely than a doomsday scenario involving human extinction or an apocalyptic collapse of industrial civilization would be "just" a future of significantly lower living standards, chronically higher risks, and the undermining of what we now consider some of our key values. Such a collapse could assume various forms, such as the worldwide spread of diseases or else of wars, triggered ultimately by scarcity of environmental resources. If this reasoning is correct, then our efforts today will determine the state of the world in which the current generation of children and young adults lives out their middle and late years.

But the seriousness of these current environmental problems is vigorously debated. Are the risks greatly exaggerated, or conversely are they underestimated? Does it stand to reason that today's human population of almost seven billion, with our potent modern technology, is causing our environment to crumble globally at a much more rapid rate than a mere few million people with stone and wooden tools already made it crumble locally in the past? Will modern technology solve our problems, or is it creating new problems faster than it solves old ones? When we deplete one resource (e.g., wood, oil, or ocean fish), can we count on being able to substitute some new resource (e.g., plastics, wind and solar energy, or farmed fish)? Isn't the rate of human population growth declining, such that we're already on course for the world's population to level off at some manageable number of people?

All of these questions illustrate why those famous collapses of past civilizations have taken on more meaning than just that of a romantic mystery. Perhaps there are some practical lessons that we could learn from all those past collapses. We know that some past societies collapsed while others didn't: what made certain societies especially vulnerable? What, exactly, were the processes by which past societies committed ecocide? Why did some past societies fail to see the messes that they were getting into, and that (one would think in retrospect) must have been obvious? Which were the solutions that succeeded in the past? If we could answer these questions, we might be able to identify which societies are now most at risk, and what measures could best help them, without waiting for more Somalia-like collapses.

But there are also differences between the modern world and its problems, and those past societies and their problems. We shouldn't be so naïve as to think that study of the past will yield simple solutions, directly transferable to our societies today. We differ from past societies in some respects that put us at lower risk than them; some of those respects often mentioned include our powerful technology (i.e., its beneficial effects), globalization, modern medicine, and greater knowledge of past societies and of distant modern societies. We also differ from past societies in some respects that put us at greater risk than them: mentioned in that connection are, again, our potent technology (i.e., its unintended destructive effects), globalization (such that now a collapse even in remote Somalia affects the U.S. and Europe), the dependence of millions (and, soon, billions) of us on modern medicine for our survival, and our much larger human population. Perhaps we can still learn from the past, but only if we think carefully about its lessons.

Efforts to understand past collapses have had to confront one major controversy and four complications. The controversy involves resistance to the idea that past peoples (some of them known to be ancestral to peoples currently alive and vocal) did things that contributed to their own decline. We are much more conscious of environmental damage now than we were a mere few decades ago. Even signs in hotel rooms now invoke love of the environment to make us feel guilty if we demand fresh towels or let the water run. To damage the environment today is considered morally culpable. . . .

As for the complications, of course it's not true that all societies are doomed to collapse because of environmental damage: in the past some societies did while others didn't; the real question is why only some societies proved fragile, and what distinguished those that collapsed from those that didn't. Some societies that I shall discuss, such as the Icelanders and Tikopians, succeeded in solving extremely difficult environmental problems, have thereby been able to persist for a long time, and are still going strong today. For example, when Norwegian colonists of Iceland first encountered an environment superficially similar to that of Norway but in reality very different, they inadvertently destroyed much of Iceland's topsoil and most of its forests. Iceland for a long time was Europe's poorest and most ecologically ravaged country. However, Icelanders eventually learned from experience, adopted rigorous measures of environmental protection, and now enjoy one of the highest per-capita national average incomes in the world. Tikopia Islanders inhabit a tiny island so far from any neighbors that they were forced to become self-sufficient in almost everything,

but they micromanaged their resources and regulated their population size so carefully that their island is still productive after 3,000 years of human occupation. Thus, this book is not an uninterrupted series of depressing stories of failure, but also includes success stories inspiring imitation and optimism.

In addition, I don't know of any case in which a society's collapse can be attributed solely to environmental damage: there are always other contributing factors. When I began to plan this book, I didn't appreciate those complications, and I naïvely thought that the book would just be about environmental damage. Eventually, I arrived at a five-point framework of possible contributing factors that I now consider in trying to understand any putative environmental collapse. Four of those sets of factors—environmental damage, climate change, hostile neighbors, and friendly trade partners—may or may not prove significant for a particular society. The fifth set of factors—the society's responses to its environmental problems—always proves significant. Let's consider these five sets of factors one by one, in a sequence not implying any primacy of cause but just convenience of presentation.

A first set of factors involves damage that people inadvertently inflict on their environment, as already discussed. The extent and reversibility of that damage depend partly on properties of people (e.g., how many trees they cut down per acre per year), and partly on properties of the environment (e.g., properties determining how many seedlings germinate per acre, and how rapidly saplings grow, per year). Those environmental properties are referred to either as fragility (susceptibility to damage) or as resilience (potential for recovery from damage), and one can talk separately of the fragility or resilience of an area's forests, its soils, its fish populations, and so on. Hence the reasons why only certain societies suffered environmental collapses might in principle involve either exceptional imprudence of their people, exceptional fragility of some aspects of their environment, or both.

A next consideration in my five-point framework is climate change, a term that today we tend to associate with global warming caused by humans. In fact, climate may become hotter or colder, wetter or drier, or more or less variable between months or between years, because of changes in natural forces that drive climate and that have nothing to do with humans. Examples of such forces include changes in the heat put out by the sun, volcanic eruptions that inject dust into the atmosphere, changes in the orientation of the Earth's axis with respect to its orbit, and changes in the distribution of land and ocean over the face of the Earth. Frequently discussed cases of natural climate change include the advance and retreat of continental ice sheets during the Ice Ages beginning over two million years ago, the so-called Little Ice Age from about A.D. 1400 to 1800, and the global cooling following the enormous volcanic eruption of Indonesia's

Mt. Tambora on April 5, 1815. That eruption injected so much dust into the upper atmosphere that the amount of sunlight reaching the ground decreased until the dust settled out, causing widespread famines even in North America and Europe due to cold temperatures and reduced crop yields in the summer of 1816 ("the year without a summer").

Climate change was even more of a problem for past societies with short human lifespans and without writing than it is today, because climate in many parts of the world tends to vary not just from year to year but also on a multi-decade time scale; e.g., several wet decades followed by a dry half-century. In many prehistoric societies the mean human generation time—average number of years between births of parents and of their children—was only a few decades. Hence towards the end of a string of wet decades, most people alive could have had no firsthand memory of the previous period of dry climate. Even today, there is a human tendency to increase production and population during good decades, forgetting (or, in the past, never realizing) that such decades were unlikely to last. When the good decades then do end, the society finds itself with more population than can be supported, or with ingrained habits unsuitable to the new climate conditions. (Just think today of the dry U.S. West and its urban or rural policies of profligate water use, often drawn up in wet decades on the tacit assumption that they were typical.) Compounding these problems of climate change, many past societies didn't have "disaster relief" mechanisms to import food surpluses from other areas with a different climate into areas developing food shortages. All of those considerations exposed past societies to increased risk from climate change.

Natural climate changes may make conditions either better or worse for any particular human society, and may benefit one society while hurting another society. (For example, we shall see that the Little Ice Age was bad for the Greenland Norse but good for the Greenland Inuit.) In many historical cases, a society that was depleting its environmental resources could absorb the losses as long as the climate was benign, but was then driven over the brink of collapse when the climate became drier, colder, hotter, wetter, or more variable. Should one then say that the collapse was caused by human environmental impact, or by climate change? Neither of those simple alternatives is correct. Instead, if the society hadn't already partly depleted its environmental resources, it might have survived the resource depletion caused by climate change. Conversely, it was able to survive its self-inflicted resource depletion until climate change produced further resource depletion. It was neither factor taken alone, but the combination of environmental impact and climate change, that proved fatal.

A third consideration is hostile neighbors. All but a few historical societies have been geographically close enough to some other societies to have had at least some contact with them. Relations with neighboring societies may be intermittently or chronically hostile. A society may be able to hold off its enemies as long as it is strong, only to succumb when it becomes weakened for any reason, including environmental damage. The proximate cause of the collapse will then be military conquest, but the ultimate cause—the factor whose change led to the collapse—will have been the factor that caused the weakening. Hence collapses for ecological or other reasons often masquerade as military defeats.

The most familiar debate about such possible masquerading involves the fall of the Western Roman Empire. Rome became increasingly beset by barbarian invasions, with the conventional date for the Empire's fall being taken somewhat arbitrarily as A.D. 476, the year in which the last emperor of the West was deposed. However, even before the rise of the Roman Empire, there had been "barbarian" tribes who lived in northern Europe and Central Asia beyond the borders of "civilized" Mediterranean Europe, and who periodically attacked civilized Europe (as well as civilized China and India). For over a thousand years, Rome successfully held off the barbarians, for instance slaughtering a large invading force of Cimbri and Teutones bent on conquering northern Italy at the Battle of Campi Raudii in 101 B.C.

Eventually, it was the barbarians rather than Romans who won the battles: what was the fundamental reason for that shift in fortune? Was it because of changes in the barbarians themselves, such that they became more numerous or better organized, acquired better weapons or more horses, or profited from climate change in the Central Asian steppes? In that case, we would say that barbarians really could be identified as the fundamental cause of Rome's fall. Or was it instead that the same old unchanged barbarians were always waiting on the Roman Empire's frontiers, and that they couldn't prevail until Rome became weakened by some combination of economic, political, environmental, and other problems? In that case we would blame Rome's fall on its own problems, with the barbarians just providing the coup de grâce. This question continues to be debated. Essentially the same question has been debated for the fall of the Khmer Empire centered on Angkor Wat in relation to invasions by Thai neighbors, for the decline in Harappan Indus Valley civilization in relation to Aryan invasions, and for the fall of Mycenean Greece and other Bronze Age Mediterranean societies in relation to invasions by Sea Peoples.

The fourth set of factors is the converse of the third set: decreased support by friendly neighbors, as opposed to increased attacks by hostile neighbors. All but a few historical societies have had friendly trade partners as well as neighboring enemies. Often, the partner and the enemy are one and the same neighbor, whose behavior shifts back and forth between friendly and hostile. Most societies depend to some extent on friendly neighbors, either for imports of essential trade goods (like U.S. imports of oil, and Japanese imports of oil, wood, and seafood, today), or else for cultural ties that lend cohesion to the society (such as Australia's cultural identity imported from Britain until recently). Hence the risk arises that, if your trade partner becomes weakened for any reason (including environmental damage) and can no longer supply the essential import or the cultural tie, your own society may become weakened as a result. This is a familiar problem today because of the First World's dependence on oil from ecologically fragile and politically troubled Third World countries that imposed an oil embargo in 1973. Similar problems arose in the past for the Greenland Norse, Pitcairn Islanders, and other societies.

The last set of factors in my five-point framework involves the ubiquitous question of the society's responses to its problems, whether those problems are environmental or not. Different societies respond differently to similar problems. For instance, problems of deforestation arose for many past societies, among which Highland New Guinea, Japan, Tikopia, and Tonga developed successful forest management and continued to prosper, while Easter Island, Mangareva, and Norse Greenland failed to develop successful forest management and collapsed as a result. How can we understand such differing outcomes? A society's responses depend on its political, economic, and social institutions and on its cultural values. Those institutions and values affect whether the society solves (or even tries to solve) its problems. In this book we shall consider this five-point framework for each past society whose collapse or persistence is discussed.

I should add, of course, that just as climate change, hostile neighbors, and trade partners may or may not contribute to a particular society's collapse, environmental damage as well may or may not contribute. It would be absurd to claim that environmental damage must be a major factor in all collapses: the collapse of the Soviet Union is a modern counter-example, and the destruction of Carthage by Rome in 146 B.C. is an ancient one. It's obviously true that military or economic factors alone may suffice. Hence a full title for this book would be "Societal collapses involving an environmental component, and in some cases also contributions of climate change, hostile neighbors, and trade partners, plus questions of societal responses." That restriction still leaves us ample modern and ancient material to consider.

*I*ssues of human environmental impacts today tend to be controversial, and opinions about them tend to fall on a spectrum between two opposite camps. One camp, usually referred to as "environmentalist" or "pro-environment," holds that our current environmental problems are serious and in urgent need of addressing, and that current rates of economic and population growth cannot be sustained. The other camp holds that environmentalists' concerns are exaggerated and unwarranted, and that continued economic and population growth is both possible and desirable. The latter camp isn't associated with an accepted short label, and so

I shall refer to it simply as "non-environmentalist." Its adherents come especially from the world of big business and economics, but the equation "non-environmentalist" = "pro-business" is imperfect; many businesspeople consider themselves environmentalists, and many people skeptical of environmentalists' claims are not in the world of big business. In writing this book, where do I stand myself with the respect to these two camps?

On the one hand, I have been a bird-watcher since I was seven years old. I trained professionally as a biologist, and I have been doing research on New Guinea rainforest birds for the past 40 years. I love birds, enjoy watching them, and enjoy being in rainforest. I also like other plants, animals, and habitats and value them for their own sakes. I've been active in many efforts to preserve species and natural environments in New Guinea and elsewhere. For the past dozen years I've been a director of the U.S. affiliate of World Wildlife Fund, one of the largest international environmentalist organizations and the one with the most cosmopolitan interests. All of those things have earned me criticism from non-environmentalists, who use phrases such as "fearmonger," "Diamond preaches gloom and doom," "exaggerates risks," and "favors endangered purple louseworts over the needs of people." But while I do love New Guinea birds, I love much more my sons, my wife, my friends, New Guineans, and other people. I'm more interested in environmental issues because of what I see as their consequences for people than because of their consequences for birds.

On the other hand, I have much experience, interest, and ongoing involvement with big businesses and other forces in our society that exploit environmental resources and are often viewed as anti-environmentalist. As a teenager, I worked on large cattle ranches in Montana, to which, as an adult and father, I now regularly take my wife and my sons for summer vacations. I had a job on a crew of Montana copper miners for one summer. I love Montana and my rancher friends, I understand and admire and sympathize with their agribusinesses and their lifestyles, and I've dedicated this book to them. In recent years I've also had much opportunity to observe and become familiar with other large extractive companies in the mining, logging, fishing, oil, and natural gas industries. For the last seven years I've been monitoring environmental impacts in Papua New Guinea's largest producing oil and natural gas field, where oil companies have engaged World Wildlife Fund to provide independent assessments of the environment. I have often been a guest of extractive businesses on their properties, I've talked a lot with their directors and employees, and I've come to understand their own perspectives and problems.

While these relationships with big businesses have given me close-up views of the devastating environmental damage that they often cause, I've also had close-up views of situations where big businesses found it in their interests to adopt environmental safeguards more draconian and effective than I've encountered even in national parks. I'm interested in what motivates these differing environmental policies of different businesses. My involvement with large oil companies in particular has brought me condemnation from some environmentalists, who use phrases such as "Diamond has sold out to big business," "He's in bed with big businesses," or "He prostitutes himself to the oil companies."

In fact, I am not hired by big businesses, and I describe frankly what I see happening on their properties even though I am visiting as their guest. On some properties I have seen oil companies and logging companies being destructive, and I have said so; on other properties I have seen them being careful, and that was what I said. My view is that, if environmentalists aren't willing to engage with big businesses, which are among the most powerful forces in the modern world, it won't be possible to solve the world's environmental problems. . . .

Critical Thinking

1. Are societies that damage their environment doomed to collapse? Is ours?
2. What does it mean, according to Jared Diamond, to say that a society collapses?
3. How do the choices of people (individually and as societies) affect whether a society collapses due to environmental damage?
4. How do Jared Diamond's points connect up with your previous readings in this book? Consider especially Selections 1, 7, 8, 11, 19, 27, 28, 29, and 40.

Acknowledgments

Chapter 1

SELECTION 1   From *Man and Nature* by George Perkins Marsh (Charles Scribner, 1864).

SELECTION 2   From *The Mountains of California* by John Muir (Houghton Mifflin, 1916).

SELECTION 3   From *The Fight for Conservation* by Gifford Pinchot (Doubleday, 1910).

SELECTION 4   From *Aldo Leopold, A Sand County Almanac: And Sketches Here and There,* Oxford University Press, 1977. Copyright © 1949, 1977 by Oxford University Press Inc. Reprinted by permission.

Chapter 2

SELECTION 5   From *Quaternary Extinctions: A Prehistoric Revolution* by Paul S. Martin and Richard G. Klein, editors. Copyright © 1984 The University of Arizona Press, 355 S. Euclid Ave. Ste. 103, Tucson, Arizona, 85719. Reprinted by permission.

SELECTION 6   From *Science,* March 10, 1967. Copyright © 1967 by American Association for the Advancement of Science. Reproduced with permission from AAAS. www.sciencemag.org

SELECTION 7   From *Science,* vol. 162, 1968. Copyright © 1968 by American Association for the Advancement of Science. Reproduced with permission from AAAS. www.sciencemag.org

SELECTION 8   From *Commonwealth Scientific and Industrial Research Organisation* (CSIRO) by Graham Turner (Commonwealth Scientific and Industrial Research Organisation (CSIRO), 2008). Copyright © 2008 by Commonwealth Scientific and Industrial Research Organisation (CSIRO). Reprinted by permission.

Chapter 3

SELECTION 9   From *Life and Death of the Salt Marsh* by John Teal and Mildred Teal (Ballantine Books, 1969). Copyright © 1969 by John and Mildred Teal. Reprinted by permission of Frances Collin, Literary Agent.

SELECTION 10   From *Science,* July 25, 1997, pp. 494–499. Copyright © 1997 by American Association for the Advancement of Science. Reproduced with permission from AAAS. www.sciencemag.org

SELECTION 11   From *Ecosystems and Human Well-Being: Synthesis* by Millennium Ecosystem Assessment (Island Press, 2005). Copyright © 2005 by World Resources Institute. Reprinted by permission.

SELECTION 12   From *Ecology,* October 1940, pp. 438–450. Copyright © 1940 by Ecological Society of America. Reprinted by permission.

Chapter 4

SELECTION 13   From *Energy and the Environment* by John M. Fowler (McGraw-Hill, 1975). Copyright © 1975 by The McGraw-Hill Companies. Reprinted by permission.

SELECTION 14   From *Huffington Post,* July 17, 2008. Copyright © 2008 by Huffington Post Inc. Reprinted by permission.

Contributors

Katia Aviles-Vazquez is currently pursuing a PhD. in Geography at the University of Texas.

David Battisti is an atmospheric sciences professor at the University of Washington.

Roger Beachy is the Director of the National Institute of Food and Agriculture in the U.S. Agriculture Department.

Lester Brown founded the World Watch Institute in Washington, D.C., in 1974. In 2001, he founded the Earth Policy Institute, which he currently serves as president and senior researcher. The *Washington Post* has called Brown "one of the world's most influential thinkers." He has authored or coauthored many highly influential books, among which are *Outgrowing the Earth: The Food Security Challenge in an Age of Falling Water Tables and Rising Temperatures* (Earthscan, 2005), *Plan B 4.0: Mobilizing to Save Civilization* (W. W. Norton, 2009), and *World on the Edge: How to Prevent Environmental and Economic Collapse* (W. W. Norton, 2011).

M. Jahi Chappell is currently a postdoctoral associate in Science & Technology Studies at Cornell University.

Joel E. Cohen is the professor of populations at the Rockefeller University and Columbia University in New York City and heads the Laboratory of Populations at Rockefeller and Columbia University. His research deals with the demography, ecology, epidemiology and social organization of human and non-human populations.

Peter Cooper is a principal scientist at the India-based International Crops Research Institute for the Semi-Arid Tropics.

Mark A. Delucchi is a research scientist at the Institute of Transportation Studies at the University of California, Davis.

Jared Diamond is Professor of Geography at the University of California, Los Angeles. He won a Pulitzer Prize for his book, *Guns, Germs, and Steel: The Fates of Human Societies* (W. W. Norton, 1997).

Nina V. Federoff is a professor of life sciences at Pennsylvania State University and Science and Technology Adviser to the Secretary of State and to the Administrator of USAID, U.S. Department of State.

David Fischhoff is Vice President Technology Strategy and Development at Monsanto Co.

Dave Foreman is the director and senior fellow of the Rewilding Institute (www.rewilding.org/), best known for promoting the restoration of the North American wilderness. His books include *Confessions of an Eco-Warrior* (Harmony, 1991) and *Rewilding North America* (Island Press, 2004).

Betsy Hartmann is the director of the Population and Development Program and professor of development studies at Hampshire College. A longstanding activist in the international women's health movement, she is the author of *Reproductive Rights and Wrongs: The Global Politics of Population Control* (South End Press, 1995), two political thrillers, and other books and articles about development, climate change, and security.

Carl Hodges is the founder of The Seawater Foundation.

Mark Z. Jacobson is Professor of Civil and Environmental Engineering at Stanford University.

Vic Knauf is the chief technical officer for Arcadia Biosciences.

David Lobell is an Assistant Professor at Stanford University in Environmental Earth System Science, and a Center Fellow in Stanford's Program on Food Security and the Environment.

Bill McKibben is an American environmentalist and writer and the founder of 350.org, an international climate campaign.

Barbara Mazur is a researcher at the DuPont Experimental Station.

Jeremy Moghtader is an instructor at the Department of Horticulture, Michigan State University.

David Molden is deputy director general for research of the Sri Lanka-based International Water Management Institute.

Ivette Perfecto is a Professor at the School of Natural Resources and Environment, University of Michigan.

Orrin H. Pilkey (b. 1934) is the James B. Duke Professor Emeritus of Geology at Duke University. A winner of the Society for Sedimentary Geology's Francis Shepard Medal for excellence in marine geology (1987), the American Geological Institute's award for outstanding contributions to public understanding of geology (1993), and the Priestley Medal, the Dickinson College Award in Memory of Joseph Priestley, for distinguished research in coastal geology and public service in policy formulation and education about America's coastal resources (2003), Pilkey has earned a reputation as one of the most outspoken critics of efforts to manage erosion along America's coastline.

Eileen Quintero is a research assistant at the School of Natural Resources and Environment, University of Michigan.

Matthew Reynolds is head of Wheat Physiology at the Mexico City-based International Maize and Wheat Improvement Center.

Pamela Ronald is Professor of Plant Pathology and Chair of the Plant Genomics Program at the University of California, Davis, and Director of Grass Genetics at the Joint Bioenergy Institute.

Mark Rosegrant is Director of the Environment and Production Technology Division at the International Food Policy Research Institute.

Andrea Samulon is currently an agribusiness campaigner with the Rainforest Action Network (http://ran.org/).

Pedro Sanchez is Director of the Tropical Agriculture and the Rural Environment Program, Senior Research Scholar and Director of the Millennium Villages Project at the Earth Institute at Columbia University.

The Secretariat of the Convention on Biological Diversity was established by the United Nations to support the goals of the Convention on Biological Diversity (CBD), signed at the Earth Summit in Rio de Janeiro, Brazil, in 1992.

Vandana Shiva is a physicist, philosopher, environmentalist, and feminist. She directs the Research Foundation for Science, Technology, and Natural Resource Policy in Dehradun, India. Among her most recent books are *Stolen Harvest: The Hijacking of the Global Food Supply* (South End Press, 1999), *Water Wars; Privatization, Pollution, and Profit* (South End Press, 2002), *Globalization's New Wars: Seed, Water and Life Forms* (Women Unlimited, New Delhi, 2005), *Earth Democracy; Justice, Sustainability, and Peace* (South End Press, 2005), *Soil Not Oil: Environmental Justice in an Age of Climate Crisis* (South End Press, 2008), and *Staying Alive: Women, Ecology and Survival in India* (Women Unlimited, New Delhi, 2009).

Paul W. Taylor is a Professor of Philosophy emeritus at the City University of New York. His work in environmental ethics has been influential; see *Respect for Nature: A Theory of Environmental Ethics* (Princeton University Press, 1986).

Avigad Vonshak is Director, J. Blaustein Institutes for Desert Research Ben-Gurion University in Israel.

The World Commission on Environment and Development (WCED) (the Brundtland Commission) was convened by the United Nations in 1983 to address growing concerns about worsening environmental deterioration and its consequences for human economies and well-being, and to propose strategies for sustainable development.

Robert S. Young is an associate professor of geology in the Department of Geosciences and Natural Resources Management of Western Carolina University.

Emily Zakem is at the School of Art and Design, University of Michigan.

Jian-Kang Zhu is Professor of Plant Cell Biology in the College of Natural and Agricultural Sciences, Botany & Plant Sciences, the University of California, Riverside.

Index